Ayahuasca Healing and Science

D1796261

Beatriz Caiuby Labate • Clancy Cavnar
Editors

Ayahuasca Healing and Science

 Springer

Editors
Beatriz Caiuby Labate
Chacruna Institute for Psychedelic Plant
Medicines
San Francisco, CA, USA

Clancy Cavnar
Chacruna Institute for Psychedelic Plant
Medicines
San Francisco, CA, USA

ISBN 978-3-030-55690-7   ISBN 978-3-030-55688-4   (eBook)
https://doi.org/10.1007/978-3-030-55688-4

© Springer Nature Switzerland AG 2021
This work is subject to copyright. All rights are reserved by the Publisher, whether the whole or part of
the material is concerned, specifically the rights of translation, reprinting, reuse of illustrations, recitation,
broadcasting, reproduction on microfilms or in any other physical way, and transmission or information
storage and retrieval, electronic adaptation, computer software, or by similar or dissimilar methodology
now known or hereafter developed.
The use of general descriptive names, registered names, trademarks, service marks, etc. in this publication
does not imply, even in the absence of a specific statement, that such names are exempt from the relevant
protective laws and regulations and therefore free for general use.
The publisher, the authors, and the editors are safe to assume that the advice and information in this book
are believed to be true and accurate at the date of publication. Neither the publisher nor the authors or the
editors give a warranty, expressed or implied, with respect to the material contained herein or for any
errors or omissions that may have been made. The publisher remains neutral with regard to jurisdictional
claims in published maps and institutional affiliations.

This Springer imprint is published by the registered company Springer Nature Switzerland AG
The registered company address is: Gewerbestrasse 11, 6330 Cham, Switzerland

Preface

The book you are now reading, *Ayahuasca Healing and Science*, is both timely and far-reaching. It covers the range of aspects of ayahuasca's legendary healing capacity, looking at the ways psychological healing is realized from consumption of a drink. Teasing apart the factors that constitute "healing" is the job of science. Here, we give deep consideration to the micro as well as the macro; from asking what the molecules in ayahuasca are and how do they change the way we think and feel, to looking at the protocols and reasons for ayahuasca-based treatment of multiple mental health disorders.

This book presents research projects where scientists use neuroimaging techniques to explore the neural correlates underlying ayahuasca's healing of problematic patterns of thought and are able to measure its ability to enhance meta-cognition and address broad patterns of dysfunction. It is not just the emotions and neurons that are affected by treatment with ayahuasca; we also take a look at the way thoughts themselves are transformed. Cognitive behavioral therapy is successful because it confronts the distorted thoughts that lead to anxiety and depression; research is revealing that ayahuasca also interferes with problematic cognitions. Its effects reveal a complex interplay between the chemical world of molecules, the world of emotions, and the world of thought, visions, and insights.

The use of this powerful agent in psychotherapy must take into account its sometimes-overwhelming and unsettling effect, especially on survivors of trauma who might not be emotionally prepared for the difficult visions and painful insights ayahuasca can provide. These considerations are important in respect to the patients who are in most need of the sublime power of ayahuasca. The problems of treating delicate emotional conditions with the sometimes-caustic cleanser that is ayahuasca are addressed herein.

The use of ayahuasca as an antidepressant, an anti-addictive agent, and as a treatment for PTSD, eating disorders, and childhood trauma points to the wide range of applications for this magical drink. Has there ever been such a multifaceted substance with such a helpful profile? Among the issues covered in this volume, we can read about ayahuasca resolving distressing mental health symptoms by weakening egocentrism, enhancing cognitive adaptability and the capacity to regulate emotions, improving mindfulness, and enhancing connection to nature. This book addresses ayahuasca's healing potential for addictions in particular, with three chapters focusing on this topic. We conclude the volume with an interview with a Shipibo healer on his perspectives on ayahuasca healing, a return to the indigenous view from whence all ayahuasca cures originate.

This book adds to the growing collection of data on this hallucinogen-religious sacrament-medicinal plant-magical potion, and we hope it encourages others to take seriously the rumors and legends circulating about the power of ayahuasca, not as a naïve believer in myths, but as a logical and discerning student of science who demands proof for such remarkable claims. The body of knowledge on this drink is expanding and we are beginning to see connections that were not apparent before, as anthropology, psychology, and the medical and biological sciences contribute their viewpoints.

Why has ayahuasca's use expanded so fast globally? What does it offer in comparison to other medicines? Why is this so popular now? Did ayahuasca come to heal us? Is it here by pure chance? Can it reveal a deeper structure to reality? For now, we may rely on the researchers and scientists who provide proof that, at the very least, ayahuasca is a potent medicine with broad applications for what ails humanity.

San Francisco, CA, USA   Clancy Cavnar
                                                                       Beatriz Caiuby Labate

Foreword: Will the Queen Save Us?

It is of no small significance that *Ayahuasca Healing and Science*, edited by the indefatigable scholars Beatriz Labate and Clancy Cavnar, will be published in 2020: the year of the pandemic, when predatory capitalism, White male supremacy, police brutality, and other aspects of patriarchy finally meet a crisis of their own making that is big enough to drown all of it. As the process unravels archaic forms of social organization, the attrition leaves behind a trail of traumas; emotional scars that reverberate psychic distress, pain, and despair. We live through an epidemic of depression and suicide. People of all walks of life need healing, and this is why this book is so timely.

Ayahuasca is a sacred Amerindian medicine whose use is spreading internationally across cultures and beliefs. In the past decade, the therapeutic use of ayahuasca received scientific support from the experts listed below, including the hardcore pioneer Jordi Riba. The plurality of views was ensured by the choice of a very diverse set of authors from a wide range of fields and countries, from neuroscience to anthropology, from psychiatry to clinical psychology, from biochemistry to holistic psychotherapy.

While the solid therapeutic effects of ayahuasca are well represented in the book, such as in the work of Brazilian researchers on the potent antidepressant effects of ayahuasca, there is a clear effort in several chapters to chart the risks and constraints of ayahuasca use. There is also a proactive determination to map the several blind spots and unknowns of this field, which include molecular, physiological, psychiatric, and cross-cultural gaps. In the coming years, many of the important questions raised in this book shall be answered by qualitative and quantitative research. It will be performed by the growing cohort of scientists dedicated to the study of this wonderful medicine, many of them authors of chapters in this book.

Ayahuasca has an exceedingly special place in the pharmacopeia of the reborn (and increasingly free) use of psychedelics for healing in Western medicine. Its head-to-head comparison with psilocybin and ketamine has yet to be performed, but it is clear that ayahuasca is a major contestant for the most potent psychedelic antidepressant. The main chemical components present in the brew, harmine and N,N-dimethyltryptamine (N,N-DMT), as well as the related compound 5-MeO DMT, present in the secretion of the toad *Bufo alvarius* and in the yopo plant snuff, have been demonstrated to induce several cellular changes related to the formation of new synapses and even new neurons entirely, as shown by the research groups of Stevens Rehen (Dakic et al., 2016; Dakic et al., 2017), Richardson Leão (Lima da Cruz et al., 2018), and David Olson (Ly et al. 2018). Congruent results were recently seen for d-LSD by a consortium of my laboratory with those of Stevens Rehen, Draulio de Araujo, Luis Fernando Tófoli, and Daniel Martins-de-Souza, with support from the Beckley Foundation (Cini et al., 2019).

The powerful synergistic effects of mixing two or more molecules able to promote synaptogenetic and neurogenetic effects, as is the case of ayahuasca, likely leads to a potent "entourage effect" that may prove as dramatic as that originally proposed by the great Raphael Mechoulam and colleagues concerning the cooperative effects of the multiple cannabinoids, which may increase clinical efficacy (Ben-Shabat et al., 1998).

It is important to note, however, that the fact that ayahuasca represents concrete hope for those in need of major neural reorganization is not only related to its biological effects. The interaction of psychedelic substances with set and setting represents a wide range of therapeutic possibilities. In this regard, ayahuasca also occupies a distinguished and strategic place due to the richness and sophistication of the Amerindian and syncretic rituals that make ayahuasca a true queen among the sacred organisms required to mitigate the ongoing mental health crisis that characterizes the twenty-first century.

Sidarta Ribeiro, Ph.D.

References

Ben-Shabat, S., Fride, E., Sheskin, T., Tamiri, T., Rhee, M. H., Vogel, Z., Bisogno, T., De Petrocellis, L., Di Marzo, V., & Mechoulam, R. (1998). An entourage effect: Inactive endogenous fatty acid glycerol esters enhance 2-arachidonoyl-glycerol cannabinoid activity. *Eur J Pharmacol, 353*(1), 23–31. https://doi.org/10.1016/s0014-2999(98)00392-6.

Cini, F. A., Ornelas, I., Marcos, E., Goto-Silva, L., Nascimento, J., Ruschi, S., Salerno, J., Karmirian, K., Costa, M., Sequerra, E., Araújo, D., Tófoli, L. F., Rennó-Costa, C., Martins-de-Souza, D., Feilding, A., Rehen, S., & Ribeiro, S. (2019) d-Lysergic acid diethylamide has major potential as a cognitive enhancer. *bioRxiv, 6*. doi: https://doi.org/10.1101/866814.

Dakic, V., Maciel, R. M., Drummond, H., Nascimento, J. M., Trindade, P., & Rehen, S. K. (2016). Harmine stimulates proliferation of human neural progenitors. *PeerJ, 4*, e2727. https://doi.org/10.7717/peerj.2727.

Dakic, V., Nascimento, J. M., Costa Sartore, R., Maciel, R. D.M., de Araujo, D. B., Ribeiro, S., Martins-de-Souza, D., & Rehen, S. K. (2017). Short term changes in the proteome of human cerebral organoids induced by 5-MeO-DMT. *Scientific reports, 7*(1), 12863. https://doi.org/10.1038/s41598-017-12779-5

Lima da Cruz, R. V., Moulin, T. C., Petiz, L. L., & Leão, R. N. (2018). A single dose of 5-MeO-DMT stimulates cell proliferation, neuronal survivability, morphological and functional changes in adult mice ventral dentate gyrus. *Frontiers in molecular neuroscience, 11*, 312. https://doi.org/10.3389/fnmol.2018.00312.

Ly, C., Greb, A. C., Cameron, L. P., Wong, J. M., Barragan, E. V., Wilson, P. C., Burbach, K. F., Soltanzadeh Zarandi, S., Sood, A., Paddy, M. R., Duim, W. C., Dennis, M. Y., McAllister, A. K., Ori-McKenney, K. M., Gray, J. A., & Olson, D. E. (2018). Psychedelics promote structural and functional neural plasticity. *Cell reports, 23*(11), 3170–3182. https://doi.org/10.1016/j.celrep.2018.05.022.

Summary

This book offers a series of perspectives on the therapeutic potential of the ritual and clinical use of the Amazonian hallucinogenic brew ayahuasca in the treatment and management of various disorders. Biomedical and anthropological data on the use of ayahuasca for treating depression, PTSD, anxiety, substance dependence, eating disorders, and its role in psychological well-being, quality of life, enhancing cognition, and for coping with grief are presented and critiqued. The use of ayahuasca associated with psychotherapy is examined, alongside the challenges of integrating plant medicines into psychiatry. Furthermore, some preliminary research with animals is introduced in which neural progenitor cells indicate that the alkaloids present in ayahuasca facilitate the formation of new neurons, suggesting that ayahuasca acts at multiple levels of neural complexity. The neurogenic effects of ayahuasca alkaloids open a new avenue of research with potential applications ranging from psychiatric disorders to brain damage and dementia. Chapters combine the review of published literature, personal experiences of the authors with the brew, and first-hand observation of the use of ayahuasca in different contexts. This book is especially timely as the growing scientific evidence of the safety and therapeutic potential of psychedelics such as MDMA and psilocybin increases the likelihood of their rescheduling and adoption into new treatment models.

Contents

About the Editors and Contributors

Editors

Dr. Beatriz Caiuby Labate (Bia Labate) is a queer Brazilian anthropologist. She has a Ph.D. in social anthropology from the State University of Campinas (UNICAMP), Brazil. Her main areas of interest are the study of plant medicines, drug policy, shamanism, ritual, and religion. She is Executive Director of the Chacruna Institute for Psychedelic Plant Medicines (http://chacruna.net). She is also Public Education and Culture Specialist at the Multidisciplinary Association for Psychedelic Studies (MAPS), and Adjunct Faculty at the East-West Psychology Program at the California Institute of Integral Studies (CIIS). She is author, co-author, and co-editor of 21 books, two special-edition journals, and several peer-reviewed articles (http://bialabate.net).

Clancy Cavnar has a doctorate in clinical psychology (Psy.D.) from John F. Kennedy University in Pleasant Hill, CA. She currently works in private practice in San Francisco, and is an associate editor at Chacruna (http://chacruna.net), a venue for publication of high-quality academic short texts on plant medicines. She is also a research associate of the Interdisciplinary Group for Psychoactive Studies (NEIP). She combines an eclectic array of interests and activities as clinical psychologist, artist, and researcher. She has a master of fine arts in painting from the San Francisco Art Institute, a master's in counseling from San Francisco State University, and she completed the Certificate in Psychedelic-Assisted Therapy program at the California Institute of Integral Studies. She is author and co-author of articles in several peer-reviewed journals and co-editor, with Beatriz Caiuby Labate, of eight books. For more information see: http://neip.info/pesquisadore/clancy-cavnar

Contributors

Adam Andros Aronovich is a psychologist, cognitive scientist, and medical anthropologist. He is an active member of the Medical Anthropology Research Center (MARC), and the Interdisciplinary Psychedelic Studies (IPS) group, both in Catalonia. He is currently working as research coordinator for the Chaikuni Institute and the Temple of the Way of Light. He is leading the qualitative part of a joint study between ICEERS, the Beckley Foundation, and the Temple of the Way of Light, assessing the therapeutic potential of ayahuasca for various Western afflictions. He has written for various publications around the world.

Ismael Apud is assistant lecturer at the faculty of psychology, Universidad de la República (UdelaR, Uruguay). He is included in the National System of Researchers, Agencia Nacional de Investigación e Innovación (ANII, Uruguay), and is a researcher at the Medical Anthropology Research Center (MARC), Universitat Rovira I Virgili (URV, Spain). He has a Ph.D. in anthropology from URV, a master's degree in the methodology of scientific research from the Universidad Nacional de Lanús (UNLa, Argentina), and degrees in psychology and in social anthropology, both from UdelaR. His research areas include mental health, addictions, religious and spiritual practices, and psychedelics. He is author and co-author of several books, chapters, and scientific articles about these topics. Some of the articles address the use of ayahuasca in spiritual, religious, and clinical settings. His theoretical approach combines medical anthropology, cognitive science of religion, and cognitive psychology.

Elena Argento, MPH, Ph.D. (c), is a research associate with the BC Centre for Excellence in HIV/AIDS, and a Ph.D. student at the University of British Columbia. Elena is passionate about investigating the social epidemiology of health among marginalized populations. Her master's project examined structural violence among sex workers in South India, and she has worked on multinational projects in Guyana and Brazil. Elena's current research focuses on drug use, criminalization, and the potential therapeutic utility of psychedelic substances for improving mental health and addiction issues. A Canadian Institutes of Health Research Doctoral Award and a Killam Pre-Doctoral Fellowship support Elena's Ph.D. research.

Rielle Capler, MHA, Ph.D. (c), is a Ph.D. candidate at the University of British Columbia in interdisciplinary studies. Rielle's research, policy, and knowledge translation work has focused on the therapeutic uses of cannabis and ayahuasca for mental health and addiction, as well as reducing barriers to access and promoting safe use. Rielle received the Governor General of Canada's Queen Elizabeth II Diamond Jubilee Medal for her work in this field, and has provided expert testimony to the federal government's cannabis legalization task force.

María Carvalho holds an undergraduate degree in psychology at the University of Porto in 1999, finished her MSc in psychology in the field of addictions in 2004, and got her PhD in psychology at the University of Porto in 2015. She is an assistant professor at the Faculty of Education and Psychology at the Catholic University of Porto, where she's been lecturing since 2005. She has been a clinical psychologist accredited by the Portuguese Bar of Psychologists since 2016. She is vice-president of the *International Center for Ethnobotanical Education Research and Service* (www.ICEERS.org), and she has coordinated Project Kosmicare at the Boom Festival since 2010. She is a founding member of the Kosmicare Association, an NGO that advocates for full-spectrum evidence-based harm reduction and psycare in recreational environments. She is currently involved in projects concerning night-life and sexual violence, and the therapeutic potential of ayahuasca.

Clancy Cavnar has a doctorate in clinical psychology (Psy.D.) from John F. Kennedy University in Pleasant Hill, CA. She currently works in private practice in San Francisco, and is associate editor at Chacruna (http://chacruna.net), a venue for publication of high-quality academic short texts on plant medicines. She is also a research associate of the Interdisciplinary Group for Psychoactive Studies (NEIP). She combines an eclectic array of interests and activities as clinical psychologist, artist, and researcher. She has a master of fine arts in painting from the San Francisco Art Institute, a master's in counseling from San Francisco State University, and she completed the Certificate in Psychedelic-Assisted Therapy program at the California Institute of Integral Studies. She is author and co-author of articles in several peer-reviewed journals and co-editor, with Beatriz Caiuby Labate, of eight books. For more information see: http://neip.info/pesquisadore/clancy-cavnar

Nicole Galvão-Coelho received her Ph.D. in psychobiology from the Federal University of Rio Grande do Norte (UFRN) in 2009, where she became assistant professor at department of physiology. She has been developing research on neuro-endocrinology of stress response, and stress-related psychopathologies, and currently coordinates the laboratory of hormonal measures and a primate center at UFRN, which houses around 200 common marmosets (Callithrix jacchus), a species that has been widely used as an animal model in neuroscience studies.

Dráulio B. Araújo is a professor of neuroimaging at the Brain Institute (UFRN), Natal, Brazil. In recent years, his research has focused on using functional neuroimaging methods (EEG and fMRI) to investigate the acute and lasting effects of ayahuasca. His research group has also been studying the antidepressant potential of ayahuasca.

Joao Paulo Maia-Oliveira graduated with a degree in medicine from the Federal University of Rio Grande do Norte in 2003. He is a psychiatrist, and received his master's and doctoral degrees in medicine, focusing on mental health, from the University of São Paulo. He is currently an assistant professor of psychiatry at the

Federal University of Rio Grande do Norte. His main research interests are schizo-phrenia, depression, and translational research.

Mauricio Diament, M.D., is a Brazilian psychiatrist from São Paulo, and a member of Interdisciplinary Cooperation for Ayahuasca Research and Outreach (ICARO) with an interest in psychedelic research on ayahuasca and its clinical implications, as well as on other psychoactive substances.

Elisabet Domínguez-Clavé holds a bachelor's degree in psychology, and master's degrees in both psychiatry and clinical research. She is a Ph.D. candidate at the Department of Pharmacology and Therapeutics of the Autonomous University of Barcelona (UAB), Spain. She is currently carrying out research on the therapeutic applications of ayahuasca under the supervision of Dr. Jordi Riba, head of the Human Neuropsychopharmacology Research Group at Sant Pau Hospital in Barcelona. She also collaborates with the Psychiatry Department at the same hospital, under the supervision of Dr. Joaquim Soler and Dr. Juan Carlos Pascual.

Matilde Elices is a psychologist with a Ph.D. in psychiatry from the Autonomous University of Barcelona (UAB), Spain. She is interested in determining the psychological mechanisms that underlie the therapeutic effects of ayahuasca and its potential therapeutic utility for the treatment of severe mental disorders. She currently works at the Mental Health Research Group at the Institute Mar of Medical Research (Spain) and collaborates with the psychiatry department at Sant Pau Hospital. Matilde has co-authored more than 20 peer-reviewed articles, most of them focused on the efficacy of psychological treatments.

Natasha Files, MSW, RSW, is an Emotion Focused Family Therapy supervisor and trainer. She is a founding board member of the International Institute for Emotion Focused Family Therapy and is co-director of Mental Health Foundations. Natasha teaches in the Social Work Department at the University of the Fraser Valley and is involved in research, with her recent research interest including the ceremonial use of ayahuasca in the treatment of eating disorders.

Jenna Fletcher, M.A., is a holistic psychotherapist based in Ottawa, Canada. She specializes in the treatment of eating disorders across the lifespan, with a focus on intergenerational healing. She conducts research on the therapeutic potential and usage of ayahuasca for the treatment of eating disorders and comorbid mental health challenges.

Bruno Ramos Gomes, M.A., is a psychologist with a master's degree in public health at São Paulo University with a research project on the use of ayahuasca to help homeless people recover in Brazil. He has been working since 2004 using harm reduction in Cracolandia, a region with hundreds of crack users in the streets of downtown São Paulo. He has also been working with ibogaine treatment since 2010, and is collaborating on ayahuasca research to treat alcohol dependence at the

Interdisciplinary Cooperation for Ayahuasca Research and Outreach (ICARO) at UNICAMP.

Débora González is a clinical psychologist with a Ph.D. in pharmacology. Her master's research and Ph.D. studies received a pre-doctoral fellowship granted by the Ministry of Health for a 36-month project at the Department of Human Pharmacology and Clinical Neurosciences and Drug Addiction Unit (IMIM— Hospital del Mar Medical Research Institute). She is co-author of several scientific papers and book chapters about ayahuasca, 2C-B, *Salvia divinorum*, and research chemicals. She is leading a longitudinal research project conducted with the Beckley Foundation at the Temple of the Way of Light in Peru on the long-term effects of ayahuasca on well-being and the psychopathological symptoms of Western users.

Kim P. C. Kuypers has been studying the acute effects of psychedelics on memory, mood, and empathy, and the underlying neurobiological mechanisms, for 15 years. Recently, she also started to look into the effects of these substances on creativity and its link with empathy. Her motivation to do this research is that by using the unique properties of these drugs, we may get a better grasp on the neurobiological mechanisms underlying these effects. We will be able to use this knowledge to enhance flexible thinking and social abilities in daily life and in pathological populations, and consequently to enhance quality of life and feelings of well-being.

Dr. Beatriz Caiuby Labate (Bia Labate) is a queer Brazilian anthropologist. She has a Ph.D. in social anthropology from the State University of Campinas (UNICAMP), Brazil. Her main areas of interest are the study of plant medicines, drug policy, shamanism, ritual, and religion. She is Executive Director of the Chacruna Institute for Psychedelic Plant Medicines (http://chacruna.net). She is also Public Education and Culture Specialist at the Multidisciplinary Association for Psychedelic Studies (MAPS), and Adjunct Faculty at the East-West Psychology Program at the California Institute of Integral Studies (CIIS). She is author, co-author, and co-editor of 21 books, two special-edition journals, and several peer-reviewed articles (http://bialabate.net).

Adele Lafrance Ph.D., C. Psych, is a psychologist and associate professor at Laurentian University in Sudbury, Ontario. Dr. Lafrance has published extensively in the field of clinical psychology, and she is the co-developer of Emotion-Focused Family Therapy. She is also studying the use of plant medicines and psychedelics in the healing of mental health issues, with a particular focus on ayahuasca and eating disorders.

Anja Loizaga-Velder, Ph.D. is a German-Mexican clinical psychologist who has been investigating the therapeutic potential of the ritual use of psychedelic plants for over 20 years. She earned a Ph.D. in medical psychology from the Heidelberg University in Germany. Anja is a researcher in health sciences at the National

Autonomous University of Mexico (UNAM) and Director of Research and Psychotherapy at the Nierika Institute for Intercultural Medicine in Mexico.

Philippe Lucas, Ph.D. (c), is vice president of Patient Research & Access at Tilray (www.tilray.ca), a federally authorized medical cannabis production, research, and distribution company based in Nanaimo, British Colombia (BC), and a graduate researcher with the Center for Addictions Research of British Columbia. His scientific research includes the therapeutic use of cannabis and psychedelics in the treatment of mental health conditions and addiction, and he has been invited to provide expert testimony before the Canadian House of Commons, the Canadian Senate, and the BC Supreme Court. Philippe first became involved with medical cannabis as a patient, and founded the Vancouver Island Compassion Society in 1999 to serve the needs of patients who might benefit from the medical use of cannabis. He is very involved in his community, and served as a Victoria City Councilor and Regional Director from 2008-2011. Philippe has received a number of accolades and awards for his work, including the Queen Elizabeth II Diamond Jubilee Medal (2013) for his work and research on medical cannabis, the University of Victoria Blue and Gold Award for academic excellence and community service (2007), and the Vancouver Island Civil Liberties Association Leadership Award (2002).

Natasha L. Mason is a Ph.D. candidate in psychopharmacology at Maastricht University. She holds an M.Sc. in cognitive and clinical neuroscience from Maastricht University, and a B.Sc. in psychology from the University of Wisconsin-Madison. Her research assesses the acute and long-term effects of recreational drugs on brain and behavior, with her main interest being the potential therapeutic mechanisms of psychedelic drugs.

Julie D. Megler is a board-certified nurse practitioner in psychiatry and family medicine. She received her master of science in nursing from the University of Miami, Florida, and post- master's certificate in psychiatry at the University of California San Francisco. She is currently in private practice in the San Francisco Bay Area. Her practice focuses on integrative mental health services for emotional and physical well-being, as well as integration of non-ordinary states of consciousness. In addition to her clinical work, Julie is on the board of directors for ERIE (Entheogenic Research, Integration, and Education). ERIE is a 501c3 nonprofit organization based out of the San Francisco Bay Area that focuses on integration and education of entheogens. She has presented on the topics of psychedelic risk reduction, integration, and therapeutic applications of ayahuasca at different conferences and has also co-authored chapters in the books *Manifesting Minds* and *The Therapeutic Use of Ayahuasca*. As an experienced clinician and activist for the psychedelic movement, Julie is dedicated to educating the community about safety and the therapeutic benefits of entheogens.

José A. Morales-García is a Spanish neuroscientist who currently works for the Spanish Research Network on Neurodegenerative Diseases (CIBERNED) in Madrid. He was awarded his Ph.D., summa cum laude, with a special doctorate mention, in neuroscience by the Autonomous University of Madrid. He is a specialist in identification and analysis of new cellular targets implicated in neurogenesis and neurodegeneration. His scientific training in different national and international universities was focused in the study *in vitro* and *in vivo* of the mechanisms underlying neurodegenerative disease, mainly Alzheimer's and Parkinsonism, in order to develop new neuroprotective, anti-inflammatory, and neurogenic compounds for the treatment of these disorders. Together with scientific research on neurodegenerative diseases, Jose A. holds a position as Associate Professor at the Medical School in Complutense University of Madrid and as Professor in the master in neuropharmacology program at the medical school in the Autonomous University of Madrid.

Sérgio Mota-Rolim received his medical degree from the Federal University of Rio Grande do Norte (UFRN) (1998–2004), with research experience in sleep, memory, and anxiety. He earned an MSc in neuroscience at the Federal University of São Paulo (2005–2007), studying the influence of biological rhythms on sleep and memory. He received a Ph.D. in neuroscience from UFRN (2008–2012), researching the epidemiological and neurophysiological aspects of lucid dreaming. In his current position as postdoctoral researcher (2013–2017) at the Brain Institute, and the Onofre Lopes University Hospital (UFRN), he is working with sleep, dreams, lucid dreaming, the neurobiological basis of music perception, and the use of DMT for the treatment of depression.

Jessica L. Nielson is a jointly appointed tenure-track assistant professor in the psychiatry department and the Institute for Health Informatics at the University of Minnesota Twin Cities. Jessica received her Ph.D. from UC Irvine in 2010 in anatomy and neurobiology, and her postdoctoral training in bioinformatics and multivariate statistics at UC San Francisco. Her research involves working with animal and human data repositories for archived and ongoing trials aimed at precision diagnosis and treatment of several neurological disorders, including spinal cord injury (SCI), traumatic brain injury (TBI), post-traumatic stress disorder (PTSD), and schizophrenia. Jessica is also currently gathering preliminary data through an anonymous online survey to assess user-reported risks and benefits of ayahuasca therapy across various contexts, specifically related to the treatment of PTSD, depression, and substance abuse.

Fernanda Palhano-Fontes is an electrical engineer with double degrees: one from the Federal University of Rio Grande do Norte (UFRN) and another from the École National Supérieure d'Electrotechnique, d'Electronique, d'Informatique (ENSEEIHT), Toulouse, France. She earned a master's and a Ph.D. in neurosciences from the Brain Institute of UFRN. In the master's program, she used functional magnetic resonance imaging (fMRI) to evaluate the acute effect of ayahuasca. In her doctoral thesis, she investigated the therapeutic potential of ayahuasca in patients with treatment-resistant depression. Her main areas of interest are psychedelics, psychiatry, and imaging techniques such as fMRI and electroencephalography.

Juan Carlos Pascual is a psychiatrist that works at Sant Pau Hospital in Barcelona. He is affiliated with the Borderline Personality Disorder Unit of the Psychiatry Department. He is also associate professor in the Department of Psychiatry and Forensic Medicine at the Autonomous University of Barcelona. He is also a researcher in the Spanish Research Network on Mental Health (CIBESAM). He has published more than 70 peer-reviewed papers related to borderline personality disorder (BPD). He has collaborated and led several financed projects on BPD targeting, genetics, neuroimaging, and pharmacological treatments.

Ana Pérez-Castillo received her Ph.D. in the field of biochemistry from the Complutense University of Madrid. Afterwards, she was awarded Fulbright and Fogarty postdoctoral fellowships to work at the Department of Medicine of the University of Minnesota. She is currently serving as a research professor at the Spanish Research Council (CSIC). She also belongs to the Spanish Research Network on Neurodegenerative Diseases (CIBERNED), an institution encompassed in the International Network of Centers of Excellence in Neurodegeneration. Her work has centered on molecular biology approaches to study pathologies of the central nervous system, specifically Parkinson's and Alzheimer's diseases. Prof. Perez-Castillo and her coworkers have identified several genes involved in these neurodegenerative disorders. She is also an expert in the field of neurogenesis. She has authored more than 150 papers, many in several of the most recognized peer-reviewed journals. She has made relevant contributions to the identification of novel therapeutic targets and treatment strategies for neurodegenerative disorders.

Daniel Perkins is a research fellow in the School of Social and Political Science at the University of Melbourne, where he is the founder and director of a global multidisciplinary research project (https://www.globalayahuascaproject.org) that is examining the use and health effects of the Amazonian psychoactive tea ayahuasca in traditional and nontraditional settings around the world. This project involves a collaboration of researchers from Australia, Brazil, Spain, Switzerland, and the Czech Republic. He is Manager of the Research, Strategy and Policy Unit in the Office of Medical Cannabis with the Victorian Department of Health and Human Services. Daniel has over 15 years of experience leading health and social research and policy projects that have included partnerships with universities, NGOs, businesses, and all levels of government. He is also a psychotherapist and neurofeedback practitioner.

Marika Renelli is currently an M.A. candidate in the applied psychology program at Laurentian University. She holds as M.Sc. in microbiology and a B.Sc. in the biological sciences from the University of Guelph. She collaborated on a project that explores the role of ceremonial ayahuasca on the healing of eating disorders.

Jordi Riba holds a Ph.D. in pharmacology. He leads the Human Neuropsychopharmacology Research Group at Sant Pau Hospital in Barcelona, and is a member of the Spanish Research Network on Mental Health (CIBESAM). He has a broad interest in psychoactive drugs, with publications on psychedelics, psychostimulants, cannabinoids, and kappa receptor agonists. He has been studying ayahuasca for over 15 years and has published nearly 40 journal articles and book chapters on the subject. He has also supervised three doctoral dissertations on the acute and long-term effects and therapeutic potential of ayahuasca in humans. He has also collaborated on the first clinical studies involving ayahuasca administration to patients with depression. His current research deals with the post-acute psychedelic "after-glow" and the use of ayahuasca in the treatment of various psychiatric conditions. He is also investigating the neuroprotective and neurogenic potential of ayahuasca alkaloids. Initial data obtained from studies in animals have revealed that several active principles present in the tea protect brain cells from hypoxia and stimulate the birth of new neurons in adult mice. These stunning results open a whole new avenue of research for ayahuasca. Potential applications of its active principles range from depression, to neurodegenerative disorders, to neural deficits associated with hypoxia and trauma.

Sidarta Tollendal Gomes Ribeiro is a Brazilian neuroscientist and Director of the Brain Institute at Universidade Federal do Rio Grande do Norte (UFRN), which he joined in 2008 as Full Professor. From 2009 to 2011, he served as Secretary of the Brazilian Society for Neuroscience and Behavior. Sidarta Ribeiro is currently the Chair of the Regional Committee in Brazil of the Pew Latin American Fellows Program in the Biomedical Sciences. He is also member of the Steering Committee of the Latin American School for Educational, Cognitive and Neural Sciences.

Emilia Sanabria, who has a Ph.D. in anthropology from the University of Cambridge, is a lecturer at Ecole Normale Supérieure de Lyon and senior research fellow at the Amsterdam Institute for Social Science Research (UVA). She has published widely on embodiment, gender, and people's relationship to drugs and pharmaceuticals. Her monograph, *Plastic Bodies* (2016), was published by Duke University Press. Her current research examines the circulation, translation, and reinvention of ayahuasca as a healing modality.

Jerome Sarris is Professor of Integrative Mental Health and Deputy Director of the National Institute for Complementary Medicine at Western Sydney University, and holds an honorary position as an associate professor in the Department of Psychiatry at the University of Melbourne. Jerome moved from clinical practice as a naturopath to academic work, and completed a doctorate at the field of psychiatry. He has particular interest in anxiety and mood disorder research pertaining to integrative medicine and psychotropic plant medicines, with over 120 publications in many eminent journals, including *The American Journal of Psychiatry, Lancet Psychiatry*, and *World Psychiatry*. Jerome is currently involved in over a dozen clinical trials and is founding Vice Chair of The International Network of Integrative Mental Health.

Milan Scheidegger holds an M.D. and completed his Ph.D. in functional and molecular neuroimaging at the Institute for Biomedical Engineering (University and ETH Zurich). As a resident physician at the Department of Psychiatry, Psychotherapy, and Psychosomatics at the University Hospital of Psychiatry in Zurich, he is currently researching the neurobiology and pharmacology of altered states of consciousness. He is member of the Swiss Society for Psycholytic Therapy (SAEPT) and investigates the potential of psychedelics such as ketamine, psilocybin, ayahuasca, and DMT to facilitate therapeutic transformation. On his ethnobotanical expeditions to Mexico, Colombia, and Brazil he explored the traditional use of psychoactive plants in shamanic rituals. In addition to empirical research, he earned an M.A. in history and philosophy of knowledge (ETH Zurich). His main interests include biosemiotics, epistemology, and phenomenology of consciousness, mindfulness, and deep ecology.

Bruno Lobão-Soares is a veterinary doctor, and has received a Ph.D. in neuroscience from the University of São Paulo. He is an assistant professor at Federal University of Rio Grande do Norte (UFRN). His current research is related to dopaminergic regulation of sleep and memory, and to new neuropharmacological tools in humans and in animal models.

Joaquim Soler is a senior clinical psychologist working in the Borderline Personality Disorder Unit of the Psychiatry Department of Sant Pau Hospital in Barcelona. He is also Associate Professor in the Department of Psychiatry and Forensic Medicine at the Autonomous University of Barcelona. He has been trained as a DBT and MBCT therapist. He has collaborated and led several publicly funded projects on borderline personality disorders (BPD), dealing with patient assessment, genetics, and the implementation of psychological and pharmacological treatments. He has published more than 60 peer-reviewed papers on BPD, depression, and mindfulness. Dr. Soler is also a researcher in the Center of Research in Mental Health Network. He is also a principal investigator of the Mindfulness Group in the Excellence Network for the Dissemination of Psychological Treatments for Mental Health Promotion in Spain. He has collaborated with the National Health Department and with international and national scientific societies. Currently, he is the president of the Spanish DBT Society.

Piera Talin is an anthropologist, junior researcher at the Amsterdam Institute for Social Science Research (UVA), and member of the ERC Advanced Grant Program "Chemical Youth: What chemicals do for youths in their everyday lives" (ERC-2012-AdvG-323646), headed by Anita Hardon at the University of Amsterdam. She holds a master's degree in cultural anthropology at the University Cà Foscari, in Venice (Italy), with a thesis on the Santo Daime community in Florianópolis (Brazil), in collaboration with the Federal University of Santa Catarina. In 2017, she, with co-author Emilia Sanabria, published the article, "Ayahuasca's Entwined Efficacy: An Ethnographic Study of Ritual Healing from 'Addiction'," in the *International Journal of Drug Policy*.

Gerald Thomas is director of Alcohol, Tobacco, Cannabis, and Gambling Prevention and Policy at the British Columbia Ministry of Health, collaborating scientist with the Centre for Addictions Research of BC, adjunct professor in the Department of Psychology at University of British Columbia, and owner and operator of Okanagan Research Consultants. He received his doctorate in political science from Colorado State University in 1998, and has worked in the area of Canadian addiction policy since 2004. He served on the secretariat of the working group that created Canada's first National Alcohol Strategy in 2007, worked on several national and provincial level projects related to substance use and addiction, and has published numerous peer-reviewed papers with leading researchers in the field.

Luis Fernando Tófoli, Ph.D., is Professor of Psychiatry at the Faculty of Medical Sciences of the University of Campinas (UNICAMP), Brazil. He heads the Laboratory of Interdisciplinary Studies on Psychoactive Substances and is a member of the State Council on Drug Policies of São Paulo. He is responsible for the Interdisciplinary Cooperation for Ayahuasca Research and Outreach (ICARO) at UNICAMP, and has recently published on the field of drug policies and the therapeutic use of psychedelics, especially ayahuasca.

Kenneth W. Tupper, Ph.D., is director of Implementation and Partnerships at the BC Centre on Substance Use and adjunct professor in the School of Population and Public Health at the University of British Columbia. His scholarly interests include: psychedelic studies; the cross-cultural and historical uses of psychoactive substances; public, professional, and school-based drug education; and creating healthy public policy for currently illegal drugs. Kenneth's doctoral research developed the concept of "entheogenic education," a theoretical frame for understanding how psychedelic plants and substances can function as cognitive tools for learning.

Marta Valle, Pharmacist with PhD in Pharmacology from the University of the Basque Country, works as Scientific Director in the Clinical Pharmacology, Modeling and Simulation Group of Parexel International. Previously, she has been Head of the Pharmacokinetic/Pharmacodynamic Modeling and Simulation Group at the Research Institut of Hospital de la Santa Creu i Sant Pau, where she collaborated closely in numerous clinical studies of human neuropsychopharmacology, with special interest in drugs of abuse such as ayahuasca. cannabis or salvinorin A. For almost 10 years she was associate professor in the Department of Pharmacology, Therapeutics and Toxicology at the Autonomous University of Barcelona and Coordinator of the Pharmacology PhD Program at the same university.

Chapter 1
Ayahuasca as a Versatile Therapeutic Agent: From Molecules to Metacognition and Back

Marta Valle, Elisabet Domínguez-Clavé, Matilde Elices, Juan Carlos Pascual, Joaquim Soler, José A. Morales-García, Ana Pérez-Castillo, and Jordi Riba

N,N-Dimethyltryptamine, β-carboline Alkaloids, and the Biochemical Effects of Ayahuasca

One of the most common versions of ayahuasca tea available globally comes from a mix of *Banisteriopsis caapi* (Malpighiaceae) and *Psychotria viridis* (Rubiaceae) (McKenna, Towers & Abbott, 1984). Based on published reports (Dos Santos et al., 2011, 2012; Riba et al., 2001, 2003; Sanches et al., 2016), this combination appears to be reasonably safe, in terms of its physiological impact, when administered to healthy individuals.

 B. caapi contains β-carboline alkaloids; mainly harmine, tetrahydroharmine (THH), and to a lesser extent, harmaline (McKenna et al. 1984; Rivier & Lindgren, 1972). *P. viridis* contains N,N-Dimethyltryptamine (DMT), which is orally inactive

M. Valle
Pharmacokinetic and Pharmacodynamic Modelling and Simulation, Hospital de la Santa Creu i Sant Pau, Barcelona, Spain
e-mail: mvallec@santpau.cat

E. Domínguez-Clavé
Department of Pharmacology, Therapeutics and Toxicology, UAB, Birmingham, AL, USA

Borderline Personality Disorder Unit, Psychiatry Service, Hospital de la Santa Creu i Sant Pau, Barcelona, Spain
e-mail: edominguezcl@santpau.cat

M. Elices
Borderline Personality Disorder Unit, Psychiatry Service, Hospital de la Santa Creu i Sant Pau, Barcelona, Spain

J. C. Pascual (✉) · J. Soler
Pharmacokinetic and Pharmacodynamic Modelling and Simulation, Hospital de la Santa Creu i Sant Pau, Barcelona, Spain

Department of Psychiatry and Forensic Medicine, UAB, Birmingham, AL, USA
e-mail: jpascual@santpau.cat; jsolerri@santpau.cat

© Springer Nature Switzerland AG 2021
B. C. Labate, C. Cavnar (eds.), *Ayahuasca Healing and Science*,
https://doi.org/10.1007/978-3-030-55688-4_1

due to its transformation by monoamine oxidase (MAO) activity in the gastrointestinal tract. Since β-carboline alkaloids are MAO inhibitors (Ott, 1999; Wang et al., 2010), they prevent DMT degradation, thereby allowing DMT to enter the general circulatory and central nervous systems (Bouso & Riba, 2014; McKenna, et al. 1984), where they can produce psychoactive effects (Szára, 1956).

The indole psychedelic DMT has long been considered to play a major role in the pharmacology of ayahuasca. DMT shows serotoninergic agonist activity at the 5-HT2A and 5-HT1A receptors sites (González-Maeso & Sealfon, 2009) and induces brief but intense modifications of the ordinary state of awareness (Strassman, Qualls, Uhlenhuth & Kellner, 1994; Riba et al., 2002). DMT also interacts with the intracellular sigma-1 receptor (S1R) (Fontanilla et al., 2009), which modulates the activity of other proteins and promotes neural plasticity (Chu & Ruoho, 2016; Tsai et al., 2009).

In mice, DMT blocks sodium channels and induces hypermobility (Fontanilla et al., 2009). DMT is also an agonist at the trace amine associated receptor (TAAR) (Bunzow et al., 2001), which has been reported to have a potential role in schizophrenia, fibromyalgia, affective disorders, addiction, drug abuse, and Parkinson's disease (Berry, Gainetdinov, Hoener & Shahid, 2017). TAAR was initially discovered through a search for novel 5HT receptors (Borowsky et al., 2001). DMT is also a substrate of the vesicle monoamine and serotonin transporters (Cozzi et al., 2009). A recent study (Szabo et al., 2016) found that DMT-mediated S1R mitigates hypoxic stress of in vitro-cultured human cortical neurons, monocyte-derived macrophages, and dendritic cells, and increases their survival. Hence, downregulation of S1R abrogates DMT-mediated effects on cellular survival and hypoxia-inducible factor 1-alpha (HIF-1a) expression in hypoxic in vitro cultures of human primary cells.

J. A. Morales-García
Center for Networked Biomedical Research on Neurodegenerative Diseases (CIBERNED), Madrid, Spain

Department of Cellular Biology, School of Medicine, Universidad Complutense de Madrid, Madrid, Spain
e-mail: jmorales@iib.uam.es

A. Pérez-Castillo
Center for Networked Biomedical Research on Neurodegenerative Diseases (CIBERNED), Madrid, Spain

Instituto de Investigaciones Biomédicas (CSIC-UAM), Madrid, Spain
e-mail: aperez@iib.uam.es

J. Riba (deceased)
Human Neuropsychopharmacology Research Group, Hospital de la Santa Creu i Sant Pau, Barcelona, Spain

Department of Neuropsychology and Psychopharmacology, Masstricht University, Maastricht, The Netherlands
e-mail: jordi.riba@maastrichtuniversity.nl

Another proposed mechanism for DMT-mediated S1R modulation is that S1R activation would, in turn, modulate Ca2+ signaling, thereby altering the function of intracellular kinases involved in cellular survival. These results suggest a novel and important role for DMT in human cellular physiology, indicating the potential value of DMT-mediated S1R modulation in future therapies targeted at hypoxia/ischemia-related pathologies.

The main constituent of the *B. caapi* vine and the principal ingredient—together with DMT—of ayahuasca preparation is the β-carboline alkaloid harmine. It is still unclear if orally ingested harmine in ayahuasca or related β-carboline alkaloids (THH and harmaline), produces psychoactive or hallucinogenic effects (Dos Santos & Hallak, 2016; Dos Santos, Bouso & Hallak, 2017). In fact, studies in humans have shown inconclusive results, describing sedative-like effects or even a lack of psychoactive effects instead of hallucinogenic (see Dos Santos, Bouso & Hallak 2017). Interestingly, a study conducted in healthy volunteers showed that the β-carbolines (including harmine) are associated with specific electroencephalographic (EEG) alterations, suggesting central/psychoactive effects (Schenberg et al., 2015).

In rodents, harmine improves memory and learning-related deficits in an object recognition task (short-term memory), the MM test (spatial memory), and a delayed match-to-sample asymmetrical 3-choice water maze task. This has been supported by improvements in several biochemical parameters in the rodent hippocampus and in hippocampal cell cultures. Besides, harmine could also induce neuroprotective effects by inhibiting DYRK1A, a protein implicated in neuronal development and apparently involved in abnormal brain development and memory and learning deficits in Alzheimer's and Parkinson's disease and in Down syndrome (Frost et al., 2011; Göckler et al., 2009). According to a recent in vitro study (Li et al., 2017), both harmaline and harmine could potentially be used to treat Alzheimer's disease, Parkinson's disease, depression, and other central nervous system diseases. However, harmine appears to be a multidrug resistance-associated protein isoform2 (MRP2) substrate (it would upregulate its expression) and would be easier metabolized than harmaline, eventually leading to low oral bioavailability (Li et al., 2017). Other preclinical studies show that β-carbolines have affinity for 5-HT2A/2 C receptors (Glennon et al., 2000).

It is not clear if the antidepressant and antiaddictive properties of ayahuasca are related to the pharmacological properties of harmine or DMT, or to a combination of these (Dos Santos, Osório, Crippa, & Hallak, 2016a; Dos Santos et al., 2016b). Future research should assess the role of other molecular mechanisms, such as S1R agonism, in the perceptual, affective, and cognitive effects of DMT and ayahuasca. Given that certain antidepressants (e.g., fluvoxamine) stimulate the S1R, it is plausible that the antidepressant effects recently reported for ayahuasca (Osório et al., 2015; Palhano-Fontes et al., 2017; Sanches et al., 2016) are mediated, at least in part, by S1R agonism.

Neurobiological Effects of Ayahuasca

Some areas implicated in cognitive control, emotion, and memory, such as the insula, the amygdala, and the hippocampus, have been reported to increase blood flow under the effects of ayahuasca (Riba et al., 2006). Other studies have identified the mediotemporal lobe (MTL), which includes the hippocampus, amygdala, and parahippocampal regions, as a target of ayahuasca-induced experience (de Araujo et al., 2012; Riba, Anderer, Jané, Saletu, & Barbanoj, 2004; Riba et al., 2006). The posterior cingulated cortex (PCC) is a key node of the default mode network (DMN) (Raichle et al., 2001). Hyperactivity in this region has been associated with psychopathology (e.g., rumination in depression) (Dutta et al., 2014). DMN is a set of functionally and structurally connected brain regions that are typically deactivated during the performance of externally oriented attention-demanding tasks. This area also exhibits high cerebral blood flow and oxygen consumption during the resting state (Lin et al., 2017).

During the acute phase of ayahuasca, studies have shown that most of the DMN exhibits decreased activity and the PCC shows reduced connectivity (Palhano-Fontes et al., 2015). The spatial brain distribution of ayahuasca-induced changes in the brain has been assessed with low-resolution electromagnetic tomography (LORETA) and EEG recordings. Riba, et al. (2004) reported a decrease of power density in the alpha-2, delta, and beta-1 frequency bands predominantly over the temporo-parieto-occipital junction, whereas theta power was reduced in the temporomedial cortex and in frontomedial regions after ayahuasca intake. These areas comprise the unimodal association cortex, multimodal association cortex, and limbic regions involved in the integration of multimodal sensory information, emotion, and memory processes (Riba et al., 2004).

Ayahuasca has also showed an excitatory effect on cortical regions involved in the processing of visual sensory information (alpha-occipital), memory-affect (MTL), and cognition-affect (theta-frontolateral and frontomedial cortex) (Valle et al., 2016). Randomized clinical trials with experienced psychedelic users suggest a prominent role for the 5-HT2A receptor in the neurophysiological and visionary effects of this substance. Ayahuasca induced significant psychedelic effects and power decreases in the delta-alpha frequency range; decreases in alpha-band oscillations were in posterior brain regions and correlated with the intensity of the visual modifications. The placebo (ketanserin) used in that study blocked these decreases and reduced this correlation.

Another study of our group (Alonso, Romero, Mananas, & Riba, 2015) assessed ayahuasca-induced changes in the dynamic interaction of brain oscillations and the directionality of drug-induced modifications using transfer entropy (TE). That study consisted of 10 healthy volunteers with previous experience in psychedelic drug use. The authors found a reduced density in the medial posterior (precuneus and cuneus) areas and in the ACC.

Similar findings have been observed using magnetoencephalography (MEG) after psilocybin administration, which modified the interaction dynamics between the higher order frontal regions and the more sensory-selective posterior areas (Muthukumaraswamy et al., 2013). 5-HT2A agonism is considered a key mechanism in modifying ayahuasca-induced brain dynamics (McKenna & Riba, 2015). Modifications in the information transfer imply that the predictability of activity in posterior areas (based on information available at anterior sites) decreases while, conversely, the predictability of activity in anterior areas increases (when information is taken into account at posterior sites). Furthermore, in the study by Alonso et al. (2015), these neurophysiological changes, which were related to a reduced top-down control and increased bottom-up information transfer in the brain, were correlated with DMT plasma concentrations and the perceived intensity of psychedelic effects.

The effects of ayahuasca on long-term ayahuasca users have also been examined. The magnetic resonance imaging (MRI) study of Bouso et al. (2015) found an association with opposite changes in brain structure in the anterior and posterior cingulate cortices. Findings included: (a) a decrease in cortical thickness in the PCC and neighboring areas, and (b) an increase in cortical thickness in the medial frontal lobes, specifically in the ACC.

Studies with radiotracer data—mainly Single Photon Emission Computed Tomography (SPECT)— have also been carried out. In a randomized double-blind clinical trial (Riba et al., 2006) increased blood perfusion in the anterior insula (bilaterally) was reported, with greater intensity in the right hemisphere, and in the anterior cingulate/frontomedial cortex (another region of the DMN) of the right hemisphere. Additional increases were observed in the left amygdala/parahippocampal gyrus. Another study (Sanches et al., 2016) found increased blood perfusion in the left nucleus accumbens, the right insula and the left subgenual area; all areas implicated in the regulation of mood and emotions. There are no reports examining the effects of ayahuasca with Positron Emission Tomography (PET) up to date.

The visual network is an area rich in 5-HT2A receptors (Savli et al., 2012). After ayahuasca intake, users are aware that the visions which disappear when opening their eyes and when attention is directed to external cues, are drug-induced (Riba et al., 2001). The visionary phenomena experienced by participants with eyes closed may be attributable to the suppression of inhibitory alpha in the visual network (Valle et al., 2016). Despite their vividness, these images clearly differ from "true hallucinations" (Riba et al., 2001), a state of awareness characterized by dream-like visions more than a pathological distortion of perception. A previous neuroimaging study found increased activity in the visual cortex under ayahuasca (de Araujo et al., 2012). According to Sampedro et al. (2017), visual areas show increased coupling with the PCC but reduced coupling with the ACC. Those authors suggested that there was a greater interplay between internally generated visual information and spontaneous mind-wandering, and a decrease in cognitive control.

Potential Therapeutic Uses of Ayahuasca

Although the full potential of ayahuasca remains unknown, a growing body of evidence accumulated over the last two decades suggests that it may have potential benefits to treat depression. This hypothesis is supported by data on the modulatory capacity of prefrontal 5-HT2A receptors on the amygdala and ACC reported in previous studies (Vollenweider & Kometer 2010). Numerous studies in animals (Aricioglu & Altunbas 2003; Farzin & Mansouri 2006; Fortunato et al., 2009, 2010a, b; Lima et al., 2007; Pic-Taylor et al., 2015; Réus et al., 2010, 2012) and humans (Barbosa Giglio & Dalgalarrondo, 2005; dos Santos, Landeira-Fernandez, Strassman, Motta & Cruz, 2007) have investigated this potential effect of ayahuasca. Several recent studies have evaluated the effects of a single dose of ayahuasca in psychiatric depressive inpatients. Osório et al. (2015) observed a significant reduction in depressive symptoms and anxiolytic effects in six resistant-to-treatment depressive patients during the first week. Those authors suggested that ayahuasca might have an earlier onset of action than traditional antidepressants that have a mean onset of therapeutic action of 2 weeks. In a subsequent study with a larger sample size (n = 17), that same group found significant decreases in depression scale scores, which occurred as soon as 80 minutes after intake up to day 21 post-intake (Sanches et al., 2016). The SPECT analysis revealed increased blood perfusion in brain regions implicated in the regulation of mood and emotional states, as mentioned before.

Recently, Palhano-Fontes et al. (2017) conducted a double-blind placebo-controlled trial (RCT) in treatment-resistant depressive patients to compare a single dosing session of ayahuasca to placebo, and found that the single dosing session showed rapid antidepressant effects. Between-group effect sizes increased from day 1 to 7, and response rates were significantly higher in the ayahuasca group at day 7. Remission rate showed a trend toward significance between groups (36% v. 7%, $p = 0.054$). As discussed above, S1R, which upregulates brain derived neurotrophic factor (BDNF) and nerve growth factor (NGF), has been implicated in depression. The regulation and expression of BDNF and NGF have been recently linked to the pathophysiology and treatment of depression (Otte et al., 2016), data that suggest that DMT-mediated S1R modulation will likely play a key role in efforts to identify future therapies to treat depression and related disorders.

Studies of anxiety symptoms in animal models (Aricioglu & Altunbas, 2003) have found that harmane (another β-carboline) diminishes anxious behaviors in the elevated plus maze (EPM) test; Hilber and Chapillon (2005) reported mixed results for harmaline in that same EPM anxiety test. Pic-Taylor et al. (2015) reported decreases in locomotor and exploratory activities in the open field and EPM tests (similar to fluoxetine). Those authors also observed increased c-fos expression in specific brain areas, confirming that ayahuasca alkaloids affect areas involved in emotional processing.

In humans, Barbosa, Giglio & Dalgalarrondo (2005) reported a decrease in anxiety associated symptoms after the first-time ritual use of ayahuasca among members of a religious group (Santo Daime) in Brazil. They also observed self-reported behavioral changes, such as increased assertiveness, vivacity, and joy. A case-control study (dos Santos et al., 2007) reported lower scores on panic- and hopelessness-related scales among individuals who took ayahuasca, but no changes in state or trait anxiety.

In recent years, promising findings about the potential effects of ayahuasca on addiction disorders have been reported. Oliveira-Lima et al. (2015) developed a substance use disorder (SUD) model in animals. Those researchers induced hyper-locomotion using ethanol, leading to locomotor sensitization. The results of that study showed that ayahuasca not only inhibited early behaviors associated with the initiation and development of ethanol addiction, but was also effective for reversing the behavioral sensitization associated with chronic ethanol administration.

In humans, effects of ayahuasca on substance use were reported in two case-control studies (Fabregas et al., 2010; Grob et al., 1996). Grob et al. (1996) reported remission of alcohol use in a sample of 15 long-term ayahuasca users compared to 15 matched controls with no prior history of ayahuasca ingestion. Fabregas et al. (2010) reported a reduction in alcohol use and cessation of drug use (except for cannabis) in two groups of jungle and urban-based ayahuasca users compared to non-ayahuasca users, with outcomes maintained at one-year follow-up. Halpern, Sherwood, Passie, Blackwell & Ruttenber (2008) reported remission of drug or alcohol abuse/dependence in a community sample of ayahuasca users (average length of membership, 6.5 years). In another case series (Thomas, Lucas, Capler, Tupper & Martin, 2013), Thomas and colleagues found statistically significant reductions in cocaine use after an ayahuasca-assisted intervention in a sample of members of a First Nations community in Canada who had no prior experience with ayahuasca. Other descriptive studies (Bouso & Riba, 2014; Doering-Silveira et al., 2005a, b; Labate, dos Santos, Strassman, Anderson & Mizumoto, 2014) have presented preliminary evidence suggesting a potentially beneficial role for ayahuasca in the treatment of SUD.

Given the growing body of evidence on the potential therapeutic use of aya-huasca in depression, anxiety and SUD, there is increasing interest in testing whether ayahuasca can be used to treat a broader range of related disorders. In this context, in a recent review (Domínguez-Clavé et al., 2016), our group postulated that ayahuasca could also be valuable in treating other impulse-related disorders, personality disorders, and even trauma.

Psychological Mechanisms Underlying the Therapeutic Effects of Ayahuasca

In recent years, our group has published research suggesting that a psychological mechanism—an increase in mindfulness-related capabilities (e.g., Soler et al., 2016)—could underlie the potential therapeutic effects of ayahuasca. In our first study, 25 individuals completed the Five Facets Mindfulness Questionnaire (FFMQ) and the Experiences Questionnaire (EQ) before and 24 h after an ayahuasca session. We found significant reductions in two facets of mindfulness, "nonjudging" and "non-reacting." A decrease in nonjudging indicates a reduced tendency to be evaluative and judgmental; that is, the person is less likely to dichotomize experiences into either "good" or "bad." Improvements in nonreacting indicate decreased reactivity to private experiences such as thoughts and feelings, regardless of whether those are pleasant or unpleasant.

The study also found significant increases in "decentering" as measured by the EQ. Decentering is considered a product of mindfulness practice. This concept has been defined as "the ability to observe one's thoughts and feelings in a detached manner, as temporary events in the mind, as neither necessarily true nor reflections of the self" (Safran & Segal, 1990). Previous reports (Bergomi, Stroehle, Michalak, Funke, & Berking, 2013; Soler et al., 2014) have found that meditation practice is beneficial for all three of the aforementioned facets (i.e., nonjudging, nonreactivity, and decentering).

However, improvement in decentering can be achieved by therapeutic interventions other than mindfulness practice, such as Acceptance and Commitment Therapy (ACT) (Hayes, Strosahl & Wilson, 1999) and Metacognitive-Based Therapy (MBT, Wells, 2009). These therapies focus on decentering, which plays a key role in contributing to their beneficial effects (Moritz et al. 2011; Van der Heiden, Muris & van der Molen, 2012; Wells et al., 2010). Franquesa et al. (2018) compared ayahuasca-naive subjects ($n = 41$) to experienced users ($n = 81$), finding that ayahuasca users scored higher on decentering measures than nonusers, even when the most and least experienced users were compared. By improving decentering, individuals gain mastery over their thoughts and emotions, minimizing or even eliminating the tendency to identify with them (Shapiro, Carlson, Astin & Freedman, 2006).

Studies have found that the capacity to "decenter" may be protective against suicidal ideation and that an individual's decentering ability is predictive of the intensity of depressive symptoms at a 6-month follow-up (Bieling et al. 2012; Hargus, Crane, Barnhofer & Williams, 2010). In this regard, the efficacy of cognitive behavior therapy (CBT) in the treatment of depression may rely on increasing this decentering capacity (Teasdale, Segal & Williams, 1995). In the context of these findings, studies have found that improving an individual's decentering capacity could directly improve depression (Bieling et al., 2012; Fresco et al., 2007a, Fresco, Segal, Buis & Kennedy, 2007b; Gecht et al. 2014; Hargus et al., 2010; Teasdale et al., 2002), generalized anxiety disorder (Hayes-Skelton, Calloway, Roemer, & Orsillo, 2015; Hoge et al., 2015), social anxiety (Hayes-Skelton & Graham, 2013), eating disorders, SUD (Shapiro, et al. 2006), and borderline personality disorders (BPD) (Soler et al., 2014). In impulsive-related disorders (such as drug abuse or BPD), an increase in the decentering capacity may diminish mood-dependent behavior by interrupting recurring maladaptive habits (Shapiro et al., 2006).

Regarding mindfulness-related capacities, ayahuasca intake seems to induce a pattern of change similar to that produced by mindfulness practice. In fact, the two main facets that improve after ayahuasca use—nonjudging and nonreacting to inner experience (Soler et al., 2016)—make up the acceptance-measuring components of the FFMQ (Baer, Smith, Hopkins, Krietemeyer & Toney, 2006).

A recently published neuroimaging study examined the effects of ayahuasca in a sample of healthy volunteers with prior experience using the substance (Sampedro et al., 2017). That study found increases in the nonjudging subscale and found that post-acute neural changes predicted sustained elevations on the nonjudging subscale at 2 months post-intake. An even more recent study (Soler et al., 2018), which compared mindfulness training to ayahuasca, found that ayahuasca induced increases in the nonjudging scale that were comparable to those of mindfulness. Based on these studies, it appears that only a few ayahuasca sessions may be as effective at improving acceptance as costly interventions. Improving this capacity allows a more detached and less judgmental stance towards potentially distressing thoughts and emotions.

In a recent observational study (Domínguez-Clavé et al., 2019), our group examined the effects of ayahuasca in a sample of volunteers participating in an ayahuasca ceremony. Of the 45 participants, 12 exhibited BPD-like traits and were allocated to a subgroup. After comparing both samples (non-BPD-like and BPD-like), we found that the BPD-trait subgroup showed significant pre-post difference in the emotion regulation (ER) subscales of Difficulties in Emotion Regulation Scale (DERS) *Lack of Control* and *Emotional Interference*. The non-BPD traits subgroup showed significant pre-post differences for those same two subscales, as well as for DERS *Emotional Non-acceptance*. The non-BPD subgroup also showed significant effects on most FFMQ subscales (*Observing*, A*cting with Awareness*, *Nonjudging* and *Nonreacting*) and in decentering (EQ).

That study was the first to examine the effects of ayahuasca on ER and the first to use a subsample of individuals with BPD-like traits. Those findings led us to believe that ayahuasca therapy could be of value in clinical populations affected by emotion dysregulation. Moreover, an exploratory study conducted by members of our team in 45 volunteers found that ayahuasca use positively influenced measures of self-compassion and self-criticism assessed 24 h before and after ayahuasca intake.

An improved understanding of this connection between ayahuasca-induced experience and mindfulness techniques could help to better characterize the therapeutic effects of ayahuasca. If ayahuasca enhances ER—or more specifically, acceptance or self-compassion—it could conceivably also exert a therapeutic effect on individuals with mental disorders by targeting ER capacities. Future studies should address the benefits of combining ayahuasca with mindfulness-related practices to provide a more focused approach to better treat specific disorders. The influence of ayahuasca on these psychological mechanisms suggests its potential to also treat trauma-related conditions and other disorders such as BPD (Bohus Dyer, Priebe, Krueger, & Steil, 2011; Bohus et al., 2013; Harner & Burgess, 2011; Harner, Budescu, Gillihan, Riley & Foa, 2015), obsessive-compulsive disorder, and phobias, in a structured, safe, and comfortable setting.

Metabolic and Connectivity Changes and Mindfulness-Related Capacities

The term "after-glow" designates the positive post-acute effects of psychedelic drugs characterized by elevated mood and openness (Pahnke, Kurland, Unger, Savage & Grof, 1970). Sampedro et al. (2017) has investigated the post-acute neurometabolic and connectivity changes induced by ayahuasca. Neuroimaging techniques (1H-magnetic resonance spectroscopy and functional connectivity) revealed post-acute reductions in glutamate+glutamine (Glx), creatine, and N-acetylaspartate+N-acetylaspartylglutamate (NAA-NAAG) in the PCC. Additionally, reductions in Glx correlated with increases in the FFMQ nonjudging subscale and were maintained after 2 months. The inverse correlation between Cr and NAA-NAAG and the scores on the Hallucinogen Rating Scale (HRS)-cognition subscale suggested a relationship between the intensity of acute effects and subsequent neurometabolic reductions. Moreover, connectivity between the ACC and PCC and between the ACC and limbic structures in the right MTL increased. The results suggest that cross-talk lingers beyond the acute stage and contributes to the "after-glow" that is reflected in enhanced mindfulness capacities. These findings confirm previous data regarding the capacity of ayahuasca to enhance mindfulness capacities, including increased "decentering" and decreased judgmental and reactive attitudes (Soler et al., 2016).

Interestingly, in a study published in 2014 (Soler et al., 2014), the authors found that ayahuasca can even increase mindfulness and self-compassion capabilities in individuals who already have high baseline capabilities. In that study, ayahuasca users had higher post-acute scores than meditators. Additionally, Sampedro and colleagues' study showed increased DMN-TPN connectivity correlated with reduced judgmental processing, inner reactivity, and increased self-kindness, providing a neurobiological basis for these modifications. These findings also points to a potential biological basis for therapeutic benefits. Conventional mindfulness training also increases this cross-talk between networks (Doll et al., 2015).

The potential of ayahuasca to influence brain dynamics at multiple levels suggests its potential to treat disorders that are highly refractory to other therapeutic interventions. Its combined effect on the psychological and neural spheres may be particularly well-suited to treating addiction disorders, where high impulsivity and self-centeredness coexist with alterations in brain function and structure.

B. Caapi, β-carbolines, and Neurogenesis

Given that β-carbolines, unlike DMT, are present in all ayahuasca brews, Morales-García et al. (2017) recently evaluated these metabolites to investigate the capacity of harmine, THH, harmaline, and harmol to induce neurogenesis in vitro using neural progenitor cells from adult mice. In the adult brain of mammals, neurogenesis—

the process of generating functional neurons from progenitor cells—is limited to specific brain regions such as the subventricular zone (SVZ) of the lateral ventricle and the subgranular zone of the dentate gyrus of the hippocampus (SGZ). Although this system in the adult brain is considered relatively robust, the number of neural stem cells (NSC) progressively decreases with age and in neurodegenerative diseases.

Impaired adult neurogenesis in neurodegenerative diseases, in addition to losing existing neurons, involves brain's endogenous loss of capacity for cell renewal and therefore the impairment and/or loss of putative function of these new neurons. Morales-García and colleagues showed that *B. caapi* alkaloids stimulate the proliferation and migration of progenitor cells and promote differentiation (predominantly) into neurons. These alkaloids increase the number and size of primary neurospheres (neural stem cell-enriched spheres), induce the loss of their undifferentiated state, and promote subsequent cell migration and differentiation into a neuronal phenotype (as indicated by the positive expression of the neuronal markers β-III-tubulin and MAP 2), as well as astrocytes. Taken together, these three effects indicate that β-carbolines have the capacity to regulate the expansion and fate of stem cell populations. The largest effects on migration have been observed for harmaline and THH. Increased migration capacity is relevant in certain conditions such as brain injury, where stem cell niches are far from the damaged area.

Neural stem cells can differentiate into neurons, astrocytes, and oligodendrocytes. In the aforementioned study, the observed increase in Tuj-1 and MAP-2 protein expression indicated differentiation, mainly towards a neuronal phenotype (Fig. 1.1). In the SVZ, both proteins were equally expressed after treatment with each of the four compounds (harmine, THH, harmaline, and harmol). However, in the SGZ, harmine administration did not influence Tuj-1 levels, a marker of immature neurons, but did significantly increase the expression of MAP-2, suggesting that harmine has a larger impact on neuronal maturation. Importantly, the magnitude of the neurogenic effects was similar for the four alkaloids, whose versatility is of interest given that, in pathological conditions, these compounds could optimize the replacement of neurons by simultaneously acting on various processes.

A likely possible explanation for the observed effects of β-carbolines on neurogenesis is the increase in monoamine levels caused by MAO inhibition. However, other studies reported neurogenesis as independent of elevated serotonin levels. A recent study (Song et al., 2016) found that serotonin depletion appears to promote, rather than decrease, hippocampal neurogenesis. Another group of researchers (Dakic et al., 2016) found that harmine—but not the MAO inhibitor pargyline—stimulated the proliferation of human neural progenitor cells in vitro. In that case, the effects of harmine were mediated through inhibition of the DYRK1A kinase rather than through MAO inhibition. These results suggest that β-carbolines regulate stem cell fate via DYRK1A or other alternative mechanisms.

The association between neurogenesis and anti-depressant activity is well-documented. Clinically effective antidepressants have been reported to stimulate this process, regardless of their specific chemical structure and mechanism of action.

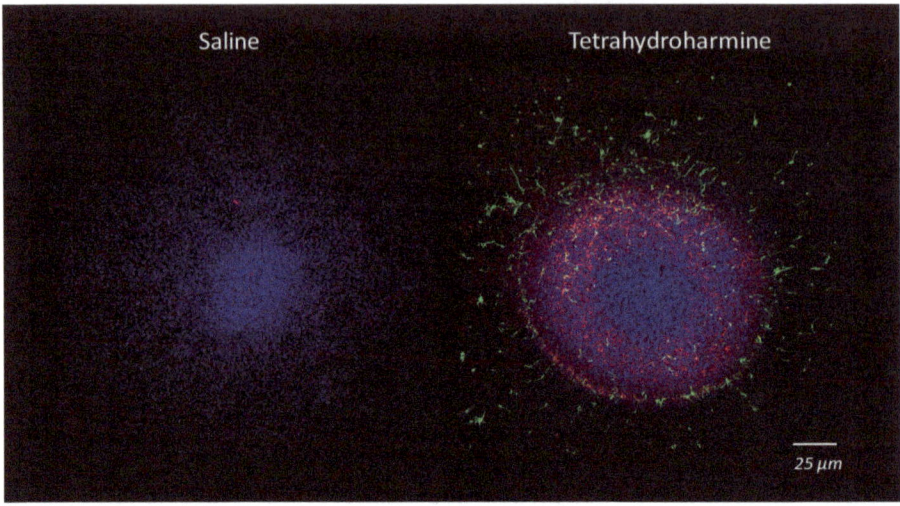

Fig. 1.1 Ayahuasca β-carboline alkaloids induce neurogenesis in vitro, promoting stem cell differentiation towards a neuronal phenotype. Rodent neural stem cells were isolated from one of the most important adult neurogenic niches, the subgranular zone of the hippocampal dentate gyrus, and cultured as free-floating neurospheres in the presence of tetrahydroharmine (THH). After 7 days, neurospheres were adhered on coated coverslips and allowed to differentiate for 3 days in the presence of THH
The figure shows triple confocal immunofluorescence images showing the expression of the neuronal markers β-III-Tubulin (TuJ-1 clone, green, early neurogenesis) and microtubule-associated protein 2 (MAP-2, red, mature neurons) in control (left) and treated (right) neurospheres. DAPI was used for nuclear staining. Scale bar = 25 μm
The left half of the figure shows results after saline (no neurogenesis)
The right half of the figure shows results after THH (evidence of early neurogenesis and presence of mature neurons)

However, it is still unclear if enhanced hippocampal neurogenesis reduces depression-like behavior (Dean & Keshavan, 2017; Hill, Sahay & Hen, 2015).

More research is needed to determine the true magnitude of the therapeutic potential of ayahuasca in mental health and to better identify the mechanisms that mediate the action of *B. caapi* alkaloids.

Conclusions

Recent years have witnessed an explosion in the number of studies on the effects of psychedelics. In this chapter, we have described the latest research on ayahuasca. Highly promising research results showing that (a) DMT can protect neurons from hypoxia; (b) the serotonin-2A receptor mediates the visual effects of ayahuasca; and (c) ayahuasca intake can diminish pathological patterns of thought and behavior, such as those observed in depression and addiction. Here, we have discussed how these modified patterns result from the enhancement of metacognitive capacities that allow individuals to observe their own thoughts and emotions in a healthier and less judgmental manner.

The recent discoveries described in this chapter about new mechanisms of action and psychological capacities of ayahuasca, such as its impact on facets of mindfulness, decentering, and emotional regulation, open new horizons in terms of potential therapeutic interventions for mental disorders. Given the important effects of ayahuasca on many regions of the brain—as shown by neuroimaging studies and neurophysiologic techniques—the optimal approach may be to combine ayahuasca with psychological interventions, as a combined treatment may yield better treatment outcomes. This chapter has also described the neural correlates underlying these psychological benefits, which have been identified through a combination of state-of-the-art neuroimaging techniques. Finally, at the cellular level, we have described how the β-carbolines present in ayahuasca stimulate the formation of new neurons in the adult mammal brain.

Acknowledgement This chapter is dedicated to the memory of Jordi Riba, whose inspiration and contribution to the work of our team has been incommensurable.

References

Alonso, J. F., Romero, S., Mananas, M. A., & Riba, J. (2015). Serotonergic psychedelics temporarily modify information transfer in humans. *International Journal of Neuropsychopharmacology, 18*, 1–9.

Aricioglu, F., & Altunbas, H. (2003). Harmane induces anxiolysis and antidepressant-like effects in rats. *Agmatine and Imidazolines: Their Novel Receptors and Enzymes, 1009*, 196–200.

Baer, R. A., Smith, G. T., Hopkins, J., Krietemeyer, J., & Toney, L. (2006). Using self-report assessment methods to explore facets of mindfulness. *Assessment, 13*, 27–45.

Barbosa, P. C. R., Giglio, J. S., & Dalgalarrondo, P. (2005). Altered states of consciousness and short-term psychological after-effects induced by the first time ritual use of ayahuasca in an urban context in Brazil. *Journal of Psychoactive Drugs, 37*(2), 193–201.

Bergomi, C., Stroehle, G., Michalak, J., Funke, F., & Berking, M. (2013). Facing the dreaded: Does mindfulness facilitate coping with distressing experiences? A moderator analysis. *Cognitive Behavior Therapy, 42*, 21–30.

Berry, M. D., Gainetdinov, R. R., Hoener, M. C., & Shahid, M. (2017). Pharmacology of human trace amine-associated receptors: Therapeutic opportunities and challenges. *Pharmacology & Therapeutics, 180*, 161–180.

Bieling, P. J., Hawley, L. L., Bloch, R. T., Corcoran, K. M., Levitan, R. D., Young, L. T., et al. (2012). Treatment-specific changes in decentering following mindfulness-based cognitive therapy versus antidepressant medication or placebo for prevention of depressive relapse. *Journal of Consulting and Clinical Psychology, 80*, 365–372.

Bohus, M., Dyer, A. S., Priebe, K., Krueger, A., & Steil, R. (2011). Dialectical behavior therapy for posttraumatic stress disorder in survivors of childhood sexual abuse. *Psychotherapie, Psychosomatik, Medizinische Psychologie, 61*, 140–147.

Bohus, M., Dyer, A. S., Priebe, K., Krueger, A., Kleindinst, N., Schmahl, C., et al. (2013). Dialectical behaviour therapy for post-traumatic stress disorder after childhood sexual abuse in patients with and without borderline personality disorder: A randomised controlled trial. *Psychotherapy and Psychosomatics, 82*, 221–233.

Borowsky, B., Adham, N., Jones, K. A., Raddaz, R., Artymyshyn, R., Ogozalek, K. L., et al. (2001). Trace amines: Identification of a family of mammalian G protein-coupled receptors. *Proceedings of the National Academy of Science, 98*(16), 8966–8971.

Bouso, J. C., & Riba, J. (2014). Ayahuasca and the treatment of drug addiction. In B. C. Labate & C. Cavnar (Eds.), *The therapeutic use of ayahuasca* (pp. 95–109). New York, NY: Springer.

Bouso, J., Palhano-Fontes, F., Rodríguez-Fornells, A., Ribeiro, S., Sanches, R., Crippa, J., et al. (2015). Long-term use of psychedelic drugs is associated with differences in brain structure and personality in humans. *European Neuropsychopharmacology, 25*, 483–492.

Bunzow, J. R., Sonders, M. S., Arttamangkul, S., Harrison, L. M., Zhang, G., Quigley, D. I., et al. (2001). Amphetamine, 3,4-methylenedioxymethamphetamine, lysergic acid diethylamide, and metabolites of the catecholamine neurotransmitters are agonists of a rat trace amine receptor. *Molecular Pharmacology, 60*(6), 1181–1188.

Chu, U. B., & Ruoho, A. E. (2016). Biochemical pharmacology of the sigma-1 receptor. *Molecular Pharmacology, 89*(1), 142–153.

Cozzi, N. V., Gopalakrishnan, A., Anderson, L. L., Feih, J. T., Shulgin, A. T., Daley, P. F., & Ruoho, A. E. (2009). Dimethyltryptamine and other hallucinogenic tryptamines exhibit substrate behavior at the serotonin uptake transporter and the vesicle monoamine transporter. *Journal of Neural Transmission, 116*(12), 1591–1599.

Dakic, V., Maciel, R. M., Drummond, H., Nascimento, J. M., Trindade, P., & Rehen, S. K. (2016). Harmine stimulates proliferation of human neural progenitors. *PeerJ, 4*, e2727.

Dean, J., & Keshavan, M. (2017). The neurobiology of depression: An integrated view. *Asian Journal of Psychiatry, 27*, 101–111. https://doi.org/10.1016/j.ajp.2017.01.025.

de Araujo, D. B., Ribeiro, S., Cecchi, G. A., Carvalho, F. M., Sanchez, T. A., Pinto, J. P., et al. (2012). Seeing with the eyes shut: Neural basis of enhanced imagery following ayahuasca ingestion. *Human Brain Mapping, 33*(11), 2550–2560. https://doi.org/10.1002/hbm.21381.

Doering-Silveira, E., Grob, C. S., de Rios, M. D., Lopez, E., Alonso, L. K., Tacla, C., & Da Silveira, D. X. (2005a). Report on psychoactive drug use among adolescents using ayahuasca within a religious context. *Journal of Psychoactive Drugs, 37*(2), 141–144.

Doering-Silveira, E., Lopez, E., Grob, C. S., de Rios, M. D., Alonso, L. K., Tacla, C., et al. (2005b). Ayahuasca in adolescence: A neuropsychological assessment. *Journal of Psychoactive Drugs, 37*, 123–128.

Doll, A., Hölzel, B. K., Boucard, C. C., Wohlschläger, A. M., & Sorg, C. (2015). Mindfulness is associated with intrinsic functional connectivity between default mode and salience networks. *Frontiers in Human Neuroscience, 9*, 461. https://doi.org/10.3389/fnhum.2015.00461.

Domínguez-Clavé, E., Soler, J., Elices, M., Pascual, J. C., Álvarez, E., de la Fuente Revenga, M., et al. (2016). Ayahuasca: Pharmacology, neuroscience, and therapeutic potential. *Brain Research Bulletin, 126*, 89–101.

Domínguez-Clavé, E., Soler, J., Pascual, J.C., Elices, M., Franquesa, A., Valle, M., et al. (2019) Ayahuasca improves emotion dysregulation in a community sample and in individuals with borderline-like traits. *Psychopharmacology (Berl), 236*(2), 573–580. https://doi.org/10.1007/s00213-018-5085-3.

Dos Santos, R. G., Bouso, J. C., & Hallak, J. E. C. (2017). Ayahuasca, dimethyltryptamine, and psychosis: A systematic review of human studies. *Therapeutic Advances in Psychopharmacology, 7*(4), 141–157.

Dos Santos, R. G., Landeira-Fernandez, J., Strassman, R. J., Motta, V., & Cruz, A. P. M. (2007). Effects of ayahuasca on psychometric measures of anxiety, panic-like and hopelessness in Santo Daime members. *Journal of Ethnopharmacology, 112*(3), 507–513.

Dos Santos, R. G., Valle, M., Bouso, J. C., Nomdedéu, J. F., Rodriguez-Espinosa, J., McIlhenny, E. H., et al. (2011). Autonomic, neuroendocrine, and immunological effects of ayahuasca: A comparative study with d-amphetamine. *Journal of Clinical Psychopharmacology, 31*(6), 717–726.

Dos Santos, R. G., Grasa, E., Valle, M., Ballester, M. R., Bouso, J. C., Nomdedéu, J. F., et al. (2012). Pharmacology of ayahuasca administered in two repeated doses. *Psychopharmacology, 219*(4), 1039–1053.

Dos Santos, R. G., & Hallak, J. E. (2016). Effects of the natural β-carboline alkaloid harmine, a main constituent of ayahuasca, in memory and in the hippocampus: A systematic literature review of preclinical studies. *Journal of Psychoactive Drugs, 49*(1), 1–10.

Dos Santos, R. G., Osório, F. L., Crippa, J. A., & Hallak, J. E. (2016a). Antidepressive and anxiolytic effects of ayahuasca: A systematic literature review of animal and human studies. *Revista Brasileira de Psiquiatria, 38*(1), 65–72.

Dos Santos, R. G., Osório, F. L., Crippa, J. A., Riba, J., Zuardi, A. W., & Hallak, J. E. (2016b). Antidepressive, anxiolytic, and antiaddictive effects of ayahuasca, psilocybin and lysergic acid diethylamide (LSD): A systematic review of clinical trials published in the last 25 years. *Therapeutic Advances in Psychopharmacology, 6*(3), 193–213.

Dutta, A., McKie, S. & Deakin, J. F. W. (2014). Resting state networks in major depressive disorder. *Psychiatry research, 224*, 139–151.

Fabregas, J. M., Gonzalez, D., Fondevila, S., Cutchet, M., Fernandez, X., Barbosa, P. C. R., et al. (2010). Assessment of addiction severity among ritual users of ayahuasca. *Drug and Alcohol Dependence, 111*(3), 257–261.

Farzin, D., & Mansouri, N. (2006). Antidepressant-like effect of harmane and other beta-carbolines in the mouse forced swim test. *European Neuropsychopharmacology, 16*(5), 324–328.

Fontanilla, D., Johannessen, M., Hajipour, A., Cozzi, N., Jackson, M., & Ruoho, A. (2009). The hallucinogen *N,N*-dimethyltryptamine (DMT) is an endogenous sigma-1 receptor regulator. *Science, 323*(5916), 934–937.

Fortunato, J. J., Reus, G. Z., Kirsch, T. R., Stringari, R. B., Stertz, L., Kapczinski, F., et al. (2009). Acute harmine administration induces antidepressive-like effects and increases BDNF levels in the rat hippocampus. *Progress in Neuro-Psychopharmacology & Biological Psychiatry, 33*, 1425–1430.

Fortunato, J. J., Reus, G. Z., Kirsch, T. R., Stringari, R. B., Fries, G. R., Kapczinski, F., et al. (2010a). Effects of beta-carboline harmine on behavioral and physiological parameters observed in the chronic mild stress model: Further evidence of antidepressant properties. *Brain Research Bulletin, 81*(4–5), 491–496.

Fortunato, J. J., Reus, G. Z., Kirsch, T. R., Stringari, R. B., Fries, G. R., Kapczinski, F., et al. (2010b). Chronic administration of harmine elicits antidepressant-like effects and increases BDNF levels in rat hippocampus. *Journal of Neural Transmission, 117*(10), 1131–1137.

Franquesa, A., Sainz-Cort, A., Gandy, S., Soler, J., Alcázar-Córcoles, M. Á., & Bouso, J. C. (2018). Psychological variables implied in the therapeutic effect of ayahuasca: A contextual approach. *Psychiatry Research, 264*, 334–339.

Fresco, D. M., Moore, M. T., van Dulmen, M. H. M., Segal, Z. V., Ma, S. H., Teasdale, J. D., & Williams, J. M. G. (2007a). Initial psychometric properties of the experiences questionnaire: Validation of a self-report measure of decentering. *Behavior Therapy, 38*(3), 234–246.

Fresco, D. M., Segal, Z. V., Buis, T., & Kennedy, S. (2007b). Relationship of posttreatment decentering and cognitive reactivity to relapse in major depression. *Journal of Consulting and Clinical Psychology, 75*(3), 447–455.

Frost, D., Meechoovet, B., Wang, T., Gately, S., Giorgetti, M., Shcherbakova, I., & Dunckley, T. (2011). β-carboline compounds, including harmine, inhibit DYRK1A and tau phosphorylation at multiple Alzheimer's disease-related sites. *PLoS One, 6*(5), e19264.

Gecht, J., Kessel, R., Forkmann, T., Gauggel, S., Drueke, B., Scherer, A., & Mainz, V. (2014). A mediation model of mindfulness and decentering: Sequential psychological constructs or one and the same? *BMC Psychology, 2*, 18. https://doi.org/10.1186/2050-7283-2-18.

Glennon, R.A., Dukat, M., Grella, B., Hong, S., Costantino, L., Teitler, M., et al. (2000). Binding of beta-carbolines and related agents at serotonin (5-HT(2) and 5-HT(1A)), dopamine (D(2)) and benzodiazepine receptors. *Drug and alcohol dependence, 60*(2), 121–32. https://doi.org/10.1016/s0376-8716(99)00148-9.

Göckler, N., Jofre, G., Papadopoulos, C., Soppa, U., Tejedor, F.J., Becker, W. (2009). Harmine specifically inhibits protein kinase DYRK1A and interferes with neurite formation. *The FEBS journal, 276*(21), 6324–37. https://doi.org/10.1111/j.1742-4658.2009.07346.x.

González-Maeso, J., & Sealfon, S. C. (2009). Agonist-trafficking and hallucinogens. *Current Medicinal Chemistry, 16*(8), 1017–1027.

Grob, C. S., McKenna, D. J., Callaway, J. C., Brito, G. S., Neves, E. S., Oberlaender, G., et al. (1996). Human psychopharmacology of hoasca, a plant hallucinogen used in ritual context in Brazil. *Journal of Nervous and Mental Disease, 184*(2), 86–94.

Halpern, J. H., Sherwood, A. R., Passie, T., Blackwell, K. C., & Ruttenber, A. J. (2008). Evidence of health and safety in American members of a religion who use a hallucinogenic sacrament. *Medical Science Monitor, 14*(8), SR15–SR22.

Hargus, E., Crane, C., Barnhofer, T., & Williams, J. M. G. (2010). Effects of mindfulness on meta-awareness and specificity of describing prodromal symptoms in suicidal depression. *Emotion, 10*(1), 34–42.

Harner, H., & Burgess, A. W. (2011). Using a trauma-informed framework to care for incarcerated women. *Journal of Obstetric, Gynecologic, and Neonatal Nursing, 40*(4), 469–476.

Harner, H. M., Budescu, M., Gillihan, S. J., Riley, S., & Foa, E. B. (2015). Posttraumatic stress disorder in incarcerated women: A call for evidence-based treatment. *Psychological Trauma, 7*(1), 58–66.

Hayes, S. C., Strosahl, K. D., & Wilson, K. G. (1999). *Acceptance and commitment therapy: An experiential approach to behavior change*. New York, NY: Guilford Press.

Hayes-Skelton, S., & Graham, J. (2013). Decentering as a common link among mindfulness, cognitive reappraisal, and social anxiety. *Behavioral and Cognitive Psychotherapy, 41*(3), 317–328.

Hayes-Skelton, S. A., Calloway, A., Roemer, L., & Orsillo, S. M. (2015). Decentering as a potential common mechanism across two therapies for generalized anxiety disorder. *Journal of Consulting and Clinical Psychology, 83*(2), 395–404.

Hilber, P., Chapillon, P. (2005). Effects of harmaline on anxiety-related behavior in mice. *Physiology and Behavior, 86*(1-2), 164–7. https://doi.org/10.1016/j.physbeh.2005.07.006.

Hill, A. S., Sahay, A., & Hen, R. (2015). Increasing adult hippocampal neurogenesis is sufficient to reduce anxiety and depression-like behaviors. *Neuropsychopharmacology, 40*(10), 2368–2378. https://doi.org/10.1038/npp.2015.85.

Hoge, E. A., Bui, E., Goetter, E., Robinaugh, D. J., Ojserkis, R. A., Fresco, D. M., & Simon, N. M. (2015). Change in decentering mediates improvement in anxiety in mindfulness-based stress reduction for generalized anxiety disorder. *Cognitive Therapy Research, 39*(2), 228–235.

Labate, B. C., dos Santos, R. G., Strassman, R., Anderson, B. T., & Mizumoto, S. (2014). Effect of Santo Daime membership on substance dependence. In B. C. Labate & C. Cavnar (Eds.), *The therapeutic use of ayahuasca* (p. 153). New York, NY: Springer.

Li, S., Zhang, Y., Deng, G., Wang, Y., Qi, S., Cheng, X., et al. (2017). Exposure characteristics of the analogous β-carboline alkaloids harmaline and harmine based on the efflux transporter of multidrug resistance protein 2. *Frontiers in Pharmacology, 8*, 541. https://doi.org/10.3389/fphar.2017.00541.

Lima, L., Ferreira, S. M., Avila, A. L., Perazzo, F. F., Schneedorf, J. M., Hinsberger, A., & Carvalho, J. C. T. (2007). Les effets de l'ayahuasca sur le systeme nerveux central: etude comportementale [Ayahuasca central nervous system effects: Behavioral study]. *Phytothérapie, 5*(5), 254–257.

Lin, P., Yang, Y., Gao, J., De Pisapia, N., Ge, S., Wang, X., et al. (2017). Dynamic default mode network across different brain states. *Scientific Reports, 7*, 46088.

McKenna, D. J., Towers, G. H., & Abbott, F. (1984). Monoamine oxidase inhibitors in South American hallucinogenic plants: Tryptamine and beta-carboline constituents of ayahuasca. *Journal of Ethnopharmacology, 10*(2), 195–223.

McKenna, D., & Riba, J. (2015). New World tryptamine hallucinogens and the neuroscience of ayahuasca. *Current Topics in Behavioral Neurosciences, 36*, 283–311.

Morales-García, J. A., de la Fuente Revenga, M., Alonso-Gil, S., Rodríguez-Franco, M. I., Feilding, A., Perez-Castillo, A., & Riba, J. (2017). The alkaloids of *Banisteriopsis caapi*, the plant source of the Amazonian hallucinogen ayahuasca, stimulate adult neurogenesis in vitro. *Scientific Reports, 7*, 5309.

Moritz, S., Kerstan, A., Veckenstedt, R., Randjbar, S., Vitzthum, F., Schmidt, C., et al. (2011). Further evidence for the efficacy of a metacognitive group training in schizophrenia. *Behaviour Research and Therapy, 49*(3), 151–157.

Muthukumaraswamy, S., Carhart-Harris, R., Moran, R., Brookes, M., Williams, T., Errtizoe, D., et al. (2013). Broadband cortical desynchronization underlies the human psychedelic state. *Journal of Neuroscience, 33*, 15171–15183.

Oliveira-Lima, A. J., dos Santos, R., Hollais, A. W., Gerardi-Junior, C. A., Baldaia, M. A., Wuo-Silva, R., et al. (2015). Effects of ayahuasca on the development of ethanol-induced behavioral sensitization and on a post-sensitization treatment in mice. *Physiology and Behavior, 142*, 28–36.

Osório, F. d. L., Sanches, R. F., Macedo, L. R., Santos, R. G., Maia-de-Oliveira, J. P., Wichert-Ana, L., et al. (2015). Antidepressant effects of a single dose of ayahuasca in patients with recurrent depression: A preliminary report. *Revista Brasileira de Psiquiatria, 37*(1), 13–20.

Ott, J. (1999). Pharmahuasca: Human pharmacology of oral DMT plus harmine. *Journal of Psychoactive Drugs, 31*(2), 171–177.

Otte, C., Gold, S. M., Penninx, B. W., Pariante, C. M., Etkin, A., Fava, M., et al. (2016). Major depressive disorder. *Nature Reviews. Disease Primers, 2*, 16065.

Pahnke, W. N., Kurland, A. A., Unger, S., Savage, C., & Grof, S. (1970). The experimental use of psychedelic (LSD) psychotherapy. *Journal of the American Medical Association, 212*(11), 1856–1863.

Palhano-Fontes, F., Andrade, K.C., Tofoli, L.F., Santos, A.C., Crippa, J.A., Hallak, J.E., et al. (2015). The psychedelic state induced by ayahuasca modulates the activity and connectivity of the default mode network. *PLoS One,10*(2):e0118143. https://doi.org/10.1371/journal.pone.0118143.

Palhano-Fontes, F., Barreto, D., Onias, H., Andrade, K. C., Novaes, M., Pessoa, J. A., et al. (2017). A randomized placebo-controlled trial on the antidepressant effects of the psychedelic ayahuasca in treatment-resistant depression. *BioRxiv*. Advance online publication. https://doi.org/10.1017/S0033291718001356.

Pic-Taylor, A., da Motta, L. G., de Morais, J. A., Junior, W. M., Santos Ade, F., Campos, L. A., et al. (2015). Behavioural and neurotoxic effects of ayahuasca infusion (*Banisteriopsis caapi* and *Psychotria viridis*) in female wistar rat. *Behavioral Processes, 118*, 102–110.

Raichle, M.E., MacLeod, A.M., Snyder, A.Z., Powers, W.J., Gusnard, D.A., Shulman, G.L. (2001). A default mode of brain function. *Proceedings of the National Academy of Sciences of the United States of America, 98*(2), 676–82. https://doi.org/10.1073/pnas.98.2.676.

Réus, G. Z., Stringari, R. B., de Souza, B., Petronilho, F., Dal-Pizzol, F., Hallak, J. E., et al. (2010). Harmine and imipramine promote antioxidant activities in prefrontal cortex and hippocampus. *Oxidative Medicine and Cellular Longevity, 3*(5), 325–331.

Réus, G. Z., Stringari, R. B., Gonçalves, C. L., Scaini, G., Carvalho-Silva, M., Jeremias, G. C., et al. (2012). Administration of harmine and imipramine alters creatine kinase and mitochondrial respiratory chain activities in the rat brain. *Depression Research and Treatment, 2012*, 987397. https://doi.org/10.1155/2012/987397.

Riba, J., Rodríguez-Fornells, A., Urbano, G., Morte, A., Antonijoan, R., Montero, M., et al. (2001). Subjective effects and tolerability of the South American psychoactive beverage ayahuasca in healthy volunteers. *Psychopharmacology, 154*(1), 85–95.

Riba, J., Anderer, P., Morte, A., Urbano, G., Jane, F., Saletu, B., & Barbanoj, M. J. (2002). Topographic pharmaco-EEG mapping of the effects of the South American psychoactive beverage ayahuasca in healthy volunteers. *British Journal of Clinical Pharmacology, 53*(6), 613–628.

Riba, J., Valle, M., Urbano, G., Yritia, M., Morte, A., & Barbanoj, M. J. (2003). Human pharmacology of ayahuasca: Subjective and cardiovascular effects, monoamine metabolite excretion, and pharmacokinetics. *Journal of Pharmacology and Experimental Therapeutics, 306*(1), 73–83.

Riba, J., Anderer, P., Jané, F., Saletu, B., & Barbanoj, M. J. (2004). Effects of the South American psychoactive beverage ayahuasca on regional brain electrical activity in humans: A functional neuroimaging study using low-resolution electromagnetic tomography. *Neuropsychobiology, 50*(1), 89–101.

Riba, J., Romero, S., Grasa, E., Mena, E., Carrió, I., & Barbanoj, M. J. (2006). Increased frontal and paralimbic activation following ayahuasca, the pan-Amazonian inebriant. *Psychopharmacology, 186*(1), 93–98.

Rivier, L., & Lindgren, J. E. (1972). "Ayahuasca," the South American hallucinogenic drink: An ethnobotanical and chemical investigation. *Economic Botany, 26*(2), 101–129.

Safran, J. D., & Segal, Z. V. (1990). *Interpersonal process in cognitive therapy*. Lanham, MD: Jason Aronson.

Sampedro, F., de la Fuente Revenga, M., Valle, M., Roberto, N., Domínguez-Clavé, E., Elices, M., et al. (2017). Assessing the psychedelic "after-glow" in ayahuasca users: Post-acute neurometabolic and functional connectivity changes are associated with enhanced mindfulness capacities. *International Journal of Neuropsychopharmacology, 20*(9), 698–711.

Sanches, R. F., de Lima Osório, F., Dos Santos, R. G., Macedo, L. R., Maia-de-Oliveira, J. P., Wichert-Ana, L., et al. (2016). Antidepressant effects of a single dose of ayahuasca in patients with recurrent depression a SPECT Study. *Journal of Clinical Psychopharmacology, 36*(1), 77–81.

Savli, M., Bauer, A., Mitterhauser, M., Ding, Y. S., Hahn, A., Kroll, T., et al. (2012). Normative database of the serotonergic system in healthy subjects using multi-tracer PET. *NeuroImage, 63*(1), 447–459.

Schenberg, E.E., Alexandre, J.F., Filev, R., Cravo, A.M., Sato, J.R., Muthukumaraswamy, S.D., et al. (2015). Acute Biphasic Effects of Ayahuasca. *PLoS one, 10*(9):e0137202. https://doi.org/10.1371/journal.pone.0137202.

Shapiro, S. L., Carlson, L. E., Astin, J. A., & Freedman, B. (2006). Mechanisms of mindfulness. *Journal of Clinical Psychology, 62*(3), 373–386.

Soler, J., Franquesa, A., Feliu-Soler, A., Cebolla, A., Garcia-Campayo, J., Tejedor, R., et al. (2014). Assessing decentering: Validation, psychometric properties, and clinical usefulness of the experiences questionnaire in a Spanish sample. *Behavior Therapy, 45*(6), 863–871.

Soler, J., Elices, M., Franquesa, A., Barker, S., Friedlander, P., Feilding, A., & Riba, J. (2016). Exploring the therapeutic potential of ayahuasca: Acute intake increases mindfulness-related capacities. *Psychopharmacology, 233*(5), 823–829.

Soler, J., Elices, M., Dominguez-Clavé, E., Pascual, J. C., Feilding, A., Navarro-Gil, M., et al. (2018). Four weekly ayahuasca sessions lead to increases in "acceptance" capacities: A comparison study with a standard 8-week mindfulness training program. *Frontiers in Pharmacology, 9*, 224.

Song, N. N., Jia, Y. F., Zhang, L., Zhang, Q., Huang, Y., Liu, X. Z., et al. (2016). Reducing central serotonin in adulthood promotes hippocampal neurogenesis. *Scientific Reports, 6*, 20338.

Strassman, R., Qualls, C., Uhlenhuth, E., & Kellner, R. (1994). Dose-response study of *N,N*-dimethyltryptamine in humans. II. Subjective effects and preliminary results of a new rating scale. *Archives of General Psychiatry, 51*(2), 98–108.

Szabo, A., Kovacs, A., Riba, J., Djurovic, S., Rajnavolgyi, E., & Frecska, E. (2016). The endogenous hallucinogen and trace amine N,N-Dimethyltryptamine (DMT) displays potent protective effects against hypoxia via Sigma-1 receptor activation in human primary iPSC-derived cortical neurons and microglia-like immune cells. *Frontiers in Neuroscience, 10*, 423. https://doi.org/10.3389/fnins.2016.00423.

Szára, S. (1956). Dimethyltryptamine: Its metabolism in man; the relation to its psychotic effect to the serotonin metabolism. *Experientia, 12*, 441–442.

Teasdale, J. D., Segal, Z., & Williams, M. J. (1995). How does cognitive therapy prevent depressive relapse and why should attentional control (mindfulness) training help? *Behavior Research and Therapy, 33*(1), 25–39.

Teasdale, J. D., Moore, R. G., Hayhurst, H., Pope, M., Williams, S., & Segal, Z. V. (2002). Metacognitive awareness and prevention of relapse in depression: Empirical evidence. *Journal of Consulting and Clinical Psychology, 70*(2), 275–287.

Thomas, G., Lucas, P., Capler, N. R., Tupper, K. W., & Martin, G. (2013). Ayahuasca-assisted therapy for addiction: Results from a preliminary observational study in Canada. *Current Drug Abuse Reviews, 6*(1), 30–42.

Tsai, S. Y., Hayashi, T., Harvey, B. K., Wang, Y., Wu, W. W., Shen, R. F., et al. (2009). Sigma-1 receptors regulate hippocampal dendritic spine formation via a free radical-sensitive mechanism involving Rac1×GTP pathway. *Proceedings of the National Academy of Sciences of the United States of America, 106*, 22468–22473.

Valle, M., Maqueda, A. E., Rabella, M., Rodríguez-Pujadas, A., Antonijoan, R. M., Romero, S., et al. (2016). Inhibition of alpha oscillations through serotonin-2A receptor activation underlies the visual effects of ayahuasca in humans. *European Neuropsychopharmacology, 26*(7), 1161–1175.

van der Heiden, C., Muris, P., & van der Molen, H. T. (2012). Randomized controlled trial on the effectiveness of metacognitive therapy and intolerance-of-uncertainty therapy for generalized anxiety disorder. *Behaviour Research and Therapy, 50*(2), 100–109.

Vollenweider, F. X., & Kometer, M. (2010). The neurobiology of psychedelic drugs: Implications for the treatment of mood disorders. *Nature Reviews in Neuroscience, 11*(9), 642–651.

Wang, Y. H., Samoylenko, V., Tekwani, B. L., Khan, I. A., Miller, L. S., Chaurasiya, N. D., et al. (2010). Composition, standardization, and chemical profiling of *Banisteriopsis caapi*, a plant for the treatment of neurodegenerative disorders relevant to Parkinson's disease. *Journal of Ethnopharmacology, 128*(3), 662–671.

Wells, Adrian. (2009). *Metacognitive Therapy for Anxiety and Depression*. Guilford Press.

Wells, A., Welford, M., King, P., Papageorgiou, C., Wisely, J., & Mendel, E. (2010). A pilot randomized trial of metacognitive therapy vs applied relaxation in the treatment of adults with generalized anxiety disorder. *Behaviour Research and Therapy, 48*(5), 429–434.

Chapter 2
Recent Evidence on the Antidepressant Effects of Ayahuasca

Fernanda Palhano-Fontes, Sérgio Mota-Rolim, Bruno Lobão-Soares, Nicole Galvão-Coelho, Joao Paulo Maia-Oliveira, and Dráulio B. Araújo

What is Major Depressive Disorder?

According to the World Health Organization, approximately 350 million people meet the clinical criteria for major depressive disorder (MDD), which is the main cause of disability worldwide and is significantly linked to suicidal risk (World Health Organization, 2017). According to the Diagnostic and Statistical Manual of Mental Disorders (DSM-V), MDD is characterized by the presence of depressed mood (irritable mood in children and adolescents) and/or anhedonia for at least 2 weeks, together with at least four of the following possible symptoms: (a) considerable change in weight, (b) insomnia or hypersomnia, (c) psychomotor agitation or retardation, (d) fatigue/loss of energy, (e) low self-esteem or inappropriate guilt, (f) diminished capacity to think, concentrate, or make decisions, (g) recurrent thoughts

F. Palhano-Fontes · S. Mota-Rolim · D. B. Araújo (✉)
Brain Institute, Federal University of Rio Grande do Norte, Natal-RN, Brazil

Onofre Lopes University Hospital, Federal University of Rio Grande do Norte, Natal-RN, Brazil
e-mail: draulio@neuro.ufrn.br

B. Lobão-Soares
Department of Biophysics and Pharmacology, Federal University of Rio Grande do Norte, Natal-RN, Brazil

N. Galvão-Coelho
Laboratory of Hormone Measurement, Department of Physiology and Behavior, Federal University of Rio Grande do Norte, Natal-RN, Brazil

J. P. Maia-Oliveira
Onofre Lopes University Hospital, Federal University of Rio Grande do Norte, Natal-RN, Brazil

Department of Clinical Medicine, Federal University of Rio Grande do Norte, Natal-RN, Brazil

© Springer Nature Switzerland AG 2021
B. C. Labate, C. Cavnar (eds.), *Ayahuasca Healing and Science*,
https://doi.org/10.1007/978-3-030-55688-4_2

21

of death, suicidal ideation, or suicide attempts (American Psychiatric Association, 2013).

Many hypotheses have been proposed to explain the physiology of depression, and it is becoming more evident that depression is a multidimensional condition (Liu et al., 2017). The most widely accepted hypothesis states that depression is a result of disturbances in serotonin (5-HT), noradrenaline, and dopamine neurotransmitters in the brain (Wong & Licinio, 2001). In fact, the use of antidepressant drugs, such as iproniazid or imipramine, has been associated with increased levels of 5-HT and noradrenaline (Wong & Licinio, 2001). Besides, low concentrations of postsynaptic serotonin 1A receptor (5-HT$_{1A}$) in the hippocampus and prefrontal cortex have been reported in patients with depression (Stahl, 2013). Changes in the 5-HT transporter were also observed in depression, as studies with monozygotic twins show that some types of polymorphism in the promoter region of its gene are more likely to develop depression (Caspi et al., 2010; Frodl et al., 2010; Starr et al., 2012).

The hypothalamic–pituitary–adrenal (HPA) axis has also been used to explain some aspects of the pathophysiology of depression, mainly based on the frequent observations of altered cortisol levels in depressive patients (Tadić et al., 2011). One view of the problem points to resistance in glucocorticoid receptor (GR), which leads to decreased negative feedback of the HPA axis, and thus result in elevated concentrations of cortisol in the plasma, urine, and cerebrospinal fluid (Zunszain et al., 2011). Although hypercortisolemia is the most common finding, hypocortisolemia has also been observed in MDD (Bremmer et al., 2007; Vreeburg et al., 2013). In this case, reduced cortisol levels would be a consequence of HPA axis fatigue, due to the recurrent depressive episodes. In fact, hypocortisolemia has proven to be dependent on depression severity, socioeconomic status, sex, age, temperament, and coping style (Moreira et al., 2016; Schuch et al., 2014; Tu et al., 2013).

Another hypothesis associates depression with changes in the immune system. Some depressive patients present increased levels of interleukin 6 (IL-6), tumor necrosis factor-alpha (TNF-α), and C-reactive protein (CRP) in the blood, showing a low-grade inflammatory profile (Leonard, 2007; Li et al., 2011; Osimo et al., 2019). It is suggested that GR resistance in immune cells leads to decreased levels of anti-inflammatory cytokines (IL-10, IL-4, examples) and enhances inflammation (Leonard, 2007; Li et al., 2011). Neurodegeneration may also be related to the pathophysiology of depression. This hypothesis is supported by evidence of reduced volume in the hippocampus and prefrontal cortex in depressive patients (Serafini, 2012). It is suggested that these volume changes would be caused by decreased neurogenesis secondary to hypercortisolemia and inflammation, which impairs the synthesis of neurotrophins like the brain-derived neurotrophic factor (BDNF) (Cai et al., 2015; Otte et al., 2016). Moreover, some studies have shown that patients with depression express low levels of BDNF (Kotan et al., 2012), although this finding is not a consensus (Elfving et al., 2012). An important observation corroborating this hypothesis is that the use of antidepressants in rats increases hippocampal volume as a consequence of increased BDNF expression (Foltran & Diaz, 2016).

Neuroimaging and Depression

Neuroimaging techniques have also helped understand the physiology of depression. Several studies have reported that major depression patients show alterations in brain anatomy and function (Dai et al., 2019). Structural magnetic resonance imaging (MRI) has consistently shown reduced hippocampal volume in MDD (Bremner et al., 2000; Lorenzetti et al., 2009; MacQueen, 2009), as well as reduced basal ganglia, caudate nucleus, and putamen (Lorenzetti et al., 2009), thalamus (Nugent et al., 2013), amygdala (Bora et al., 2012), insula (Soriano-Mas et al., 2011), and anterior cingulate cortex (Bora et al., 2012). Deficits in both global and local white matter integrity are also observed (Shen et al., 2017).

Functional brain changes have involved particularly the limbic system, a set of brain regions fundamental to emotion processing and mood regulation (Davidson et al., 2002; Groenewold et al., 2013; Joormann & Stanton, 2016). The amygdala, for instance, is a brain structure highly sensitive to negative emotional stimuli that have been repeatedly implicated in depression. Amygdala hyperactivity has been correlated with depressive symptoms and its regulation has been associated with antidepressant effects (Godlewska et al., 2012; Murphy et al., 2009; Sheline et al., 2001; Williams et al., 2015). For instance, selective serotonin reuptake inhibitors (SSRI), such as citalopram, lead to reduced amygdala response to fearful facial expressions compared to placebo (Godlewska et al., 2012; Murphy et al., 2009).

Another structure of the limbic system, the anterior cingulate cortex (ACC), appears to be involved in depression. Positron emission tomography (PET) has found the decreased metabolic activity of the subgenual portion of the ACC in patients with depression when compared to healthy controls (Drevets et al., 1997; Videbech, 2000). This region has also been an indicator of the antidepressant response to different interventions, such as antidepressants, psychotherapy, and deep brain stimulation (Delaveau et al., 2011; Siegle et al., 2012). For example, increased functional connectivity between the ACC and the limbic system has been reported after successful antidepressant treatment with the SSRI sertraline (Anand et al. 2005).

Other brain regions, such as the ventromedial prefrontal cortex (vmPFC), the posterior cingulate cortex (PCC), and the inferior parietal cortex, also seem to be important in depression. Together with the medial prefrontal cortex (mPFC) and the ACC, these regions form a brain network known as the default mode network (DMN), which is characterized by greater activity during periods of rest than during a goal-directed task (Raichle et al., 2001). The DMN has been used as a consistent marker for self-referential processes, including spontaneous thoughts, autobiographical memories, mental simulations, and mind-wandering (Buckner et al., 2008; Northoff et al., 2006). MDD has been associated with hyperconnectivity within the DMN and hyperconnectivity between frontoparietal control systems and regions of the DMN (Kaiser et al., 2015). These results are consistent with clinical symptoms, such as rumination, in which MDD patients are impulsively focused on the past and future negative consequences, rather than reflecting on the present moment (Papageorgiou & Wells, 2004). Furthermore, evidence suggests that patients with depression fail to modulate different regions of the DMN during an emotional regulation task (Sheline et al., 2009).

Other brain networks have also been implicated in depression (Kaiser et al., 2015). Typically, patients present reduced functional connectivity within the fronto-parietal network, involved in cognitive control of attention and emotion regulation, and decreased connectivity between frontoparietal and the dorsal attention networks (Kaiser et al., 2015).

Sleep and Depression

Sleep disturbances are observed in about 75% of patients with depression (Minkel et al., 2017). Changes in sleep may be the first sign of a depressive episode and, for many patients, it persists even after treatment. Besides, patients with sleep distur-bances have poorer outcomes than those without (Murphy & Peterson, 2015). Sleep complaints in MDD patients include difficulty falling asleep, frequent nocturnal awakenings, early-morning awakening, nonrestorative sleep, and recurrent night-mares (Minkel et al., 2017).

Most consistently, studies have found the following differences when comparing MDD patients with healthy controls (Minkel et al., 2017):

1. Prolonged sleep onset latency, increased wake time during sleep, increased num-ber of awakenings during sleep (i.e., fragmentation or discontinuity of sleep), increased early morning wake-up time, and decreased total sleep time;
2. Reduced rapid eye movement sleep (REMS) latency (i.e., shorter interval between sleep onset and the first REMS episode), increased amount of REMS, longer first REMS period, and increased REM density (i.e., higher number of eye movements during REMS);
3. Decreased amount of deep sleep (DS) and of DS percentage over total sleep (Minkel et al., 2017; Murphy & Peterson, 2015).

Reduced REMS latency has been the most robust polysomnographic finding in depression (Minkel et al., 2017; Murphy & Peterson, 2015). Some suggest that these changes are due to low levels of serotonin (Siegel, 2017), somewhat related to the secretion of cortisol at night (Poland et al., 1992). In fact, an inverse correlation between cortisol level and REMS latency has been observed, as subjects with shorter REMS latency are the ones with higher HPA activity (Asnis et al., 1983), while SSRI treatment may stabilize HPA functioning (Hinkelmann et al., 2012).

A recent study suggests that cortisol changes may be a physiological target that links sleep disturbance and depression (Santiago et al., 2020). The authors found that treatment-resistant depressive patients had lower salivary cortisol awakening response and sleep duration, with higher latency to sleep than healthy volunteers with a good sleep profile. Furthermore, a low cortisol profile was associated with more severe depressive symptoms and worse sleep quality. However, although

BDNF has been proposed as a biomarker of good sleep quality, no difference in serum BDNF levels was observed (Santiago et al., 2020).

Sleep changes in MDD have been treated with psychotherapy, medication, or both (Minkel et al., 2017; Murphy & Peterson, 2015). Psychotherapy includes cognitive behavioral therapy for insomnia and sleep hygiene, such as (a) using the bed only for sleep and sex, (b) going to sleep only when really tired, (c) leaving the bed after 15 min if unable to fall asleep, (d) establishing regular sleep and wake up time, and (e) avoid napping during the day (Perlis et al., 2010). Pharmacological strategies include hypnotics and sedating agents, such as non-benzodiazepine GABA agonists, and melatonin. Antidepressants are also useful, especially tricycles and MAO inhibitors (Schutte-Rodin et al., 2008).

Ayahuasca and Depression

Thus far, a few clinical trials have been designed to test the antidepressant effects of ayahuasca. In a first open-label trial, 17 patients with refractory depression participated in a single dosing session with ayahuasca (Osório et al., 2015; Sanches et al., 2016). Depressive symptoms were monitored by clinical psychiatric scales (Carneiro et al., 2015). Assessments occurred before the session (baseline), during the session, and following 1, 2, 7, 14, and 21 days after the session. A significant decrease in depressive symptoms was observed already during the ayahuasca session (40 min after ingestion), which remained significantly reduced for 21 days (Osório et al., 2015; Sanches et al., 2016). Subacute changes in cerebral blood flow were also assessed 8 h after the session, and significant increased cerebral blood flow was observed in the nucleus accumbens, right insula, and the subgenual portion of the anterior cingulate cortex (sgACC); brain areas recurrently involved in depression (Sanches et al., 2016).

Although promising, this study did not control for the placebo effect, which is high and can benefit up to 40% of the patients in clinical trials for depression (Sonawalla & Rosenbaum, 2002). The follow-up randomized placebo-controlled trial (RCT) with ayahuasca for treatment-resistant depression was conducted from January 2014 to June 2016 (registered at http://clinicaltrials.gov/NCT02914769). The trial was a parallel two-arm design, in which half of the patients were dosed with ayahuasca and the other half with placebo (Palhano-Fontes et al., 2019).

We assessed 218 patients for eligibility, and 35 met clinical criteria for the trial. All patients had used at least two different antidepressant medications unsuccessfully (referred to as "treatment-resistant depression") and were all naïve to ayahuasca. Six patients had to be excluded during the washout period,[1] leaving 29 patients in the trial: 14 in the ayahuasca group and 15 in the placebo group. We

---

[1] Before the dosing session, patients remained abstinent from any antidepressant medication, for a period of 2 weeks, typically.

assessed depression severity with the Montgomery-Åsberg Depression Rating Scale (MADRS), and the Hamilton Depression Scale (HAM-D), this time at baseline, and followed-up at 1, 2, 7, 14 days and monthly for the next 6 months after the session.

The substance used as a placebo was a brownish liquid, with a bitter and sour taste that came from water, beer yeast, citric acid, zinc sulfate, and caramel colorant. It was designed to mimic some of the characteristics of ayahuasca, including modest gastrointestinal distress due to the presence of zinc sulfate. We used a single batch of ayahuasca, prepared by our friend and leader of a Barquinha church, Edilsom Fernandes da Silva, from Ji-Paraná, Rondônia, Brazil. Patients received a single dose of 1 ml/kg of placebo or ayahuasca at an adjusted dose of 0.36 mg/kg of N,N-DMT (Palhano-Fontes et al., 2019).

All participants gave written informed consent before the trial, which was conducted in accordance with the Declaration of Helsinki that guides ethical principles for medical research, and with the Consolidated Standards of Reporting Trials (CONSORT) (Schulz et al., 2011; World Medical Association, 2001). The trial was approved by the Ethics Committee of the Hospital Universitário Onofre Lopes (HUOL), from the Universidade Federal do Rio Grande do Norte (UFRN), in Natal, Brazil.

Prior to participation they were informed about different aspects of the trial, including the 50% possibility of taking ayahuasca and 50% of taking the placebo. Possible effects were extensively presented and discussed.

Patients then underwent a washout period of about 2 weeks prior to the dosing session to reduce interaction between their medications and ayahuasca. During this period, patients were followed by our psychiatrists for medication weaning.

Our trial was also designed to explore the nature of the subacute effects of ayahuasca. In other words, we were interested in understanding changes happening 1 day after the session. For that, we made a number of assessments 1 day before and 1 day after the session with ayahuasca or placebo.

Patients arrived at the hospital on Tuesday at around 2 p.m. They were always received by a member of our crew, and had their baseline psychiatric evaluation, which included depression severity checked with the HAM-D and MADRS scales (Palhano-Fontes et al., 2019). In the same afternoon, patients had a neuropsychological evaluation, followed by a magnetic resonance imaging (MRI) scanning session for functional MRI (fMRI). We also used linguistic analysis based on participants' free reports on their thoughts during the fMRI session, as well as on four stories created based on images with different emotional contents (Mota et al., 2017). They had dinner, and at around 9 pm, they were prepared with the EEG cap system, and then allowed to go to sleep. Brain (EEG), heart (ECG), and muscle (EMG) activities were recorded continuously throughout the night, and were followed by two health professionals: a nurse and a physician.

On the next day (Wednesday), patients were awakened at 6 am to collect saliva, and blood was collected about 45 min later, both at fasting and at rest. After a light breakfast, participants were prepared for the dosing session, when they would receive either ayahuasca or placebo. Each session lasted approximately 9 h, usually going from 7 am to 4 pm. Sessions took place in a bedroom-like environment

Fig. 2.1 Setting where dosing sessions took place consisted of a bedroom-like hospital site equipped with a recliner, a bed, and a desk, controlled temperature, natural and dimmed light

specifically designed for the study, equipped with a recliner, a bed, and a desk, controlled temperature, and natural and dimmed light (Fig. 2.1). Throughout the dosing session, at least two members of our team accompanied the participant, giving assistance when necessary.

At around 10 am, participants were interviewed by a trained psychiatrist for clinical evaluation and another session of clinical scales, this time including evaluating for dissociative, mania-like, and psychotomimetic symptoms, with three scales: Clinician Administered Dissociative States Scale (CADSS), Brief Psychiatric Rating Scale (BPRS), and Young Mania Rating Scale (YMRS) (Palhano-Fontes et al., 2019). Before dosing, we reinforced that they could drink ayahuasca and feel little, or drink placebo and feel something. The transitory nature of the effects was also reinforced. We also mentioned a set of helpful attitudes and strategies to be taken, such as maintaining an open attitude toward the experience. Participants were asked to remain silent, calm, with their eyes closed, while maintaining attention on their body, thoughts, and emotions. We told them we were going to ask for every detail they experienced; "so, please, pay attention."

Ayahuasca or placebo was taken at around 10:30 a.m. Participants were monitored throughout the session with EEG, ECG, and EMG. They listened to music during most of the session, and the soundtrack included Brazilian singers. Instrumental, classical, and Andean music were also present. A link to the adapted playlists, created by L. F. Tofoli, is in the footnote.[2] Saliva and psychiatric scales were also applied at three time points along the session. These assessments occurred 1h40, 2h40, and 4h after intake. At these assessments, psychiatrists monitored depressive, dissociative, psychotic, and mania-like symptoms.

When the acute effects had ceased, at around 3 pm, patients described their experiences in detail (recorded in video). They also responded to three questionnaires, two of which are sensitive to the effects of classical psychedelics, the Hallucinogenic Rating Scale (HRS) (Mizumoto et al., 2011) and the Mystical Experience Questionnaire (MEQ30) (Maclean et al., 2012), and one to assess changes in spontaneous thinking, the Amsterdam Resting State Questionnaire (ARSQ) (Diaz et al.,

---

[2] https://open.spotify.com/user/lftofoli/playlist/6sRsIxM4oNEpCPSFDQxAJB?si=THVcUA_tRPSIFIgZmMSvtg https://open.spotify.com/user/lftofoli/playlist/5wAyiC2RvBpbZUZeLMERza?si=H999tZ1LSJ6FyUCFgrSnJw

2013). After psychiatric evaluation, they were allowed to go home at around 4:00 pm. To get a feeling of how the session was, access: https://vimeo.com/246973683.

On the following afternoon (Thursday), they returned for psychiatric evaluation (MADRS), neuropsychological, MRI/fMRI, and another sleep EEG, from Thursday to Friday. The next morning (Friday), they repeated blood and saliva exams. After breakfast, they retrospectively described their experience, reported any perceived changes following dosing, and, at around 9 am, they were discharged by a psychiatrist. Patients were asked to return for follow-up assessments at 7 days (D7), 14 days (D14), 1 month (M1) and from then on, every month, for 6 months (M2, M3, M4, M5, M6). Also, 7 days after the session (following Tuesday), they were introduced to a new treatment scheme based on a different commercially available antidepressant.

During the session with ayahuasca, patients showed slight transient changes in dissociative and psychotomimetic symptoms, and we did not observe significant increases in mania-like symptoms. We also observed significant transient nausea (ayahuasca = 71%, placebo = 26%), vomiting (ayahuasca = 57%, placebo = 0%), transient anxiety (ayahuasca = 50%, placebo = 73%), and transient headache (ayahuasca = 42%, placebo = 53%).

Figure 2.2 shows changes in depression severity over time. We observed significantly reduced depressive symptoms 1 day (D1) after the session with ayahuasca that persisted for at least 7 days (D7) when compared to placebo (Palhano-Fontes et al., 2019). We did not find significant differences in the follow-up assessments, possibly due to the small number of patients who returned to the hospital for the psychiatric evaluations (Fig. 2.2). Nonetheless, it is worth mentioning that other recent studies using psychedelics found reduced symptoms of depression and anxiety for longer periods after the psychedelic session (Gasser et al., 2014; Griffiths et al., 2016; Ross et al., 2016).

One day after the session, response rates[3] were 50% in the ayahuasca group and 46% in the placebo group. Two days after dosing, the response rate was 77% in the ayahuasca group and 64% in the placebo. A significant between-groups difference was observed 7 days after dosing, with 64% of responders in the ayahuasca group, and 27% in the placebo (Palhano-Fontes et al., 2019). Also, 7 days after dosing, remission rate (MADRS≤10) was statistically different between-groups, with 36% of remitters in the ayahuasca group and 7% in the placebo (Palhano-Fontes et al., 2019).

Our results are comparable with trials that used ketamine, although the dynamic is different. While ketamine effect size is largest 1 day after the session (Cohen's d = 0.89) and reduces toward 1 week after the session (Cohen's d = 0.41), the effect sizes observed for ayahuasca is smallest and compatible to ketamine at day 1 (Cohen's d = 0.84), but it increases and is largest 7 days after the session (Cohen's d = 1.49) (Palhano-Fontes et al., 2019). Furthermore, our results suggest that the

---

[3] Response was defined as a reduction of 50% or more in MADRS baseline scores.

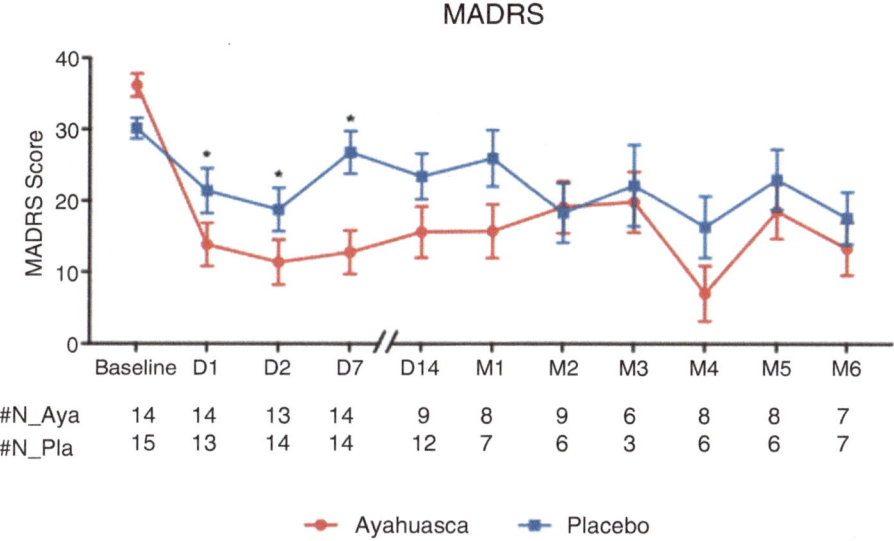

MADRS

	Baseline	D1	D2	D7	D14	M1	M2	M3	M4	M5	M6
#N_Aya	14	14	13	14	9	8	9	6	8	8	7
#N_Pla	15	13	14	14	12	7	6	3	6	6	7

— Ayahuasca   — Placebo

Fig. 2.2 Depression severity as a function of time. Significant differences are observed between ayahuasca (red circles) and placebo (blue squares) at D1 ($p = 0.04$), D2 ($p = 0.04$) and D7 ($p < 0.0001$). No differences were found after D7. #N_Aya and #N_pla are the numbers of patients in ayahuasca and placebo groups who attended psychiatric evaluation at that time point. Values are (mean ± SEM). MADRS scores: mild depression (11–19), moderate (20–34), severe (≥ 35) *$p < 0.05$

ayahuasca session significantly reduced suicidality (i.e., suicide attempts, suicide planning, and suicidal ideation), from baseline to 1, 2, and 7 days after the dosing session (Zeifman et al., 2019).

Neurobiological Bases Supporting the Antidepressant Effects of Ayahuasca

During the last few decades, the scientific exploration of ayahuasca and its effects has allowed insights about its biological mechanisms, safety limits, and potential therapeutic effects. Ayahuasca preparations contain N,N-dimethyltryptamine (N,N-DMT), a serotonin and sigma-1 receptors agonist (Carbonaro & Gatch, 2016), both of which have been implicated in depression. Furthermore, the MAOI present in ayahuasca directly affects the monoaminergic systems, increasing the viability of monoamines, such as serotonin, noradrenaline, and dopamine in the synaptic cleft, and tetrahydroharmine, one of these MAOI, is also a serotonin reuptake inhibitor (McKenna et al., 1984).

Cellular and systemic levels studies have been developed using models of neural cells on a dish, human stem cells, and cerebral organoids. Using human neural cells and cerebral organoids derived from stem cells, a recent study has observed that harmine is associated with proliferation of neural cells (neurogenesis), linked to the inhibition of an enzyme called DYRK1A (dual specificity tyrosine-phosphorylation-regulated kinase) (Dakic et al., 2016). In another study, harmine, tetrahydroharmine, and harmaline were found to stimulate neural stem cell proliferation, migration, and differentiation into adult neurons, using brain progenitor cells from adult mice (Morales-García et al., 2017).

Changes in protein expression in brain organoids has also been observed after being exposed to 5-MeO-DMT, an analog of N,N-DMT (Dakic et al., 2017), naturally occurring in plants and in the Sonoran Desert toad (*Incilius alvarius*) (Davis et al., 2019). A cascade of proteins was changed that were involved in processes of novel dendritic spine formation, neuronal plasticity, and memory (Dakic et al., 2017). Furthermore, 5-MeO-DMT was also shown to induce increased dendritic complexity and to change electrophysiological properties of these newborn neurons (Lima da Cruz et al., 2018).

Animal models have also allowed for a more comprehensive exploration of the antidepressant effects of ayahuasca. These models have been extensively used as a tools to understand the biological bases of neuropsychiatric conditions as well as the effects of new treatments (Cardoso et al., 2004; Coimbra et al., 2017; Figueiredo et al., 2011; França et al., 2015; Ramaker & Dulawa, 2017; Willner & Belzung, 2015). Such studies allow for a deeper knowledge not accessible in human individuals and are conducted according to national and international regulations of animal welfare and ethical considerations.

Studies with ayahuasca and its components have shown promise in decreasing anxiety and depressive-like behavior in different animal models, such as in zebrafish, mice, and nonhuman primates (Cameron et al., 2018; Fortunato et al., 2010a; Fortunato et al., 2009, 2010b; Savoldi et al., 2017; Winne et al., 2020).

Studies in nonhuman primates have also been important, since they show similar brain morphology, function, and social organization with humans, fitting better as a translational animal model to psychiatric conditions (Galvão-Coelho et al., 2008; Galvão-Coelho et al., 2017). For instance, young common marmosets subjected to social isolation were used as a model of depression, upon which the effects of ayahuasca were tested. Results suggest reduced depressive-like behaviors and fecal cortisol levels improvements after a single session with ayahuasca compared to placebo. The effects were faster and more robust than in animals treated with the antidepressant nortriptyline (da Silva et al., 2018).

In humans, ayahuasca also seems to have a significant impact on the endocrine and immune systems. Evidence suggests increased prolactin and cortisol levels, and reduced lymphocytes CD3 and CD4, 2 h after a single ayahuasca intake in healthy volunteers (Dos Santos et al., 2011). Patients from our recent RCT with ayahuasca presented hypocortisolemia and blunted salivary cortisol awakening response at baseline. We found that patients treated with ayahuasca, but not with placebo, stabilized their salivary cortisol awakening response to the same level found in the healthy participants, 2 days after treatment (Galvao et al., 2018). These same patients treated with ayahuasca (but not with placebo) also presented an increase in serum BDNF levels, which was correlated with lower depressive symptoms (MADRS scores) (rho = −0.55). This finding suggests that, most probably, ayahuasca affects the fast expression of this critical molecule, as is suggested for the new fast-acting antidepressants, such as ketamine (Ly et al., 2018).

Data from our RCT revealed changes in blood inflammatory biomarkers: C-reactive protein (CRP) and IL-6. At baseline, patients displayed higher CRP levels than healthy controls. Two days after ayahuasca ingestion, CRP levels decreased in both patients and healthy individuals treated with ayahuasca (but not with placebo). In addition, we found a significant positive correlation (rho = 0.70) between decreased inflammation and depressive symptoms (MADRS). No relevant variation was detected in IL-6 levels. Although usually, IL-6 and CRP are physiologically linked, some trials have shown isolated increases in CRP, which can be stimulated independently of IL-6 (Harris et al., 1999; Saito et al., 2016; Więdłocha et al., 2018). Since depression is also considered a pro-inflammatory disease, this work indicates that psychedelics can also have direct or indirect impact on the immune system involved in depressive processes, and also opens a new road for testing psychedelics on other inflammatory diseases (Galvão-Coelho et al. 2020).

The effects of ayahuasca on sleep also corroborate with its antidepressant effects. Barbanoj et al. (2008) investigated the impact of ayahuasca on sleep in the night following a single diurnal ayahuasca session. Results indicate that ayahuasca decreases REMS duration, and tends to increase REMS latency (Barbanoj et al., 2008). As mentioned before, increased amount of REMS and reduced REMS latency are common findings in depression (Minkel et al., 2017; Riemann et al., 2001; Tsuno et al., 2005).

The default mode network (DMN) has also been implicated in the pathophysiology of depression and changes in its activity and connectivity have been consistently observed during the effects of psychedelics. A previous fMRI study conducted by our group suggests that ayahuasca reduces the activity of most DMN nodes (Palhano-Fontes et al., 2015), while patients with depression tend to demonstrate increased DMN activity and functional connectivity in this region (Sheline et al., 2009).

The DMN has also been playing center stage in a recent entropic brain theory, which holds that the altered state of consciousness induced by psychedelics is characterized by higher entropy of the brain's functional connectivity (Carhart-Harris, 2018; Carhart-Harris et al., 2014). Entropy is a physical quantity associated with the level of uncertainty of a system. According to this theory, while in an ordinary state of consciousness, the DMN would act by restricting and maintaining the efficient functioning of the brain; under the psychedelic influence, the decreased activity and connectivity within the DMN would reduce such restrictions. With these constraints temporarily gone, the more flexible experiences under the influence of psychedelics would be represented by a state of increased entropy. In line with this theory, previous studies have observed increased entropy of brain connections while under the influence of psychedelics (Petri et al., 2014; Viol et al., 2017; Viol et al., 2019).

More recently, we have suggested that ayahuasca may exert its long-lasting effects on mood by modulating neural networks supporting interoceptive, affective, and self-referential functions, such as the DMN and the salience networks (Pasquini et al., 2020). We found increased functional connectivity of the salience network and decreases in the DMN 1 day after the dosing session in healthy individuals (Pasquini et al., 2020), consistent with our previous observation that the activity and connectivity within the salience network is increased during the acute effects of ayahuasca (Fig. 2.3), together with changes in the DMN (Palhano-Fontes et al., 2015). Similar increased functional connectivity within the DMN was also observed following psychedelic psilocybin in an open-label trial for depression (Carhart-Harris et al., 2017). Altogether, these studies suggest that brain changes could be a mechanism influencing the therapeutic response we observe in the use of psychedelics for depression.

The therapeutic benefit of psychedelics also seems to depend on the quality of psychedelic experience, such as the peak experience (Bogenschutz et al., 2015; Garcia-Romeu et al., 2015; Griffiths et al., 2016; Majić et al., 2015; Ross et al., 2016). An open-label trial with psilocybin for tobacco addiction found a significant correlation between cessation outcomes and mystical experience measures (Garcia-Romeu et al., 2015). The same was found in another trial with psilocybin for alcohol dependence, where mystical experiences were correlated with decreased craving and increased abstinence (Bogenschutz et al., 2015). Furthermore, recent RCT with psilocybin for depression and anxiety in patients with life-threatening cancer showed that mystical experiences mediate the therapeutic effects (Griffiths et al., 2016; Ross et al., 2016).

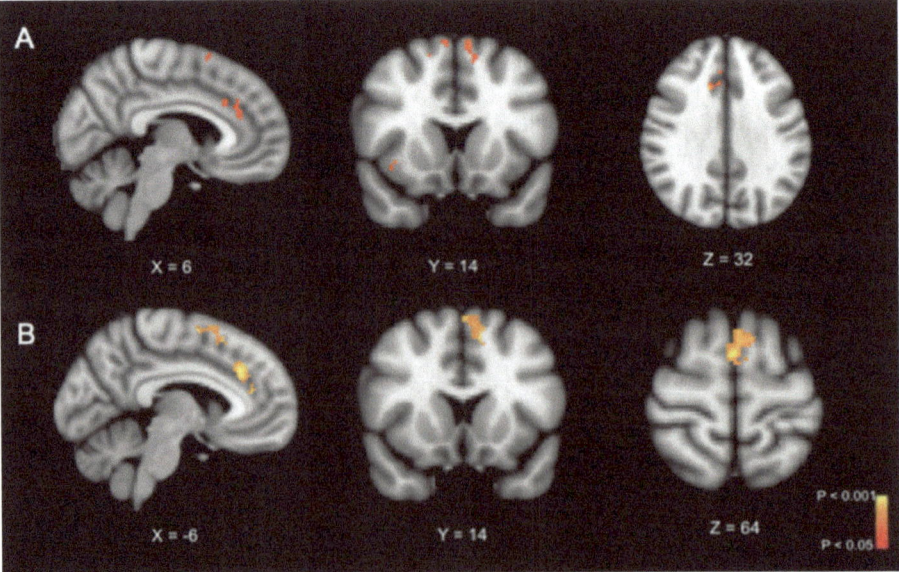

Fig. 2.3 Activity (panel A) and functional connectivity (panel B) increases within the salience network during acute effects of ayahuasca

In our recent RCT with ayahuasca for depression, out of the 29 patients, 27 patients responded to the HRS and 15 patients responded to the MEQ. We found that the HRS dimension that assesses changes in visual perception was the only one to show relationship with the antidepressant outcome (Palhano-Fontes et al., 2019). In fact, it has been argued that visions may play a crucial role in the therapeutic benefits of ayahuasca, as a mechanism to provide personal insights (Frecska et al., 2016). Our results give support to this hypothesis. In addition to a previous fMRI study that found increased activity in vision-related cortices during the acute effects of ayahuasca (de Araujo et al., 2012), we also found a negative correlation between changes in depressive symptoms after the ayahuasca session and the MEQ factor related to perception of time and space (Palhano-Fontes et al., 2019).

Final Remarks

Psychedelic substances from different plants and fungi have been used for a long time for healing purposes, and form a central pillar of different ancient cultures (Schultes & Hofmann, 1979). Despite the presence of these substances throughout nature and history, they have attracted little scientific attention until the 1940s, with the discovery of lysergic acid diethylamide (LSD) (Hofmann, 1979). During that period, LSD was regarded with great interest by psychiatry due to its therapeutic potential (Oram, 2014). Between the 1950s and mid-1960s, more than 1000 articles had been published, with more than 40,000 subjects having participated in clinical trials with LSD (Bakalar & Grinspoon, 1997). Due to the Drug War policy under Richard Nixon's presidency in the 1970s, all research with psychedelics was practically shut down, and psychedelics became schedule I substances, defined as drugs with severe safety concerns, no medical use, and high risk of abuse. Back then, there were no controlled trials with psychedelics for any human condition.

After a few decades of discontinuation, research involving psychedelics has once again gained momentum. Recent studies suggest that it is safe when used in appropriate set and setting. In the case of ayahuasca, its clinical potential has been tested in recent studies on anxiety and depression (dos Santos et al., 2016; Osório et al., 2015; Palhano-Fontes et al., 2019; Sanches et al., 2016), and alcohol addiction (Thomas et al., 2013).

Our recent study was the first RCT to test a classic psychedelic against treatment-resistant depression (Palhano-Fontes et al., 2019). One point worth mentioning is the rapid nature of the observed antidepressant effects of ayahuasca. Currently available antidepressants, such SSRIs and tricyclics, take around 2 weeks for the onset of the desired therapeutic response (Cai et al., 2015; Otte et al., 2016), while our study suggests significant antidepressant effects of ayahuasca already 1 day after dosing.

The placebo effect observed in our study was higher than usually seen in clinical trials for depression, including those with ketamine (Palhano-Fontes et al., 2019; Romeo et al., 2015). While we found a placebo response rate of 46% 1 day after dosing, and 26% 7 days after, studies with ketamine found placebo responses on the order of 0–6% at 1 day and 0–11% 7 days after dosing.

Several factors contribute to the high placebo effects observed in our study. First, most of our patients come from low socioeconomic status, and most live under intense and constant stress, in insecure environments, under intense financial and psychosocial stressors. During our trial, they stayed in a very comfortable and supportive environment for a period of 4 days. It seems as though this is more of a care effect than a typical placebo effect. For several of them, our laboratory was like staying in a five-star hotel.

Most of our patients (76%) had concurrent personality disorder, most of them from cluster B (borderline and histrionic), and other studies have related personality disorder to higher placebo responses (Palhano-Fontes et al., 2019). This was the first clinical trial with psychedelics to report including individuals with personality disorder. No serious adverse events were observed, although the session tended to be more challenging for these individuals. The number of individuals with personality disorder included in our trial was small, but the preliminary results found are promising, which warrants additional research exploring the safety and clinical utility of ayahuasca as an intervention for personality disorder (Zeifman et al., 2019).

For many years, indigenous knowledge and ayahuasca communities have been pointing out the therapeutic benefits of ayahuasca. This wisdom is now being explored with scientific methods and new interpretations are being made in light of new findings. Preliminary evidence supports many claims and previous anecdotal reports. The evidence reviewed here suggests good support for the use of ayahuasca for treatment-resistant depression. The antidepressant effects we observed were accompanied by changes in a variety of systems, including hormonal, inflammatory, and immune systems, and brain functions related to perception, memory, attention, emotion, and cognition. Such observations reveal that the nature of the effects of ayahuasca are indeed multifactorial, and spans from gross biochemical changes to subtle cognitive processes.

References

American Psychiatric Association. (2013). *Diagnostic and statistical manual of mental disorders* (DSM-5). Washington, DC: Author.

Anand, A., Li, Y., Wang, Y., Wu, J., Gao, S., Bukhari, L., ... & Lowe, M. J. (2005). Activity and connectivity of brain mood regulating circuit in depression: a functional magnetic resonance study. *Biological psychiatry, 57*(10), 1079–1088.

Asnis, G. M., Halbreich, U., Sachar, E. J., Nathan, R. S., Ostrow, L. C., Novacenko, H., Davis, M., Endicott, J., & Puig-Antich, J. (1983). Plasma cortisol secretion and REM period latency in adult endogenous depression. *The American Journal of Psychiatry, 140*(6), 750–753.

Bakalar, J. B., & Grinspoon, L. (1997). *Psychedelic drugs reconsidered*. New York City, NY: The Lindesmith Center.

Barbanoj, M. J., Riba, J., Clos, S., Giménez, S., Grasa, E., Romero, S., Gimenez, S., Grasa, E., & Romero, S. (2008). Daytime ayahuasca administration modulates REM and slow-wave sleep in healthy volunteers. *Psychopharmacology, 196*(2), 315–326.

Bogenschutz, M. P., Forcehimes, A. A., Pommy, J. A., Wilcox, C. E., Barbosa, P., & Strassman, R. J. (2015). Psilocybin-assisted treatment for alcohol dependence: A proof-of-concept study. *Journal of Psychopharmacology, 29*(3), 289–299.

Bora, E., Fornito, A., Pantelis, C., & Yücel, M. (2012). Gray matter abnormalities in major depressive disorder: A meta-analysis of voxel based morphometry studies. *Journal of Affective Disorders, 138*(1–2), 9–18.

Bremmer, M. A., Deeg, D. J. H., Beekman, A. T. F., Penninx, B. W. J. H., Lips, P., & Hoogendijk, W. J. G. (2007). Major depression in late life is associated with both hypo- and hypercortisolemia. *Biological Psychiatry, 62*(5), 479–486.

Bremner, J. D., Narayan, M., Anderson, E. R., Staib, L. H., Miller, H. L., & Charney, D. S. (2000). Hippocampal volume reduction in major depression. *The American Journal of Psychiatry, 157*(1), 115–118.

Buckner, R. L., Andrews-Hanna, J. R., & Schacter, D. L. (2008). The brain's default network: Anatomy, function, and relevance to disease. *Annals of the New York Academy of Sciences, 1124*, 1–38.

Cai, S., Huang, S., & Hao, W. (2015). New hypothesis and treatment targets of depression: An integrated view of key findings. *Neuroscience Bulletin, 31*(1), 61–74.

Cameron, L. P., Benson, C. J., Dunlap, L. E., & Olson, D. E. (2018). Effects of N,N-dimethyltryptamine on rat behaviors relevant to anxiety and depression. *ACS Chemical Neuroscience, 9*(7), 1582–1590. https://doi.org/10.1021/acschemneuro.8b00134.

Carbonaro, T. M., & Gatch, M. B. (2016). Neuropharmacology of N,N-dimethyltryptamine. *Brain Research Bulletin, 126*, 74–88. https://doi.org/10.1016/j.brainresbull.2016.04.016.

Cardoso, R. C., Lobão-Soares, B., Bianchin, M. M., Carlotti, C. G., Walz, R., Alvarez-Silva, M., Trentin, A. G., & Nicolau, M. (2004). Enhancement of blood-tumor barrier permeability by Sar-[D-Phe8] des-Arg9BK, a metabolically resistant bradykinin B1 agonist, in a rat C6 glioma model. *BMC Neuroscience, 5*(1), 38.

Carhart-Harris, R. L. (2018). The entropic brain – Revisited. *Neuropharmacology, 142*, 167–178.

Carhart-Harris, R. L., Leech, R., Hellyer, P. J., Shanahan, M., Feilding, A., Tagliazucchi, E., Chialvo, D. R., & Nutt, D. (2014). The entropic brain: A theory of conscious states informed by neuroimaging research with psychedelic drugs. *Frontiers in Human Neuroscience, 8*, 20.

Carhart-Harris, R. L., Roseman, L., Bolstridge, M., Demetriou, L., Pannekoek, J. N., Wall, M. B., Tanner, M., Kaelen, M., McGonigle, J., Murphy, K., Leech, R., Curran, H. V., & Nutt, D. J. (2017). Psilocybin for treatment-resistant depression: fMRI-measured brain mechanisms. *Scientific Reports, 7*(1), 13187.

Carneiro, A. M., Fernandes, F., & Moreno, R. A. (2015). Hamilton depression rating scale and Montgomery-Asberg depression rating scale in depressed and bipolar I patients: Psychometric properties in a Brazilian sample. *Health and Quality of Life Outcomes, 13*(1), 42.

Caspi, A., Hariri, A. R., Holmes, A., Uher, R., & Moffitt, T. E. (2010). Genetic sensitivity to the environment: The case of the serotonin transporter gene and its implications for studying complex diseases and traits. *The American Journal of Psychiatry, 167*(5), 509–527.

Coimbra, N. C., Paschoalin-Maurin, T., Bassi, G. S., Kanashiro, A., Biagioni, A. F., Felippotti, T. T., Elias-Filho, D. H., Mendes-Gomes, J., Cysne-Coimbra, J. P., Almada, R. C., & Others. (2017). Critical neuropsychobiological analysis of panic attack-and anticipatory anxiety-like behaviors in rodents confronted with snakes in polygonal arenas and complex labyrinths: A comparison to the elevated plus-and T-maze behavioral tests. *Brazilian Journal of Psychiatry, 39*(1), 72–83.

Dai, L., Zhou, H., Xu, X., & Zuo, Z. (2019). Brain structural and functional changes in patients with major depressive disorder: A literature review. *PeerJ, 7*, e8170.

Dakic, V., Maciel, R. d. M., Drummond, H., Nascimento, J. M., Trindade, P., & Rehen, S. K. (2016). Harmine stimulates proliferation of human neural progenitors. *PeerJ, 4*, e2727.

Dakic, V., Minardi Nascimento, J., Costa Sartore, R., Maciel, R. d. M., de Araujo, D. B., Ribeiro, S., Martins-de-Souza, D., & Rehen, S. K. (2017). Short term changes in the proteome of human cerebral organoids induced by 5-MeO-DMT. *Scientific Reports, 7*(1), 12863.

da Silva, F. S., Silva, E. A. S., de Sousa, G. M., Jr., Maia-de-Oliveira, J. P., Soares-Rachetti, V. d. P., de Araujo, D. B., Sousa, M. B. C., Lobão-Soares, B., Hallak, J., & Galvão-Coelho, N. L. (2018). Acute effects of ayahuasca in a juvenile non-human primate model of depression. *bioRxiv*, 254268. https://doi.org/10.1101/254268.

Davidson, R. J., Pizzagalli, D., Nitschke, J. B., & Putnam, K. (2002). Depression: Perspectives from affective neuroscience. *Annual Review of Psychology, 53*, 545–574.

Davis, A. K., So, S., Lancelotta, R., Barsuglia, J. P., & Griffiths, R. R. (2019). 5-methoxy-N,N-dimethyltryptamine (5-MeO-DMT) used in a naturalistic group setting is associated with unintended improvements in depression and anxiety. *The American Journal of Drug and Alcohol Abuse, 45*(2), 161–169.

de Araujo, D. B., Ribeiro, S., Cecchi, G. A., Carvalho, F. M., Sanchez, T. A., Pinto, J. P., de Martinis, B. S., Crippa, J. A., Hallak, J. E. C., & Santos, A. C. (2012). Seeing with the eyes shut: Neural basis of enhanced imagery following ayahuasca ingestion. *Human Brain Mapping, 33*(11), 2550–2560.

Delaveau, P., Jabourian, M., Lemogne, C., Guionnet, S., Bergouignan, L., & Fossati, P. (2011). Brain effects of antidepressants in major depression: A meta-analysis of emotional processing studies. *Journal of Affective Disorders, 130*(1–2), 66–74.

Diaz, B. A., Van Der Sluis, S., Moens, S., Benjamins, J. S., Migliorati, F., Stoffers, D., Den Braber, A., Poil, S.-S. S., Hardstone, R., Van't Ent, D., Boomsma, D. I., De Geus, E., Mansvelder, H. D., Van Someren, E. J. W., & Linkenkaer-Hansen, K. (2013). The Amsterdam Resting-State Questionnaire reveals multiple phenotypes of resting-state cognition. *Frontiers in Human Neuroscience, 7*, 446. https://doi.org/10.3389/fnhum.2013.00446.

dos Santos, R. G., Osório, F. L., Crippa, J. A. S., & Hallak, J. E. C. (2016). Antidepressive and anxiolytic effects of ayahuasca: A systematic literature review of animal and human studies. *Revista Brasileira de Psiquiatria, 38*(1), 65–72.

Dos Santos, R. G., Valle, M., Bouso, J. C., Nomdedéu, J. F., Rodríguez-Espinosa, J., McIlhenny, E. H., Barker, S. A., Barbanoj, M. J., & Riba, J. (2011). Autonomic, neuroendocrine, and immunological effects of ayahuasca: A comparative study with d-amphetamine. *Journal of Clinical Psychopharmacology, 31*(6), 717–726.

Drevets, W. C., Price, J. L., Simpson, J. R., Jr., Todd, R. D., Reich, T., Vannier, M., & Raichle, M. E. (1997). Subgenual prefrontal cortex abnormalities in mood disorders. *Nature, 386*(6627), 824–827. https://doi.org/10.1038/386824a0.

Elfving, B., Buttenschøn, H. N., Foldager, L., Poulsen, P. H. P., Andersen, J. H., Grynderup, M. B., Hansen, Å. M., Kolstad, H. A., Kaerlev, L., Mikkelsen, S., Thomsen, J. F., Børglum, A. D., Wegener, G., & Mors, O. (2012). Depression, the Val66Met polymorphism, age, and gender influence the serum BDNF level. *Journal of Psychiatric Research, 46*(9), 1118–1125.

Figueiredo, C. P., Antunes, V. L. S., Moreira, E. L. G., De Mello, N., Medeiros, R., Di Giunta, G., Lobão-Soares, B., Linhares, M., Lin, K., Mazzuco, T. L., Prediger, R. D. S., & Walz, R. (2011). Glucose-dependent insulinotropic peptide receptor expression in the hippocampus and neocortex of mesial temporal lobe epilepsy patients and rats undergoing pilocarpine induced status epilepticus. *Peptides, 32*(4), 781–789.

Foltran, R. B., & Diaz, S. L. (2016). BDNF isoforms: A round trip ticket between neurogenesis and serotonin? *Journal of Neurochemistry, 138*(2), 204–221.

Fortunato, J. J., Réus, G. Z., Kirsch, T. R., Stringari, R. B., Fries, G. R., Kapczinski, F., Hallak, J. E., Zuardi, A. W., Crippa, J. A., & Quevedo, J. (2010a). Effects of beta-carboline harmine on behavioral and physiological parameters observed in the chronic mild stress model: Further evidence of antidepressant properties. *Brain Research Bulletin, 81*(4–5), 491–496.

Fortunato, J. J., Réus, G. Z., Kirsch, T. R., Stringari, R. B., Fries, G. R., Kapczinski, F., Hallak, J. E., Zuardi, A. W., Crippa, J. A., & Quevedo, J. (2010b). Chronic administration of harmine elicits antidepressant-like effects and increases BDNF levels in rat hippocampus. *Journal of Neural Transmission, 117*(10), 1131–1137.

Fortunato, J. J., Réus, G. Z., Kirsch, T. R., Stringari, R. B., Stertz, L., Kapczinski, F., Pinto, J. P., Hallak, J. E., Zuardi, A. W., Crippa, J. A., & Quevedo, J. (2009). Acute harmine administration induces antidepressive-like effects and increases BDNF levels in the rat hippocampus. *Progress in Neuro-Psychopharmacology & Biological Psychiatry, 33*(8), 1425–1430.

França, A. S. C., Lobão-Soares, B., Muratori, L., Nascimento, G., Winne, J., Pereira, C. M., Jeronimo, S. M. B., & Ribeiro, S. (2015). D2 dopamine receptor regulation of learning, sleep and plasticity. *European Neuropsychopharmacology, 25*(4), 493–504.

Frecska, E., Bokor, P., & Winkelman, M. (2016). The therapeutic potentials of ayahuasca: Possible effects against various diseases of civilization. *Frontiers in Pharmacology, 7*, 35.

Frodl, T., Reinhold, E., Koutsouleris, N., Donohoe, G., Bondy, B., Reiser, M., Möller, H.-J., & Meisenzahl, E. M. (2010). Childhood stress, serotonin transporter gene and brain structures in major depression. *Neuropsychopharmacology, 35*(6), 1383–1390.

Galvao, A. C. M., Almeida, R. N., dos Santos Silva, E. A., de Morais Freire, F. A., Palhano-Fontes, F., Onias, H., Arcoverdee, E., Maia-de-Oliveira, J. P., Araujo, D., Lobao-Soares, B., & Galvao-Coelho, N. L. (2018). A single dose of ayahuasca modulates salivary cortisol in treatment-resistant depression. *bioRxiv*, 257238. https://doi.org/10.1101/257238.

Galvão-Coelho, N. L., Galvão, A. C. d. M., da Silva, F. S., & de Sousa, M. B. C. (2017). Common marmosets: A potential translational animal model of juvenile depression. *Frontiers in Psychiatry, 8*, 175.

Galvão-Coelho, N. L., Silva, H. P. A., Leão, A. d. C., & de Sousa, M. B. C. (2008). Common marmosets (Callithrix jacchus) as a potential animal model for studying psychological disorders associated with high and low responsiveness of the hypothalamic-pituitary-adrenal axis. *Reviews in the Neurosciences, 19*(2–3), 187–201.

Galvão-Coelho, N. L., de Menezes Galvão, A. C., de Almeida, R. N., Palhano-Fontes, F., Campos Braga, I., Lobão Soares, B., ... & de Araujo, D. B. (2020). Changes in inflammatory biomarkers are related to the antidepressant effects of ayahuasca. *Journal of Psychopharmacology, 34*(10), 1125–1133.

Garcia-Romeu, A., Griffiths, R., & Johnson, M. (2015). Psilocybin-occasioned mystical experiences in the treatment of tobacco addiction. *Current Drug Abuse Reviews, 7*(3), 157–164.

Gasser, P., Holstein, D., Michel, Y., Doblin, R., Yazar-Klosinski, B., Passie, T., & Brenneisen, R. (2014). Safety and efficacy of lysergic acid diethylamide-assisted psychotherapy for anxiety associated with life-threatening diseases. *The Journal of Nervous and Mental Disease, 202*(7), 513–520.

Godlewska, B. R., Norbury, R., Selvaraj, S., Cowen, P. J., & Harmer, C. J. (2012). Short-term SSRI treatment normalises amygdala hyperactivity in depressed patients. *Psychological Medicine, 42*(12), 2609–2617.

Griffiths, R. R., Johnson, M. W., Carducci, M. A., Umbricht, A., Richards, W. A., Richards, B. D., Cosimano, M. P., & Klinedinst, M. A. (2016). Psilocybin produces substantial and sustained decreases in depression and anxiety in patients with life-threatening cancer: A randomized double-blind trial. *Journal of Psychopharmacology, 30*(12), 1181–1197.

Groenewold, N. A., Opmeer, E. M., de Jonge, P., Aleman, A., & Costafreda, S. G. (2013). Emotional valence modulates brain functional abnormalities in depression: Evidence from a meta-analysis of fMRI studies. *Neuroscience and Biobehavioral Reviews, 37*(2), 152–163.

Harris, T. B., Ferrucci, L., Tracy, R. P., Corti, M. C., Wacholder, S., Ettinger, W. H., Jr., Heimovitz, H., Cohen, H. J., & Wallace, R. (1999). Associations of elevated interleukin-6 and C-reactive protein levels with mortality in the elderly. *The American Journal of Medicine, 106*(5), 506–512.

Hinkelmann, K., Moritz, S., Botzenhardt, J., Muhtz, C., Wiedemann, K., Kellner, M., & Otte, C. (2012). Changes in cortisol secretion during antidepressive treatment and cognitive improvement in patients with major depression: A longitudinal study. *Psychoneuroendocrinology, 37*(5), 685–692.

Hofmann, A. (1979). How LSD originated. *Journal of Psychedelic Drugs, 11*(1–2), 53–60.

Joormann, J., & Stanton, C. H. (2016). Examining emotion regulation in depression: A review and future directions. *Behaviour Research and Therapy, 86*, 35–49.

Kaiser, R. H., Andrews-Hanna, J. R., Wager, T. D., & Pizzagalli, D. A. (2015). Large-scale network dysfunction in major depressive disorder: A meta-analysis of resting-state functional connectivity. *JAMA Psychiatry, 72*(6), 603–611.

Kotan, Z., Sarandöl, E., Kırhan, E., Ozkaya, G., & Kırlı, S. (2012). Serum brain-derived neurotrophic factor, vascular endothelial growth factor and leptin levels in patients with a diagnosis of severe major depressive disorder with melancholic features. *Therapeutic Advances in
Psychopharmacology, 2*(2), 65–74.

Leonard, B. E. (2007). Inflammation, depression and dementia: Are they connected? *Neurochemical
Research, 32*(10), 1749–1756.

Lima da Cruz, R. V., Moulin, T. C., Petiz, L. L., & Leão, R. N. (2018). A single dose of 5-meo-DMT
stimulates cell proliferation, neuronal survivability, morphological and functional changes in
adult mice ventral dentate gyrus. *Frontiers in Molecular Neuroscience, 11*, 312.

Li, M., Soczynska, J. K., & Kennedy, S. H. (2011). Inflammatory biomarkers in depression: An
opportunity for novel therapeutic interventions. *Current Psychiatry Reports, 13*(5), 316–320.

Liu, B., Liu, J., Wang, M., Zhang, Y., & Li, L. (2017). From serotonin to neuroplasticity: Evolvement
of theories for major depressive disorder. *Frontiers in Cellular Neuroscience, 11*, 305.

Lorenzetti, V., Allen, N. B., Fornito, A., & Yucel, M. (2009). Structural brain abnormalities in
major depressive disorder: A selective review of recent MRI studies. *Journal of Affective
Disorders, 117*(1–2), 1–17.

Ly, C., Greb, A. C., Cameron, L. P., Wong, J. M., Barragan, E. V., Wilson, P. C., Burbach, K. F.,
Soltanzadeh Zarandi, S., Sood, A., Paddy, M. R., Duim, W. C., Dennis, M. Y., McAllister,
A. K., Ori-McKenney, K. M., Gray, J. A., & Olson, D. E. (2018). Psychedelics promote struc-
tural and functional neural plasticity. *Cell Reports, 23*(11), 3170–3182.

Maclean, K. A., Leoutsakos, J.-M. S., Johnson, M. W., & Griffiths, R. R. (2012). Factor analysis of
the mystical experience questionnaire: A study of experiences occasioned by the hallucinogen
psilocybin. *Journal for the Scientific Study of Religion, 51*(4), 721–737.

MacQueen, G. M. (2009). Magnetic resonance imaging and prediction of outcome in patients with
major depressive disorder. *Journal of Psychiatry & Neuroscience, 34*(5), 343–349.

Majić, T., Schmidt, T. T., & Gallinat, J. (2015). Peak experiences and the afterglow phenomenon:
When and how do therapeutic effects of hallucinogens depend on psychedelic experiences?
Journal of Psychopharmacology, 29(3), 241–253.

McKenna, D. J., Towers, G. H., & Abbott, F. (1984). Monoamine oxidase inhibitors in South
American hallucinogenic plants: Tryptamine and beta-carboline constituents of ayahuasca.
Journal of Ethnopharmacology, 10(2), 195–223.

Minkel, J. D., Krystal, A. D., & Benca, R. M. (2017). Unipolar major depression. In M. H. Kryger,
T. Roth, & W. C. Dement (Eds.), *Principles and practice of sleep medicine* (6th ed.,
pp. 1352–1362). Philadelphia, PA: Elsevier.

Mizumoto, S., da Silveira, D. X., Barbosa, P. C. R., & Strassman, R. J. (2011). Hallucinogen
Rating Scale (HRS) – Versão brasileira: tradução e adaptação transcultural [Hallucinogen
Rating Scale (HRS) – Brazilian version: Translation and cross-cultural adaptation]. *European
Archives of Psychiatry and Clinical Neuroscience, 38*, 231–237.

Morales-García, J. A., de la Fuente Revenga, M., Alonso-Gil, S., Rodríguez-Franco, M. I.,
Feilding, A., Perez-Castillo, A., & Riba, J. (2017). The alkaloids of *Banisteriopsis caapi*, the
plant source of the Amazonian hallucinogen ayahuasca, stimulate adult neurogenesis in vitro.
Scientific Reports, 7(1), 5309.

Moreira, M. A., Guerra, R. O., do Nascimento Falcão Freire, A., Dos Santos Gomes, C., & Maciel,
Á. C. C. (2016). Depressive symptomatology and cortisol concentrations in elderly community
residents: A cross-sectional study. *Aging Clinical and Experimental Research, 28*(1), 131–137.

Mota, N. B., Copelli, M., & Ribeiro, S. (2017). Thought disorder measured as random speech
structure classifies negative symptoms and schizophrenia diagnosis 6 months in advance. *NPJ
Schizophrenia, 3*, 18.

Murphy, M. J., & Peterson, M. J. (2015). Sleep disturbances in depression. *Sleep Medicine Clinics,
10*(1), 17–23.

Murphy, S. E., Norbury, R., O'Sullivan, U., Cowen, P. J., & Harmer, C. J. (2009). Effect of a single
dose of citalopram on amygdala response to emotional faces. *British Journal of Psychiatry,
194*(6), 535–540.

Northoff, G., Heinzel, A., de Greck, M., Bermpohl, F., Dobrowolny, H., & Panksepp, J. (2006). Self-referential processing in our brain – A meta-analysis of imaging studies on the self. *NeuroImage, 31*(1), 440–457.

Nugent, A. C., Davis, R. M., Zarate, C. A., Jr., & Drevets, W. C. (2013). Reduced thalamic volumes in major depressive disorder. *Psychiatry Research, 213*(3), 179–185.

Oram, M. (2014). Efficacy and enlightenment: LSD psychotherapy and the Drug Amendments of 1962. *Journal of the History of Medicine and Allied Sciences, 69*(2), 221–250.

Osimo, E. F., Baxter, L. J., Lewis, G., Jones, P. B., & Khandaker, G. M. (2019). Prevalence of low-grade inflammation in depression: A systematic review and meta-analysis of CRP levels. *Psychological Medicine, 49*(12), 1958–1970.

Osório, F. d. L., Sanches, R. F., Macedo, L. R., dos Santos, R. G., Maia-de-Oliveira, J. P., Wichert-Ana, L., de Araujo, D. B., Riba, J., Crippa, J. A., & Hallak, J. E. (2015). Antidepressant effects of a single dose of ayahuasca in patients with recurrent depression: A preliminary report. *Revista Brasileira de Psiquiatria, 37*(1), 13–20.

Otte, C., Gold, S. M., Penninx, B. W., Pariante, C. M., Etkin, A., Fava, M., Mohr, D. C., & Schatzberg, A. F. (2016). Major depressive disorder. *Nature Reviews. Disease Primers, 2*, 16065.

Palhano-Fontes, F., Andrade, K. C., Tofoli, L. F., Jose, A. C. S., Crippa, A. S., Hallak, J. E. C., Ribeiro, S., & De Araujo, D. B. (2015). The psychedelic state induced by ayahuasca modulates the activity and connectivity of the default mode network. *PLoS One, 10*(2), e0118143.

Palhano-Fontes, F., Barreto, D., Onias, H., Andrade, K. C., Novaes, M. M., Pessoa, J. A., Mota-Rolim, S. A., Osório, F. L., Sanches, R., Dos Santos, R. G., Tófoli, L. F., de Oliveira Silveira, G., Yonamine, M., Riba, J., Santos, F. R., Silva-Junior, A. A., Alchieri, J. C., Galvão-Coelho, N. L., Lobão-Soares, B., et al. (2019). Rapid antidepressant effects of the psychedelic ayahuasca in treatment-resistant depression: A randomized placebo-controlled trial. *Psychological Medicine, 49*(4), 655–663.

Papageorgiou, C., & Wells, A. (2004). Nature, functions, and beliefs about depressive rumination. In C. Papageorgio & A. Wells (Eds.), *Depressive rumination: Nature, theory and treatment*. Hoboken, NJ: Wiley.

Pasquini, L., Palhano-Fontes, F., & Araujo, D. B. (2020). Subacute effects of the psychedelic ayahuasca on the salience and default mode networks. *Journal of Psychopharmacology, 34*(6), 623–635. https://doi.org/10.1177/0269881120909409.

Perlis, M. L., Smith, M. T., Jungquist, C., Nowakowski, S., Orff, H., & Soeffing, J. (2010). Cognitive-behavioral therapy for insomnia. In H. Attarian (Ed.), *Clinical handbook of insomnia* (pp. 281–296). Totowa, NJ: Humana Press.

Petri, G., Expert, P., Turkheimer, F., Carhart-Harris, R., Nutt, D., Hellyer, P. J., & Vaccarino, F. (2014). Homological scaffolds of brain functional networks. *Journal of the Royal Society, Interface, 11*(101), 20140873. https://doi.org/10.1098/rsif.2014.0873.

Poland, R. E., McCracken, J. T., Lutchmansingh, P., & Tondo, L. (1992). Relationship between REM sleep latency and nocturnal cortisol concentrations in depressed patients. *Journal of Sleep Research, 1*(1), 54–57.

Raichle, M. E., MacLeod, A. M., Snyder, A. Z., Powers, W. J., Gusnard, D. A., & Shulman, G. L. (2001). A default mode of brain function. *Proceedings of the National Academy of Sciences of the United States of America, 98*(2), 676–682.

Ramaker, M. J., & Dulawa, S. C. (2017). Identifying fast-onset antidepressants using rodent models. *Molecular Psychiatry, 22*(5), 656–665.

Riemann, D., Berger, M., & Voderholzer, U. (2001). Sleep and depression – Results from psychobiological studies: An overview. *Biological Psychology, 57*(1–3), 67–103.

Romeo, B., Choucha, W., Fossati, P., & Rotge, J.-Y. (2015). Meta-analysis of short- and mid-term efficacy of ketamine in unipolar and bipolar depression. *Psychiatry Research, 230*(2), 682–688.

Ross, S., Bossis, A., Guss, J., Agin-Liebes, G., Malone, T., Cohen, B., Mennenga, S. E., Belser, A., Kalliontzi, K., Babb, J., Su, Z., Corby, P., & Schmidt, B. L. (2016). Rapid and sustained symptom reduction following psilocybin treatment for anxiety and depression in patients with life-threatening cancer: A randomized controlled trial. *Journal of Psychopharmacology, 30*(12), 1165–1180.

Saito, J., Shibasaki, J., Shimokaze, T., Kishigami, M., Ohyama, M., Hoshino, R., Toyoshima, K., & Itani, Y. (2016). Temporal relationship between serum levels of interleukin-6 and C-reactive protein in therapeutic hypothermia for neonatal hypoxic-ischemic encephalopathy. *American Journal of Perinatology, 33*(14), 1401–1406.

Sanches, R. F., de Lima Osório, F., Dos Santos, R. G., Macedo, L. R. H., Maia-de-Oliveira, J. P., Wichert-Ana, L., de Araujo, D. B., Riba, J., Crippa, J. A. S., & Hallak, J. E. C. (2016). Antidepressant effects of a single dose of ayahuasca in patients with recurrent depression: A SPECT study. *Journal of Clinical Psychopharmacology, 36*(1), 77–81.

Santiago, G. T. P., de Menezes Galvão, A. C., de Almeida, R. N., Mota-Rolim, S. A., Palhano-Fontes, F., Maia-de-Oliveira, J. P., de Araújo, D. B., Lobão-Soares, B., & Galvão-Coelho, N. L. (2020). Changes in cortisol but not in brain-derived neurotrophic factor modulate the association between sleep disturbances and major depression. *Frontiers in Behavioral Neuroscience, 14*, 44.

Savoldi, R., Polari, D., Pinheiro-da-Silva, J., Silva, P. F., Lobao-Soares, B., Yonamine, M., Freire, F. A. M., & Luchiari, A. C. (2017). Behavioral changes over time following ayahuasca exposure in zebrafish. *Frontiers in Behavioral Neuroscience, 11*, 139.

Schuch, J. J. J., Roest, A. M., Nolen, W. A., Penninx, B. W. J. H., & de Jonge, P. (2014). Gender differences in major depressive disorder: Results from the Netherlands study of depression and anxiety. *Journal of Affective Disorders, 156*, 156–163.

Schultes, R. E., & Hofmann, A. (1979). *Plants of the gods: Origins of hallucinogenic use*. London, UK: Hutchinson.

Schulz, K. F., Altman, D. G., Moher, D., & CONSORT Group. (2011). CONSORT 2010 statement: Updated guidelines for reporting parallel group randomised trials. *International Journal of Surgery, 9*(8), 672–677.

Schutte-Rodin, S., Broch, L., Buysse, D., Dorsey, C., & Sateia, M. (2008). Clinical guideline for the evaluation and management of chronic insomnia in adults. *Journal of Clinical Sleep Medicine, 4*(5), 487–504.

Serafini, G. (2012). Neuroplasticity and major depression, the role of modern antidepressant drugs. *World Journal of Psychiatry, 2*(3), 49–57.

Sheline, Y. I., Barch, D. M., Donnelly, J. M., Ollinger, J. M., Snyder, A. Z., & Mintun, M. A. (2001). Increased amygdala response to masked emotional faces in depressed subjects resolves with antidepressant treatment: An fMRI study. *Biological Psychiatry, 50*(9), 651–658.

Sheline, Y. I., Barch, D. M., Price, J. L., Rundle, M. M., Vaishnavi, S. N., Snyder, A. Z., Mintun, M. A., Wang, S., Coalson, R. S., & Raichle, M. E. (2009). The default mode network and self-referential processes in depression. *Proceedings of the National Academy of Sciences of the United States of America, 106*(6), 1942–1947.

Shen, X., Reus, L. M., Cox, S. R., Adams, M. J., Liewald, D. C., Bastin, M. E., Smith, D. J., Deary, I. J., Whalley, H. C., & McIntosh, A. M. (2017). Subcortical volume and white matter integrity abnormalities in major depressive disorder: Findings from UK Biobank imaging data. *Scientific Reports, 7*(1), 5547.

Siegel, J. M. (2017). Rapid eye movement sleep. In M. Kryger, T. Roth, & W. Dement (Eds.), *Principles and practice of sleep medicine* (pp. 78–95). London, UK: Elsevier.

Siegle, G. J., Thompson, W. K., Collier, A., Berman, S. R., Feldmiller, J., Thase, M. E., & Friedman, E. S. (2012). Toward clinically useful neuroimaging in depression treatment. *Archives of General Psychiatry, 69*(9), 913.

Sonawalla, S. B., & Rosenbaum, J. F. (2002). Placebo response in depression. *Dialogues in Clinical Neuroscience, 4*(1), 105–113.

Soriano-Mas, C., Hernández-Ribas, R., Pujol, J., Urretavizcaya, M., Deus, J., Harrison, B. J., Ortiz, H., López-Solà, M., Menchón, J. M., & Cardoner, N. (2011). Cross-sectional and longitudinal assessment of structural brain alterations in melancholic depression. *Biological Psychiatry, 69*(4), 318–325.

Stahl, S. M. (2013). *Stahl's essential psychopharmacology: Neuroscientific basis and practical applications.* Cambridge, UK: Cambridge University Press.

Starr, L. R., Hammen, C., Brennan, P. A., & Najman, J. M. (2012). Serotonin transporter gene as a predictor of stress generation in depression. *Journal of Abnormal Psychology, 121*(4), 810–818.

Tadić, A., Wagner, S., Gorbulev, S., Dahmen, N., Hiemke, C., Braus, D. F., & Lieb, K. (2011). Peripheral blood and neuropsychological markers for the onset of action of antidepressant drugs in patients with major depressive disorder. *BMC Psychiatry, 11*, 16.

Thomas, G., Lucas, P., Capler, N. R., Tupper, K. W., & Martin, G. (2013). Ayahuasca-assisted therapy for addiction: Results from a preliminary observational study in Canada. *Current Drug Abuse Reviews, 6*(1), 30–42.

Tsuno, N., Besset, A., & Ritchie, K. (2005). Sleep and depression. *The Journal of Clinical Psychiatry, 66*(10), 1254–1269.

Tu, M. T., Zunzunegui, M.-V., Guerra, R., Alvarado, B., & Guralnik, J. M. (2013). Cortisol profile and depressive symptoms in older adults residing in Brazil and in Canada. *Aging Clinical and Experimental Research, 25*(5), 527–537.

Videbech, P. (2000). PET measurements of brain glucose metabolism and blood flow in major depressive disorder: A critical review. *Acta Psychiatrica Scandinavica, 101*(1), 11–20.

Viol, A., Palhano-Fontes, F., Onias, H., de Araujo, D. B., Hövel, P., & Viswanathan, G. M. (2019). Characterizing complex networks using entropy-degree diagrams: Unveiling changes in functional brain connectivity induced by ayahuasca. *Entropy, 21*(2), 128.

Viol, A., Palhano-Fontes, F., Onias, H., de Araujo, D. B., & Viswanathan, G. M. (2017). Shannon entropy of brain functional complex networks under the influence of the psychedelic ayahuasca. *Scientific Reports, 7*(1), 7388.

Vreeburg, S. A., Hoogendijk, W. J. G., DeRijk, R. H., van Dyck, R., Smit, J. H., Zitman, F. G., & Penninx, B. W. J. H. (2013). Salivary cortisol levels and the 2-year course of depressive and anxiety disorders. *Psychoneuroendocrinology, 38*(9), 1494–1502.

Więdłocha, M., Marcinowicz, P., Krupa, R., Janoska-Jaździk, M., Janus, M., Dębowska, W., Mosiołek, A., Waszkiewicz, N., & Szulc, A. (2018). Effect of antidepressant treatment on peripheral inflammation markers – A meta-analysis. *Progress in Neuro-Psychopharmacology & Biological Psychiatry, 80*, 217–226.

Williams, L. M., Korgaonkar, M. S., Song, Y. C., Paton, R., Eagles, S., Goldstein-Piekarski, A., Grieve, S. M., Harris, A. W. F., Usherwood, T., & Etkin, A. (2015). Amygdala reactivity to emotional faces in the prediction of general and medication-specific responses to antidepressant treatment in the randomized iSPOT-D trial. *Neuropsychopharmacology, 40*(10), 2398–2408.

Willner, P., & Belzung, C. (2015). Treatment-resistant depression: Are animal models of depression fit for purpose? *Psychopharmacology, 232*(19), 3473–3495.

Winne, J., Boerner, B. C., Malfatti, T., Brisa, E., Doerl, J., Nogueira, I., Leão, K. E., & Leão, R. N. (2020). Anxiety-like behavior induced by salicylate depends on age and can be prevented by a single dose of 5-MeO-DMT. *Experimental Neurology, 326*, 113175.

Wong, M. L., & Licinio, J. (2001). Research and treatment approaches to depression. *Nature Reviews Neuroscience, 2*(5), 343–351.

World Health Organization. (2017, February 22). *Depression* [Fact sheet]. World Health Organization. http://www.who.int/mediacentre/factsheets/fs369/en/

World Medical Association. (2001). World Medical Association Declaration of Helsinki. Ethical principles for medical research involving human subjects. *Bulletin of the World Health Organization, 79*(4), 373–374.

Zeifman, R. J., Palhano-Fontes, F., Hallak, J., Arcoverde, E., Maia-Oliveira, J. P., & Araujo, D. B. (2019). The impact of ayahuasca on suicidality: Results from a randomized controlled trial. *Frontiers in Pharmacology, 10*, 1325.

Zunszain, P. A., Anacker, C., Cattaneo, A., Carvalho, L. A., & Pariante, C. M. (2011). Glucocorticoids, cytokines and brain abnormalities in depression. *Progress in Neuro-Psychopharmacology & Biological Psychiatry, 35*(3), 722–729.

Chapter 3
Psychedelic Medicines: A Paradigm Shift from Pharmacological Substitution Towards Transformation-Based Psychiatry

Milan Scheidegger

Psychedelic Medicines Reveal Novel Drug Targets for Mental Health Disorders

Scientific interest into the neurobiology and phenomenology of drug-induced altered states of consciousness and their potential to facilitate therapeutic transformation has resumed. Promising directions include the rediscovery of psychoactive substances for the treatment of various mental health disorders (Nichols, Johnson & Nichols, 2017). During the last decade, the discovery of the antidepressant effect of the dissociative anesthetic ketamine fueled the search for novel drug targets for mood disorders (Murrough, Abdallah & Mathew, 2017). Ketamine was proposed to represent a novel class of rapid-acting glutamatergic antidepressants due to its immediate, albeit transient, antidepressant effect.

Likewise, the therapeutic potential of psilocybin was assessed for various conditions, including end-of-life anxiety, depression, obsessive-compulsive disorder, and nicotine and alcohol addiction, with promising preliminary results (Bogenschutz et al., 2015; Carhart-Harris et al., 2016; Johnson, Garcia-Romeu, Cosimano, & Griffiths, 2014; Moreno, Wiegand, Taitano & Delgado, 2006). Compared to the transient effects of ketamine, psilocybin showed more sustained improvements in depressive symptoms, even several weeks after a single administration, which continues to inspire the search for rapid-acting serotonergic drug targets.

Recently, due to its therapeutic potential, scientific interest in the N,N-dimethyltryptamine (DMT)-containing Amazonian plant medicine ayahuasca has started to grow (Frecska, Bokor & Winkelman, 2016). In South America, ayahuasca is used in ritualistic settings as traditional medicine for curative and religious purposes by indigenous and mestizo people. Currently, this plant medicine is spreading

M. Scheidegger (✉)
Department of Psychiatry, Psychotherapy and Psychosomatics, University Hospital of Psychiatry Zurich, Zurich, Switzerland
e-mail: milan.scheidegger@bli.uzh.ch

© Springer Nature Switzerland AG 2021

B. C. Labate, C. Cavnar (eds.), *Ayahuasca Healing and Science*,
https://doi.org/10.1007/978-3-030-55688-4_3

all over the world with increasing numbers of Westerners seeking ayahuasca (Labate & Cavnar, 2014; Labate & Jungaberle, 2011; Tupper, 2008).

This rapid dissemination points (a) to an unmet need for increased psychological wellbeing through rapid and sustainable relief from various mental health problems and (b) to the observational evidence that ayahuasca facilitates transformational processes with beneficial health outcomes. Indeed, ayahuasca users reported greater wellbeing compared to both general psychedelic and nonpsychedelic drug users in a recent global online survey (Lawn et al., 2017).

First evidence from preclinical and human studies suggest that ayahuasca has the potential to reduce symptoms of anxiety and depression (Santos et al., 2016). In an open-label clinical trial in depressed patients, a single dose of ayahuasca showed immediate improvements in depressive symptoms for up to 3 weeks (Osório et al., 2015). Recent data from a randomized controlled trial confirm rapid antidepressant effects of ayahuasca in treatment-resistant patients (Palhano-Fontes et al., 2018, in this volume). However, more systematic research is needed to move beyond observational evidence and further verify therapeutic efficacy and biomechanisms of ayahuasca under controlled conditions.

An Emerging Paradigm Shift from Substitution- to Transformation-Based Therapy

Compared to conventional psychotropic drugs, psychedelic medicines represent a novel class of rapid-acting serotonergic compounds with different mechanisms of action. While ketamine has been mostly used as stand-alone drug treatment with a more biochemically driven mechanism of action, other psychoactive substances such as LSD, MDMA, or psilocybin are preferably studied as adjuncts to psychotherapy, involving an element of guidance, introspection, and self-discovery associated with their positive clinical outcomes. Accordingly, psychedelics administered without psychological support or in less suitable settings without guidance may have limited or even adverse effects. Because psychedelics are increasingly recognized as facilitators of psychotherapeutic processes, rather than just molecular drug targets, it is very likely that we are facing an emerging paradigm shift from pharmacological substitution towards transformation-based psychiatric treatment.

The standard treatment paradigm with pharmaceutical psychotropic drugs is essentially substitution-oriented. Accordingly, mental illness is primarily treated as a disorder caused by certain deficits in brain metabolism, such as neurotransmitter deficiencies of serotonin, norepinephrine, and dopamine that are associated with specific symptoms such as anxious and depressed mood or lack of motivation and energy (Nutt, 2008). Consequently, successful symptomatic treatment includes pharmacological substitution of these neurotransmitter deficits daily, e.g., with an SSRI antidepressant that increases serotonin levels in the brain.

In contrast, the psychedelic-assisted treatment approach as proposed here follows the distinct model of *transformation-based therapy*; psychological illness is not primarily resulting from a deficient brain metabolism that needs to be substituted, but follows from misguided bio-psycho-social processes that await transformation. In contrast to substitution-oriented treatment approaches, the goal of transformation-based therapy is to move the patient from a suboptimal state into a more adaptive state of consciousness that needs to be further stabilized by concurrent psychotherapy. Although maladaptive processes, including stress-related changes in brain chemistry and function, may be associated with deficits in serotonergic or glutamatergic neurotransmission (Popoli, Yan, McEwen, & Sanacora, 2012), these neurotransmitter deficiencies are not necessarily treated as the root cause of psychological suffering. Accordingly, the psychedelic-assisted therapy approach is less based on a long-term pharmacological substitution of neurotransmitters than on a rapid change of dysfunctional neuronal regulatory circuits that allow for a more sustainable process of bio-psycho-social adaptation. The concept of transformation-based therapy, as introduced here, is therefore less focused on the resolution of specific symptoms and their substitution by means of psychotropic drugs, than on empowering the innate dynamic capacity of the patient to regain sustained access to more adaptive states of consciousness that are naturally followed by symptom resolution on many levels.

In clinical settings, psychedelics are hypothesized to work through a biphasic mechanism of action. They act primarily as nonspecific "catalysts" that facilitate the potential for transformative change that has to be specifically guided by psychotherapeutic processes including (1) identification of the dysfunctional states of mental stagnation, (2) setting of conscious intentions towards transformation, (3) providing psychological support during drug-induced altered states of consciousness, followed by (4) posttreatment psychotherapeutic integration to further stabilize adaptive states and prevent relapse into previous dysfunctional states. Secondarily, in the medium and long term, catalytic exposure to psychedelics parallels the effects of adaptogens on the human body and mind. Conceptually, adaptogens are stress response modifiers that nonspecifically increase an organism's resistance to various adverse influences, thereby promoting adaptation to the environment (Panossian, 2017). As outlined below, the psychophysiological effects of ayahuasca are particularly aligned with the adaptogenic activity of specific herbal medicines that exhibit polyvalent beneficial effects on physical and mental health.

Adaptogenic Effects of Ayahuasca on Physical and Mental Health

Current uses of adaptogens in pharmacotherapy are related to their potential for treating stress-related physical, mental, and behavioral disorders. Although the mechanisms of action of adaptogens are specifically related to stress protection and environmental adaptation, it is very unlikely that the pharmacological activity of phytochemicals is associated with only one specific molecular target. Instead, the metabolic regulation of homeostasis by adaptogens at the cellular and systems levels is associated with a multitude of targets, which requires a holistic network pharmacology approach (Panossian, 2017).

The adaptogenic activity of the ayahuasca brew is predominantly related to its main active ingredients: the psychotropic indole alkaloid N,N-dimethyltryptamine (DMT) and different β-carboline alkaloids (e.g., harmine, harmaline, and tetrahydroharmine). In order for DMT to become bioavailable, oral formulations usually contain plant-based sources of DMT (e.g., from *Psychotria viridis*) combined with β-carbolines (e.g., from *Banisteriopsis caapi*) that act as selective reversible MAO-A inhibitors to prevent degradation of DMT in the body (Callaway et al., 1996). DMT is a structural analogue of serotonin and is widely found in nature, including plants, mammalian organisms, and in human brains and body fluids. DMT interacts with a variety of serotonin receptors, mainly 5HT2A, 5HT2C, and 5HT1A, and also with glutamate, dopamine, acetylcholine, sigma-1, and trace amine-associated receptors (Carbonaro & Gatch, 2016; Valle et al., in this volume).

The widespread changes in neuronal excitability resulting from 5-HT2A receptor activation induce marked psychotropic effects on mood and cognition (Vollenweider & Kometer, 2010). Compared to the rapid onset and experiential intensity of intravenous DMT (Strassman, Qualls, Uhlenhuth & Kellner, 1994), the subjective effects of orally activated DMT in ayahuasca preparations appear in a slower, progressive way and disappear after 4–6 h (Riba et al., 2003), which is much more amenable for psychotherapeutic applications. Similar to other adaptogenic phytochemicals, regular ayahuasca use is considerably safe and does not produce pharmacological tolerance even after repeated administration (Barbosa et al., 2012; Santos et al., 2012).

Traditionally, herbal medicines were used to treat the symptoms of several disorders, which is in line with the broad range of beneficial effects of the β-carboline constituents in ayahuasca and their immunomodulatory, neurotrophic, neuroprotective, cognitive-enhancing, and antidepressant effects, among many others (Moloudizargari, Mikaili, Aghajanshakeri, Asghari, & Shayegh, 2013; Morales-García et al., 2017; Santos & Hallak, 2016). Most notably, recent evidence shows that also DMT is neuroprotective by increasing neuronal cell survival under oxidative stress, suggesting its potential adaptogenic role in counteracting the adverse effects of stress in the body (Szabo et al., 2016). This is of crucial relevance, since stress-related disturbances in brain function are associated with the pathogenesis of many psychiatric disorders. Hence, the pharmacology of ayahuasca is a typical example of the multitarget action of adaptogens and may be particularly well suited for the systemic treatment of complex diseases, chronic conditions, and psychiatric syndromes, where conventional reductionist approaches have often been disappointing.

Adaptogenic Effects of Ayahuasca on Brain Dynamics

Understanding how specific drugs affect the neurodynamics of adaptive and mal-adaptive behavioral states represents a current research priority in biomedical science. It was recently discovered that resting-state brain activity is organized into multiple large-scale functional connectivity networks that enable efficient neuronal communication between different brain hubs that subserve complex mental functions (Fox, Snyder, Corbetta, Van Essen & Raichle, 2005). Dysfunctional brain network dynamics was found to correlate with symptom severity in various psychiatric conditions, with recovery of connectivity observed following successful treatment (Fox & Greicius, 2010). Hence, psychedelic medicines have been proposed to provide promising tools to enhance therapeutic transformation by resetting the neurocircuits that underlie maladaptive neurobehavioral states (Nichols, Johnson & Nichols, 2017).

The effects produced by classical psychedelics are principally mediated through the serotonergic 5-HT1A/2A receptors (Nichols, 2016). Serotonergic psychedelics, such as psilocybin, LSD, and DMT, share the common biomechanism of increased excitability of the prefrontal cortex via 5HT2A receptor agonism (Vollenweider & Kometer, 2010). Due to widespread changes in neuronal excitability, both short-lived and persisting connectivity networks emerge between brain regions during the psychedelic state that normally do not show significant functional association (Petri et al., 2014; Tagliazucchi, Carhart-Harris, Leech, Nutt & Chialvo, 2014).

Drug-induced increases in global brain connectivity allow for better integration of information, while increased metastability and repertoire of connectivity motives accelerate phase transitions and transformation processes (Gallimore, 2015). The breakdown between conventional pathways of neuronal communication and the formation of novel functional associations is in line with a recent hypothesis that psychedelics lead to entropic activity within the brain (Carhart-Harris et al., 2014). During the drug-induced high entropy state, brain activity is less constrained and allows for a larger repertoire of states that succeed in a more random fashion.

The observed increase in entropy may be related to the temporary removal of some of the restrictions that are necessary for sustaining ordinary consciousness resulting in a more flexible mental state. Consistent with this hypothesis, increased brain entropy and changes in global network integration were found subsequent to ayahuasca ingestion (Viol, Palhano-Fontes, Onias, Araujo & Viswanathan, 2017). Phenomenologically, this slightly higher entropy state is perceived as experientially richer, but not as informative as normal waking consciousness: The way meaning is attributed to things is broadened, expanding the sense of reality, including experiential novelty or even bizarreness. This higher entropy state is associated with increased cognitive flexibility, creativity, and imagination (Gallimore, 2015). From a therapeutic perspective, such broadband alterations in functional network connectivity may increase global brain plasticity, whereby the maladaptive patterns that are linked to dysfunctional emotions, thought patterns, and behaviors can be transformed.

Depressive symptomatology is thought to arise from a distorted sense of self, one's relation to others, and the external world, as reflected in disrupted brain connectivity in self-referential information processing networks (Nejad, Fossati & Lemogne, 2013). If depression is redefined as the tendency to enter into, and the inability to disengage from, a negative mood state, rather than the mood state per se (Holtzheimer & Mayberg, 2011), then targeting maladaptive forms of rumination and their underlying neural correlates may become key for effective treatment. Accordingly, active pharmacological disruption of pathological connectivity in core networks relevant to depression would be expected to mediate adaptive changes in brain network dynamics within the plasticity window that opens after drug administration.

Recently, the default mode network (DMN) was proposed as a target for such network-based interventions, because it shows pathological hyperactivity and connectivity in depression and is linked to maladaptive patterns of ruminative self-referential information processing (Sheline, Price, Yan & Mintun, 2010). Interestingly, rapid-acting glutamatergic antidepressants such as ketamine were found to decrease connectivity within key brain networks such as the DMN (Lehmann et al., 2016; Scheidegger et al., 2012). A similar pattern of attenuated DMN activity or connectivity was found for serotonergic drugs such as psilocybin (Carhart-Harris et al., 2012), ayahuasca (Palhano-Fontes et al., 2015; Valle et al., in this volume), and following successful antidepressant treatment (Nejad, Fossati & Lemogne, 2013). This supports the notion that rapid antidepressant action relies on attenuation of self-referential information processing systems such as the DMN. In conclusion, these findings suggest that the pharmacological modulation of glutamate- and serotonin-responsive cerebral circuits that are associated with a disruption of the habitual brain network architecture and an overall increase in brain entropy, represent early biomechanisms to break out of the inflexible and self-referential modes of thinking that characterize depression.

Adaptogenic Effects of Ayahuasca on Mental Functioning

Besides the biological effects, there is a wide range of drug-induced psychedelic experiences that are valued as personally meaningful and deeply transformative. The psychotropic effects of ayahuasca typically include a heightened state of introspection characterized by increased awareness, dreamlike visions and geometric patterns, evocation of intense emotions, and recollection of autobiographical memories. Occasionally, ayahuasca visions also include archetypal or psycho-spiritual themes (Shanon, 2002). Accordingly, cerebral areas associated with episodic memories, internal sensations, and conscious experience of emotions were found to be activated following ayahuasca intake (Riba et al., 2006; Valle et al., in this volume), thereby allowing for improved cognitive-emotional integration.

Insight into maladaptive behavioral, emotional, and cognitive patterns reflects the general action of ayahuasca as a psycho-integrative medicine: brain activity that is normally driven by prefrontal cognitive control mechanisms is shifted towards intense emotionally laden discharges from the lower limbic areas of the brain (Frecska, Bokor & Winkelman, 2016). Contrary to other psychotropic drugs such as opioids, benzodiazepines, or antipsychotics that tend to cloud and narrow the field of consciousness, ayahuasca usually increases the level of mental clarity and emotional awareness compared to everyday consciousness. Thus, dysfunctional thoughts and emotions can be better recognized, restructured, and integrated, which accelerates and deepens the process of psychotherapeutic transformation. This may benefit patients suffering from unresolved emotional traumas and conflicts that await conscious resolution and reframing.

Many psychiatric symptoms result from early childhood trauma and attachment disorders, which limit the ability to relate and increase vulnerability due to lower stress resilience and reduced interpersonal coping strategies. Since early traumas affect the preverbal phases of human development, they are often difficult to resolve through cognitive-behavioral therapy, which is more palliative by providing basic skills to alleviate symptoms. For patients suffering from emotional neglect and attachment disorders, psychedelic-assisted therapy could resolve interpersonal deficits in more profound and sustainable ways. In supportive environments, psychedelics promote feelings of trust, empathy, bonding, closeness, tenderness, forgiveness, acceptance, and connectedness (Belser et al., 2017; Watts Day, Krzanowski, Nutt & Carhart-Harris, 2017). These empathogenic effects can serve in developing increased levels of relational embeddedness, self-acceptance, and self-care through direct experience of corrective emotions. With increased levels of connectedness to the self, to others, and to the environment (Carhart-Harris, Erritzoe, Haijen, Kaelen & Watts, 2017; Forstmann & Sagioglou, 2017), self-destructive behavioral patterns can be attenuated and general improvement in relationship homeostasis may be reached. This facilitates posttraumatic growth by contributing to the resolution of behavioral consequences of traumas by releasing the person from dysfunctional habits that underlie the dynamics of self-destructive and addictive behaviors (Argento, Capler, Thomas, Lucas & Tupper, in this volume). Indeed, a recent observational study highlights the therapeutic value of ayahuasca in reducing eating disorder symptoms by experiencing greater self-acceptance, increased capacity to regulate painful emotions, and gaining insights about the root causes of the illness (Lafrance et al., 2017; in this volume). Through enhanced awareness of the likely personal consequences of maladaptive behaviors, ayahuasca may provide a motivation for more sustained changes in attitudes and behaviors.

Another way of understanding psychedelic-assisted transformation is to compare it with the behavioral technique of exposure therapy: Anxiety is usually best resolved through repeated exposure to the feared object or context. Comparably, psychedelic experiences share the same sequence of exposing the patient to the entire possible spectrum of intense feelings and mental states. Ayahuasca experiences are often described as going through an initial state of confronting innermost fears: fear of insanity, fear of death, paranoid thoughts, or the despair of loneliness and rejection (Kjellgren, Eriksson & Norlander, 2009). If the subject is able to surrender, this challenging initial phase is usually followed by a sudden transformation of the experience into emotional relief, insightful reflections, states of self-transcendence, and changed worldview and outlook on life (Frecska, Bokor, & Winkelman, 2016). In analogy to behavioral exposure therapy that aims at decreasing fear responses, a state of nonspecific stress resistance could be achieved by adaptogens such as ayahuasca. Through repeated exposures to the psychedelic state, the mind trains its intrinsic capability to deal with difficult emotions that may finally lead to a more relaxed and accepting attitude towards general human suffering. Indeed, a recent observational study of people suffering from the death of a loved one found lower levels of grief in ayahuasca users, suggesting its potential to increase the capacity to deal with difficult emotions (González, Carvalho, Cantillo, Aixalá & Farré, 2017).

From a psychological perspective, enhanced cognitive flexibility, creativity, and imagination are generally found in psychedelic states, with long-term increases in creative problem-solving abilities and the personality trait of openness (Lebedev et al., 2016; Maclean, Johnson & Griffiths, 2011; Sweat, Bates & Hendricks, 2016). The adaptogenic potential of ayahuasca might therefore be specifically related to its ability to increase cognitive flexibility, e.g., through divergent thinking (Kuypers et al., 2016; Mason and Kuypers, in this volume). Ayahuasca was also found to elevate levels of mindfulness, self-compassion, and "decentering," which refers to the capacity to observe thoughts and emotions as transitory mental events without being trapped by them (Sampedro et al., 2017; Soler et al., 2016; Valle et al., in this volume). Changes in cognitive flexibility and mindfulness-related capabilities strongly support the implementation of concomitant psychotherapy to train patients how to stabilize and maintain more balanced states of mind that allow for a better adaptation to their environment. Indeed, longitudinal observational studies reported that ayahuasca users are well adapted and integrated in their social, working, and familiar environments and that ayahuasca is used as a tool for personal development (Bouso et al., 2012).

Converging Roles of Mindfulness and Psychedelics for Transformational Mental Healthcare

From a phenomenological perspective, some drug-induced peak experiences share a striking resemblance to mystical experiences reached in deep meditation. In these states, the usual ego boundaries are dissolved to a degree where the subject of experience feels universally connected with all existence. Such transformative experiences can be achieved by various consciousness-altering techniques, such as artistic creativity, ecstasy, psychedelics, trance, meditation, and relaxation.

In his cartography of ecstatic and meditative states (see Fig. 3.1), Roland Fischer considered the varieties of human experience on (1) the perception-hallucination continuum of increasing *ergotropic arousal* (e.g., creative, psychotropic, and ecstatic experiences) and (2) along the perception-meditation continuum of

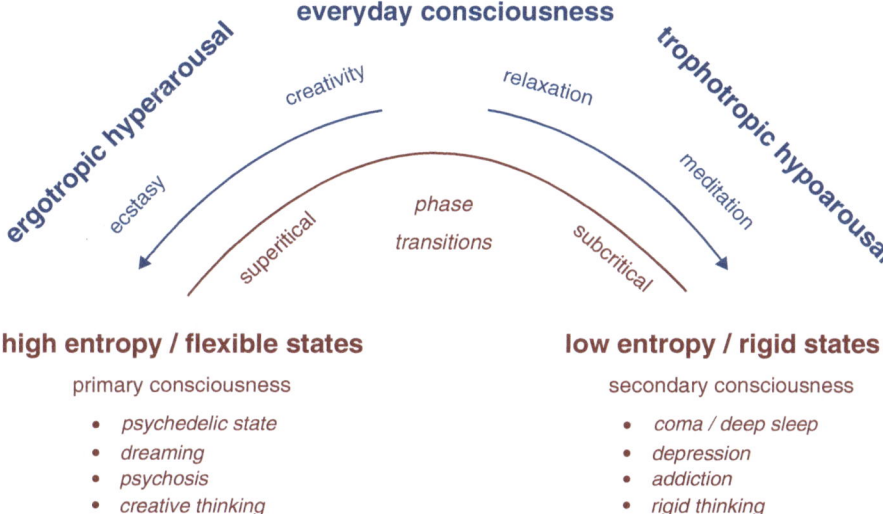

Fig. 3.1 Phenomenological map of altered states of consciousness. A continuum of ergotropic vs. trophotropic arousal (top spectrum; adapted from Fischer, 1971) is contrasted with high entropy primary consciousness vs. low entropy secondary consciousness (bottom spectrum; adapted from Carhart-Harris et al., 2014)

increasing *trophotropic arousal* (e.g., hypo-aroused meditative states) (Fischer, 1971). Both directions represent changes in self-referential processing in terms of mutually exclusive psychophysiological alterations of everyday consciousness. While the creative person or meditation expert has learned to purposefully navigate through different states of consciousness, mental illness is characterized as the loss of freedom to skillfully navigate towards more adaptive mental states.

With progression towards peak or breakthrough states of ego dissolution and self-transcendence, therapeutic transformation, or "unlearning" of previously conditioned self-referential behaviors and attitudes, is facilitated. But, due to their mutually exclusive psychophysiology, ordinary and altered states of consciousness cannot occur at the same time, which raises questions of transfer and stabilization: How can brief transformative moments of altered perception be integrated within the boundaries of everyday consciousness? To further develop and maintain mindfulness-related capabilities, sustainable therapeutic transformation strongly relies on concomitant mental training, otherwise short-term beneficial effects are likely to fade away.

Fischer's two-sided phenomenological map of altered states of consciousness parallels contemporary adoptions of Freudian ideas into the entropic brain hypothesis (Carhart-Harris et al., 2014). Freud divided the psyche into two fundamentally distinct modes of activity: the primary process and the secondary process (Freud, 1995). Dreaming and psychosis typify a primitive style of thinking, directly animated by the drives, that is dominant in infancy and characterized by disorder, conceptual paradox, symbolic imagery, intense emotions, and animistic thinking. The entropic brain hypothesis refers to this mode of cognition as "primary consciousness," which is characterized by unconstrained thinking and less ordered higher brain entropy dynamics.

Healthy psychological life results from an adaptive equilibrium between *primary* and *secondary processes*, which presuppose the binding of free-flowing energy, by intervening as a system of control characterized by order, precision, conceptual consistency, controlled emotions, and rational thinking. This constrained style of cognition is dominant in normal waking consciousness, referred to as "secondary consciousness," which is associated with more ordered neurodynamics.

Following this model, the phenomenology of psychedelic states is characterized by phase transitions from more rigid and inflexible secondary states towards less-constrained primary states characterized by elevated levels of brain entropy and uncertainty, corresponding to the ergotropic spectrum of consciousness (see Fig. 3.1). This psychedelic-induced regression into primary consciousness through temporarily relaxed neurophysiological constraints explains psychedelic phenomena such as expanded perception, vivid imagination, emotional intensity, conceptual paradox, and loss of usual self-boundaries.

Interestingly, drug-induced altered states (e.g., ketamine, psilocybin or ayahuasca) share very similar neuronal biosignatures as altered states reached by non-pharmacological methods (e.g., meditation): The expanded state of consciousness is generally associated with reduced activity or connectivity in cortical midline areas that mediate self-referential information processing, such as the default mode network (DMN) (Brewer et al., 2011; Carhart-Harris et al., 2012; Palhano-Fontes et al., 2015; Scheidegger et al., 2012). Such states of attenuated self-referential processing are of particular clinical importance. While the field of consciousness is usually narrowed in depressive, anxious, obsessive-compulsive, or addictive states, it can be broadened towards more flexible and less biased modes of conscious awareness as a result of psychedelic-assisted therapy. Negative emotional biases related to critical self-referential processing are particularly well-resolved through mindfulness training that develops skills such as meta-awareness, self-regulation, and a positive relationship between self and other that transcends self-focused needs and increases pro-social characteristics (Vago & Silbersweig, 2012). Likewise, improvements in mindfulness-related capabilities may be enhanced not only by meditation practices but also through distinctive features of transformative experiences induced by ayahuasca or psilocybin (Griffiths et al., 2017; Soler et al., 2016).

In settings that are supportive for cultivating mindfulness, drug-induced mystical-type experiences facilitate enduring trait-level increases in pro-social attitudes and behaviors such as interpersonal closeness, gratefulness, forgiveness, and purposefulness, thereby supporting psycho-spiritual development (Griffiths et al., 2017). Most notably, the intensity of drug-induced mystical-type experiences was found to be predictive of positive long-term outcomes in several clinical studies (Bogenschutz et al., 2015; Garcia-Romeu et al., 2014; Griffiths et al., 2016; Roseman et al., 2018; Ross et al., 2016). Through expanded states of consciousness, patients may become increasingly aware of how to navigate their own consciousness and break out from maladaptive states through increased meta-awareness and cognitive flexibility. The induction of transformative experiences and mindfulness-related capabilities is therefore expected to contribute to the noted increases in wellbeing and mood following exposure to psychedelics.

From a conceptual perspective, it makes sense to assume that through attenuation of ego boundaries and expansion of consciousness towards states of self-transcendence, a transient state of release from clinical symptoms is induced. This follows logically from the notion that symptoms arise at the interface between the individual and its environment, signaling a homeostatic imbalance with a need for behavioral adjustment or removal of noxious influences. When ego boundaries are dissolved to a degree where the conceptual distinction between external and internal world is suspended, there is no interface for the generation of symptoms anymore. With progressive reorientation in body, space, and time, the constructivist and process-related dynamics of symptom formation and possible avenues for their resolution become more evident to the patient. From this clinical epistemological perspective, it can be argued that drug-induced states of self-transcendence provide unique chances to gain metacognitive awareness into the intrinsic dynamic of root causes and processes underlying the formation of symptoms and open a psychotherapeutic window for adaptation, transformation, and personal growth.

Transformational Psychotherapy as an Emerging New Treatment Paradigm

The transformation-based psychotherapy framework, as proposed here, is very aligned with process-oriented adaptations of third wave cognitive behavioral therapies (CBT). While standard CBT is more outcome-oriented, modifying psychiatric symptoms as directly as possible, the third wave of CBT emphasizes a set of new behavioral and cognitive approaches such as emotions, relationships, values, mindfulness, acceptance, and meta-cognition. Established therapeutic approaches include acceptance and commitment therapy (ACT), dialectical behavior therapy (DBT), mindfulness-based cognitive therapy (MBCT), meta-cognitive therapy (MCT), compassion-focused therapy (CFT), and many others (Hayes & Hofmann, 2017). As reviewed above, the adaptogenic effects of ayahuasca on psychological functioning are particularly aligned with mindfulness-based approaches (Soler et al., 2016; Valle et al., in this volume), that aim to increase acceptance, self-compassion, present-moment and meta-cognitive awareness, and psychological flexibility in the service of achieving core life values (Hayes, Levin, Plumb-Vilardaga, Villatte & Pistorello, 2013).

Adaptive changes in value orientation are likely to follow from recognition of maladaptive orientations (e.g., self-centeredness, addictive cravings, experiential avoidance). Accordingly, a process-oriented view on therapeutic transformation is encouraged, starting with the assessment of the current state of stagnation and identifying the direction for adaptive change by encouraging an awareness of the patient's potential for self-direction. Ayahuasca may become particularly useful to support this process of self-discovery, creation of meaning, and enriching the patient's self-awareness towards the attainment of insight, sense of agency, identity, self-esteem, emotional awareness, and increased psychological wellbeing.

Although psychedelic medicines show considerable promise as tools for psychotherapeutic transformation, they also pose considerable challenges on currently established treatment paradigms. In the case of depression, a significant number of patients treated according to the substitution-based approach with an SSRI antidepressant fail to respond adequately (Gaynes, 2009) or relapse shortly after initial therapeutic progress. Are these the patients that might benefit from psychedelic-assisted therapy instead?

Possible answers are provided by the bipartite model of brain serotonin function that was proposed recently to contrast different therapeutic approaches to emotional processing (Carhart-Harris & Nutt, 2017). The chronic antidepressant action of SSRIs reduces limbic responsiveness and increases emotional resilience to environmental stress, likely via postsynaptic 5-HT1A receptor signaling. This contrasts with the more important role of 5-HT2A receptor signaling for psychedelics, which promotes emotional release and active coping, and requires willingness of the patient to confront and work through sources of stress during the window of plasticity opening after drug administration.

Psychedelic-assisted therapy might therefore be particularly well suited for patients with so-called psychoneurotic disorders to loosen otherwise fixed maladaptive patterns of cognition and behavior so they might overcome symptoms that are usually more challenging to resolve within the substitution-oriented paradigm. But what about the treatment-resistant patients with more serious psychiatric illnesses or complex traumatic histories, personality disorders, and other comorbidities? The clinical decision whether a patient is better cared for in a palliative or substitution-oriented framework or whether he might benefit from psychedelic-assisted transformational psychotherapy necessitates a careful medical screening, indication, and risk-benefit analysis, which requires a high level of clinical expertise.

Another limiting factor is that pretreatment with serotonergic antidepressants attenuates the effects of serotonergic psychedelics due to altered 5HT2A-receptor function, therefore substitution- and transformation-based approaches cannot be combined at the same time (Bonson, Buckholtz & Murphy, 1996). Hence, if psychedelic-assisted therapy is becoming an established treatment approach in the near future, strategies to stratify patient populations to substitution- versus transformation-based treatments become crucial to ensure clinical benefit and safety.

The Ritual Use of Psychedelic Medicines in Nonclinical Settings

The effects of psychedelic medicines as nonspecific amplifiers of subjective experience are highly context-dependent. In nonclinical indigenous or psychospiritual settings, altered states of consciousness are traditionally experienced in group rituals or ceremonies. The ritual itself facilitates transitions between different states of consciousness through heightening levels of awareness, synchronizing bio-psycho-social rhythms, and shifting mental processes towards more creative-imaginative and emotional directions.

During the process of loosening of ego boundaries, the ritual provides a safe space for the loss of conscious control. When a group of individuals moves from a solid self-centered to a more liquid state of group dynamics, individuals sometimes feel like merging into a coherent group organism, accompanied by a sense of solidarity and transcendence that is perceived as deeply joyful, cathartic, purposeful, or even spiritual (Kent, 2010). Through setting predefined personal intentions before the ritual, an attitude of positive expectancy for individual or collective transformation is generated and supports the broadening of conscious perspectives from the habitual towards the realm of new possibilities.

Besides encouraging insight, meaning, compassion, and connectedness, rituals can also provoke challenging cathartic experiences, strengthening acceptance of unpleasant feelings, loss of control, and general human suffering. Hereby the whole spectrum of experiences that are part of human existence, including states of ego dissolution and self-transcendence, are safely contained within ritual practices that encourage the opportunity for mutual social support and emotional bonding among participants, motivating lasting behavioral changes. Further anthropological and ethnographical investigations about the traditional ritualistic use of psychedelic medicines (Labate & Cavnar, 2014; Labate Cavnar & Gearin, 2016) and the importance of contextual factors could be informative to design prospective therapeutic environments.

Future Directions and Challenges

The challenge of how a traditional indigenous practice can be translated into the Western world remains. Taking a pragmatic stance might be useful to resolve the tensions between different belief systems that naturally create separation and dissent. If the commonly valued goal of medicinal use of ayahuasca is to reduce human suffering, then it seems legitimate to introduce a traditional Amazonian plant medicine into contemporary clinical practice if there is evidence-based proof for its therapeutic efficacy.

A pragmatic knowledge transfer does not necessarily require adopting the entire ceremonial context with an indigenous shaman and associated spiritual belief systems for the medicine to work properly. Pragmatism primarily motivates a careful analysis of the functional role of the medicine in a given context. What is the functional role of ayahuasca in the traditional indigenous context? How is therapeutic transformation facilitated, and what are the necessary conditions to secure participants safety and beneficial outcomes? The same pragmatic questions apply to other contexts: How does ayahuasca work in a way that makes sense to the Western mind? What are the essential functional conditions that need to be met to achieve a desired therapeutic outcome? How should the traditional setting be adapted and which belief systems do we create to make sense of the ayahuasca experience? The answers might look very different depending on context, because pragmatism is not oriented towards finding absolute truths, but is more focused on revealing what works best in a given context. While traditionalists might emphasize the need for strict rules and orthodoxy to preserve a specific tradition that evolved around the ingestion of ayahuasca, I am advocating here for the possibility of change and cultural evolution. If we allow different cultures and approaches to merge, there is true chance for discovery and innovation by means of pragmatic knowledge transfer.

The same reasoning applies to questions about whether it is necessary to work with the whole molecular plant matrix that is contained in the original ayahuasca plant species, or if a scientific version of "pharmahuasca" will have similar experiential and therapeutic properties, if only a limited number of active ingredients is extracted and standardized according to the protocols of pharmaceutical science and tested under empirically controlled laboratory conditions.

If science and medicine are ready to adopt ayahuasca into their pharmacopoeia of approved medicinal herbs, the future will teach us how pharmahuasca may become an evidence-based treatment for various mental health problems, and how the indigenous wisdom that evolved around the traditional use of ayahuasca may contribute to meet the challenges of intercultural knowledge transfer.

References

Argento, E., Capler, R., Thomas, G., Lucas, P., & Tupper, K. (this volume). Perspectives on healing and recovery from addiction with ayahuasca-assisted therapy among members of an Indigenous community in Canada. In B. C. Labate & C. Cavnar (Eds.), *Ayahuasca healing and science*. Cham: Springer.

Barbosa, P. C. R., Mizumoto, S., Bogenschutz, M. P., & Strassman, R. J. (2012). Health status of ayahuasca users. *Drug Testing and Analysis, 4*(7–8), 601–609. https://doi.org/10.1002/dta.1383.

Belser, A. B., Agin-Liebes, G., Swift, T. C., Terrana, S., Devenot, N., Friedman, H. L., et al. (2017). Patient experiences of psilocybin-assisted psychotherapy: An interpretative phenomenological analysis. *Journal of Humanistic Psychology, 57*(4), 354–388. https://doi.org/10.1177/0022167817706884.

Bogenschutz, M. P., Forcehimes, A. A., Pommy, J. A., Wilcox, C. E., Barbosa, P. C. R., & Strassman, R. J. (2015). Psilocybin-assisted treatment for alcohol dependence: A proof-of-concept study. *Journal of Psychopharmacology (Oxford, England), 29*(3), 289–299. https://doi.org/10.1177/0269881114565144.

Bonson, K. R., Buckholtz, J. W., & Murphy, D. L. (1996). Chronic administration of serotonergic antidepressants attenuates the subjective effects of LSD in humans. *Neuropsychopharmacology: Official Publication of the American College of Neuropsychopharmacology, 14*(6), 425–436. https://doi.org/10.1016/0893-133X(95)00145-4.

Bouso, J. C., González, D., Fondevila, S., Cutchet, M., Fernández, X., Ribeiro Barbosa, P. C., et al. (2012). Personality, psychopathology, life attitudes and neuropsychological performance among ritual users of ayahuasca: A longitudinal study. *PLoS One, 7*(8), e42421. https://doi.org/10.1371/journal.pone.0042421.

Brewer, J. A., Worhunsky, P. D., Gray, J. R., Tang, Y.-Y., Weber, J., & Kober, H. (2011). Meditation experience is associated with differences in default mode network activity and connectivity. *Proceedings of the National Academy of Sciences, 108*(50), 20254–20259. https://doi. org/10.1073/pnas.1112029108.

Callaway, J. C., Raymon, L. P., Hearn, W. L., McKenna, D. J., Grob, C. S., Brito, G. S., & Mash, D. C. (1996). Quantitation of N,N-dimethyltryptamine and harmala alkaloids in human plasma after oral dosing with ayahuasca. *Journal of Analytical Toxicology, 20*(6), 492–497.

Carbonaro, T. M., & Gatch, M. B. (2016). Neuropharmacology of N,N-dimethyltryptamine. *Brain Research Bulletin, 126*(Pt 1), 74–88. https://doi.org/10.1016/j.brainresbull.2016.04.016.

Carhart-Harris, R. L., & Nutt, D. J. (2017). Serotonin and brain function: A tale of two receptors. *Journal of Psychopharmacology, 33*(9), 1091–1120. https://doi. org/10.1177/0269881117725915.

Carhart-Harris, R. L., Bolstridge, M., Rucker, J., Day, C. M. J., Erritzoe, D., Kaelen, M., et al. (2016). Psilocybin with psychological support for treatment-resistant depression: An open-label feasibility study. *The Lancet Psychiatry, 3*(7), 619–627. https://doi.org/10.1016/S2215-0366(16)30065-7.

Carhart-Harris, R. L., Erritzoe, D., Haijen, E., Kaelen, M., & Watts, R. (2017). Psychedelics and connectedness. *Psychopharmacology, 29*(4), 1–4. https://doi.org/10.1007/s00213-017-4701-y.

Carhart-Harris, R. L., Erritzoe, D., Williams, T., Stone, J. M., Reed, L. J., Colasanti, A., et al. (2012). Neural correlates of the psychedelic state as determined by fMRI studies with psilocybin. *Proceedings of the National Academy of Sciences, 109*(6), 2138–2143. https://doi. org/10.1073/pnas.1119598109.

Carhart-Harris, R. L., Leech, R., Hellyer, P. J., Shanahan, M., Feilding, A., Tagliazucchi, E., et al. (2014). The entropic brain: A theory of conscious states informed by neuroimaging research with psychedelic drugs. *Frontiers in Human Neuroscience, 8*, 20. https://doi.org/10.3389/fnhum.2014.00020.

Fischer, R. (1971). A cartography of the ecstatic and meditative states. *Science, 174*(4012), 897–904.

Forstmann, M., & Sagioglou, C. (2017). Lifetime experience with (classic) psychedelics predicts pro-environmental behavior through an increase in nature relatedness. *Journal of Psychopharmacology, 31*(8), 975–988. https://doi.org/10.1177/0269881117714049.

Fox, M. D., & Greicius, M. (2010). Clinical applications of resting state functional connectivity. *Frontiers in Systems Neuroscience, 4*, 19. https://doi.org/10.3389/fnsys.2010.00019.

Fox, M. D., Snyder, A. Z., Corbetta, M., Van Essen, D. C., & Raichle, M. E. (2005). The human brain is intrinsically organized into dynamic, anticorrelated functional networks. *Proceedings of the National Academy of Sciences of the United States of America, 102*(27), 9673–9678. https://doi.org/10.1073/pnas.0504136102.

Frecska, E., Bokor, P., & Winkelman, M. (2016). The therapeutic potentials of ayahuasca: Possible effects against various diseases of civilization. *Frontiers in Pharmacology, 7*(e42421), 35–17. https://doi.org/10.3389/fphar.2016.00035.

Freud, S. (1995). *The standard edition of the complete psychological works of Sigmund Freud.* London: Hogarth Press and the Institute of Psycho-Analysis.

Gallimore, A. R. (2015). Restructuring consciousness – The psychedelic state in light of integrated information theory. *Frontiers in Human Neuroscience, 9*, 346. https://doi.org/10.3389/fnhum.2015.00346.

Garcia-Romeu, A., Griffiths, R. R., & Johnson, M. W. (2014). Psilocybin-occasioned mystical experiences in the treatment of tobacco addiction. *Current Drug Abuse Reviews, 7*(3), 157–164.

Gaynes, B. N. (2009). Identifying difficult-to-treat depression: Differential diagnosis, subtypes, and comorbidities. *The Journal of Clinical Psychiatry, 70*(Suppl 6), 10–15. https://doi. org/10.4088/JCP.8133su1c.02.

González, D., Carvalho, M., Cantillo, J., Aixalá, M., & Farré, M. (2017). Potential use of ayahuasca in grief therapy. *Omega, 68*(4), 30222817710879. https://doi.org/10.1177/0030222817710879.

Griffiths, R. R., Johnson, M. W., Carducci, M. A., Umbricht, A., Richards, W. A., Richards, B. D., et al. (2016). Psilocybin produces substantial and sustained decreases in depression and anxiety in patients with life-threatening cancer: A randomized double-blind trial. *Journal of Psychopharmacology, 30*(12), 1181–1197. https://doi.org/10.1177/0269881116675513.

Griffiths, R. R., Johnson, M. W., Richards, W. A., Richards, B. D., Jesse, R., Maclean, K. A., & Klinedinst, M. A. (2017). Psilocybin-occasioned mystical-type experience in combination with meditation and other spiritual practices produces enduring positive changes in psychological functioning and in trait measures of prosocial attitudes and behaviors. *Journal of Psychopharmacology (Oxford, England), 5,* 269881117731279. https://doi.org/10.1177/0269881117731279.

Hayes, S. C., & Hofmann, S. G. (2017). The third wave of cognitive behavioral therapy and the rise of process-based care. *World Psychiatry, 16*(3), 245–246. https://doi.org/10.1002/wps.20442.

Hayes, S. C., Levin, M. E., Plumb-Vilardaga, J., Villatte, J. L., & Pistorello, J. (2013). Acceptance and commitment therapy and contextual behavioral science: Examining the progress of a distinctive model of behavioral and cognitive therapy. *Behavior Therapy, 44*(2), 180–198. https://doi.org/10.1016/j.beth.2009.08.002.

Holtzheimer, P. E., & Mayberg, H. S. (2011). Stuck in a rut: Rethinking depression and its treatment. *Trends in Neurosciences, 34*(1), 1–9. https://doi.org/10.1016/j.tins.2010.10.004.

Johnson, M. W., Garcia-Romeu, A., Cosimano, M. P., & Griffiths, R. R. (2014). Pilot study of the 5-HT2AR agonist psilocybin in the treatment of tobacco addiction. *Journal of Psychopharmacology (Oxford, England), 28*(11), 983–992. https://doi.org/10.1177/0269881114548296.

Kent, J. L. (2010). *Psychedelic information theory: Shamanism in the age of reason.* Seattle, WA: PIT Press.

Kjellgren, A., Eriksson, A., & Norlander, T. (2009). Experiences of encounters with ayahuasca – "The vine of the soul". *Journal of Psychoactive Drugs, 41*(4), 309–315. https://doi.org/10.1080/02791072.2009.10399767.

Kuypers, K. P. C., Riba, J., la Fuente Revenga, de, M, Barker, S., Theunissen, E. L., & Ramaekers, J. G. (2016). Ayahuasca enhances creative divergent thinking while decreasing conventional convergent thinking. *Psychopharmacology, 233*(18), 3395–3403. https://doi.org/10.1007/s00213-016-4377-8.

Labate, B. C., & Cavnar, C. (2014). *Ayahuasca shamanism in the Amazon and beyond.* New York City, NY: Oxford University Press.

Labate, B. C., & Jungaberle, H. (2011). *The internationalization of ayahuasca.* Münster: LIT Verlag.

Labate, B. C., Cavnar, C., & Gearin, A. K. (2016). *The world ayahuasca diaspora.* New York City, NY: Routledge.

Lafrance, A., Loizaga-Velder, A., Fletcher, J., Renelli, M., Files, N., & Tupper, K. W. (2017). Nourishing the spirit: Exploratory research on ayahuasca experiences along the continuum of recovery from eating disorders. *Journal of Psychoactive Drugs, 49*(5), 427–435. https://doi.org/10.1080/02791072.2017.1361559.

Lafrance, A., Renelli, M., Fletcher, J., Files, N., Tupper, K., & Loizaga-Velder, A. (this volume). Ayahuasca as a healing tool along the continuum of recovery from eating disorders. In B. C. Labate & C. Cavnar (Eds.), *Ayahuasca healing and science.* Cham: Springer.

Lawn, W., Hallak, J. E., Crippa, J. A., Santos, dos, R, Porffy, L., Barratt, M. J., et al. (2017). Well-being, problematic alcohol consumption and acute subjective drug effects in past-year ayahuasca users: A large, international, self-selecting online survey. *Scientific Reports, 7*(1), 15201. https://doi.org/10.1038/s41598-017-14700-6.

Lebedev, A. V., Kaelen, M., Lövdén, M., Nilsson, J., Feilding, A., Nutt, D. J., & Carhart-Harris, R. L. (2016). LSD-induced entropic brain activity predicts subsequent personality change. *Human Brain Mapping, 37*(9), 3203–3213. https://doi.org/10.1002/hbm.23234.

Lehmann, M., Seifritz, E., Henning, A., Walter, M., Böker, H., Scheidegger, M., & Grimm, S. (2016). Differential effects of rumination and distraction on ketamine induced modulation of resting state functional connectivity and reactivity of regions within the default-mode network. *Social Cognitive and Affective Neuroscience, 11*(8), 1227–1235. https://doi.org/10.1093/scan/nsw034.

Maclean, K. A., Johnson, M. W., & Griffiths, R. R. (2011). Mystical experiences occasioned by the hallucinogen psilocybin lead to increases in the personality domain of openness. *Journal of Psychopharmacology, 25*(11), 1453–1461. https://doi.org/10.1177/0269881111420188.

Mason, N. L., & Kuypers, K. (this volume). Acute and long-term effects of ayahuasca on higher-order cognitive processes. In B. C. Labate & C. Cavnar (Eds.), *Ayahuasca healing and science.* Cham: Springer.

Moloudizargari, M., Mikaili, P., Aghajanshakeri, S., Asghari, M., & Shayegh, J. (2013). Pharmacological and therapeutic effects of *Peganum harmala* and its main alkaloids. *Pharmacognosy Reviews, 7*(14), 199–218. https://doi.org/10.4103/0973-7847.120524.

Morales-García, J. A., la Fuente Revenga, de, M, Alonso-Gil, S., Rodríguez-Franco, M. I., Feilding, A., Perez-Castillo, A., & Riba, J. (2017). The alkaloids of *Banisteriopsis caapi,* the plant source of the Amazonian hallucinogen ayahuasca, stimulate adult neurogenesis in vitro. *Scientific Reports, 7*(1), 5309. https://doi.org/10.1038/s41598-017-05407-9.

Moreno, F. A., Wiegand, C. B., Taitano, E. K., & Delgado, P. L. (2006). Safety, tolerability, and efficacy of psilocybin in 9 patients with obsessive-compulsive disorder. *The Journal of Clinical Psychiatry, 67*(11), 1735–1740.

Murrough, J. W., Abdallah, C. G., & Mathew, S. J. (2017). Targeting glutamate signalling in depression: Progress and prospects. *Nature Reviews Drug Discovery, 16*(7), 472–486. https://doi.org/10.1038/nrd.2017.16.

Nejad, A. B., Fossati, P., & Lemogne, C. (2013). Self-referential processing, rumination, and cortical midline structures in major depression. *Frontiers in Human Neuroscience, 7,* 666. https://doi.org/10.3389/fnhum.2013.00666.

Nichols, D. E. (2016). Psychedelics. *Pharmacological Reviews, 68*(2), 264–355. https://doi.org/10.1124/pr.115.011478.

Nichols, D. E., Johnson, M. W., & Nichols, C. D. (2017). Psychedelics as medicines: An emerging new paradigm. *Clinical Pharmacology & Therapeutics, 101*(2), 209–219. https://doi.org/10.1002/cpt.557.

Nutt, D. J. (2008). Relationship of neurotransmitters to the symptoms of major depressive disorder. *The Journal of Clinical Psychiatry, 69*(Suppl E1), 4–7.

Osório, F. de L, Sanches, R. F., Macedo, L. R., Santos, Dos, R. G, Maia-de-Oliveira, J. P., Wichert-Ana, L., et al. (2015). Antidepressant effects of a single dose of ayahuasca in patients with recurrent depression: A preliminary report. *Revista Brasileira de Psiquiatria, 37*(1), 13–20. https://doi.org/10.1590/1516-4446-2014-1496.

Palhano-Fontes, F., Andrade, K. C., Tofoli, L. F., Santos, A. C., Crippa, J. A. S., Hallak, J. E. C., et al. (2015). The psychedelic state induced by ayahuasca modulates the activity and connectivity of the default mode network. *PLoS One, 10*(2), e0118143. https://doi.org/10.1371/journal.pone.0118143.

Palhano-Fontes, F., Barreto, D., Onias, H., Andrade, K. C., Novaes, M. M., Pessoa, J. A., et al. (2018). Rapid antidepressant effects of the psychedelic ayahuasca in treatment-resistant depression: A randomized placebo-controlled trial. *Psychological Medicine, 7,* 1–9. https://doi.org/10.1017/S0033291718001356.

Palhano-Fontes, F., Mota-Rolim, S., Lobão-Soares, B., Galvão-Coelho, N., Maia-Oliveira, J. P., & Araújo, D. B. (this volume). Recent evidence on the antidepressant effects of ayahuasca. In B. C. Labate & C. Cavnar (Eds.), *Ayahuasca healing and science.* Cham: Springer.

Panossian, A. (2017). Understanding adaptogenic activity: Specificity of the pharmacological action of adaptogens and other phytochemicals. *Annals of the New York Academy of Sciences, 1401*(1), 49–64. https://doi.org/10.1111/nyas.13399.

Petri, G., Expert, P., Turkheimer, F., Carhart-Harris, R., Nutt, D., Hellyer, P. J., & Vaccarino, F. (2014). Homological scaffolds of brain functional networks. *Journal of the Royal Society, 11*(101), 20140873–20140873. https://doi.org/10.1098/rsif.2014.0873.

Popoli, M., Yan, Z., McEwen, B. S., & Sanacora, G. (2012). The stressed synapse: The impact of stress and glucocorticoids on glutamate transmission. *Nature Reviews Neuroscience, 13*(1), 22–37. https://doi.org/10.1038/nrn3138.

Riba, J., Romero, S., Grasa, E., Mena, E., Carrió, I., & Barbanoj, M. J. (2006). Increased frontal and paralimbic activation following ayahuasca, the pan-Amazonian inebriant. *Psychopharmacology, 186*(1), 93–98. https://doi.org/10.1007/s00213-006-0358-7.

Riba, J., Valle, M., Urbano, G., Yritia, M., Morte, A., & Barbanoj, M. J. (2003). Human pharmacology of ayahuasca: Subjective and cardiovascular effects, monoamine metabolite excretion, and pharmacokinetics. *The Journal of Pharmacology and Experimental Therapeutics, 306*(1), 73–83. https://doi.org/10.1124/jpet.103.049882.

Roseman, L., Nutt, D. J., & Carhart-Harris, R. L. (2018). Quality of acute psychedelic experience predicts therapeutic efficacy of psilocybin for treatment-resistant depression. *Frontiers in Pharmacology, 8*, 51–10. https://doi.org/10.3389/fphar.2017.00974.

Ross, S., Bossis, A., Guss, J., Agin-Liebes, G., Malone, T., Cohen, B., et al. (2016). Rapid and sustained symptom reduction following psilocybin treatment for anxiety and depression in patients with life-threatening cancer: A randomized controlled trial. *Journal of Psychopharmacology, 30*(12), 1165–1180. https://doi.org/10.1177/0269881116675512.

Sampedro, F., la Fuente Revenga, de, M, Valle, M., Roberto, N., Domínguez-Clavé, E., Elices, M., et al. (2017). Assessing the psychedelic "after-glow" in ayahuasca users: Post-acute neurometabolic and functional connectivity changes are associated with enhanced mindfulness capacities. *The International Journal of Neuropsychopharmacology, 20*(9), 698–711. https://doi.org/10.1093/ijnp/pyx036.

Santos, Dos, R. G, & Hallak, J. E. C. (2016). Effects of the natural β-carboline alkaloid harmine, a main constituent of ayahuasca, in memory and in the hippocampus: A systematic literature review of preclinical studies. *Journal of Psychoactive Drugs, 49*(1), 1–10. https://doi.org/10.1080/02791072.2016.1260189.

Santos, Dos, R. G, Grasa, E., Valle, M., Ballester, M. R., Bouso, J. C., Nomdedéu, J. F., et al. (2012). Pharmacology of ayahuasca administered in two repeated doses. *Psychopharmacology, 219*(4), 1039–1053. https://doi.org/10.1007/s00213-011-2434-x.

Santos, Dos, R. G, Osório, F. L., Crippa, J. A. S., & Hallak, J. E. C. (2016). Antidepressive and anxiolytic effects of ayahuasca: A systematic literature review of animal and human studies. *Revista Brasileira de Psiquiatria, 38*(1), 65–72. https://doi.org/10.1590/1516-4446-2015-1701.

Scheidegger, M., Walter, M., Lehmann, M., Metzger, C., Grimm, S., Boeker, H., et al. (2012). Ketamine decreases resting state functional network connectivity in healthy subjects: Implications for antidepressant drug action. *PLoS One, 7*(9), e44799. https://doi.org/10.1371/journal.pone.0044799.

Shanon, B. (2002). *The antipodes of the mind.* New York City, NY: Oxford University Press.

Sheline, Y. I., Price, J. L., Yan, Z., & Mintun, M. A. (2010). Resting-state functional MRI in depression unmasks increased connectivity between networks via the dorsal nexus. *Proceedings of the National Academy of Sciences, 107*(24), 11020–11025. https://doi.org/10.1073/pnas.1000446107.

Soler, J., Elices, M., Franquesa, A., Barker, S., Friedlander, P., Feilding, A., et al. (2016). Exploring the therapeutic potential of ayahuasca: Acute intake increases mindfulness-related capacities. *Psychopharmacology, 233*(5), 823–829. https://doi.org/10.1007/s00213-015-4162-0.

Strassman, R. J., Qualls, C. R., Uhlenhuth, E. H., & Kellner, R. (1994). Dose-response study of N,N-dimethyltryptamine in humans. II. Subjective effects and preliminary results of a new rating scale. *Archives of General Psychiatry, 51*(2), 98–108.

Sweat, N. W., Bates, L. W., & Hendricks, P. S. (2016). The associations of naturalistic classic psychedelic use, mystical experience, and creative problem solving. *Journal of Psychoactive Drugs, 48*(5), 344–350. https://doi.org/10.1080/02791072.2016.1234090.

Szabo, A., Kovacs, A., Riba, J., Djurovic, S., Rajnavolgyi, E., & Frecska, E. (2016). The endogenous hallucinogen and trace amine N,N-dimethyltryptamine (DMT) displays potent protective effects against hypoxia via sigma-1 receptor activation in human primary IPSC-derived cortical neurons and microglia-like immune cells. *Frontiers in Neuroscience, 10*(35), 423–411. https://doi.org/10.3389/fnins.2016.00423.

Tagliazucchi, E., Carhart-Harris, R., Leech, R., Nutt, D., & Chialvo, D. R. (2014). Enhanced repertoire of brain dynamical states during the psychedelic experience. *Human Brain Mapping, 35*(11), 5442–5456. https://doi.org/10.1002/hbm.22562.

Tupper, K. W. (2008). The globalization of ayahuasca: Harm reduction or benefit maximization? *The International Journal of Drug Policy, 19*(4), 297–303. https://doi.org/10.1016/j.drugpo.2006.11.001.

Vago, D. R., & Silbersweig, D. A. (2012). Self-awareness, self-regulation, and self-transcendence (S-ART): A framework for understanding the neurobiological mechanisms of mindfulness. *Frontiers in Human Neuroscience, 6*, 296. https://doi.org/10.3389/fnhum.2012.00296.

Valle, M., Domínguez-Clavé, E., Elices, M., Pascual, J. C., Soler, J., Morales-García, J. A., Pérez-Castillo, A., & Riba, J. (this volume). Ayahuasca as an unusually versatile therapeutic agent: from molecules to meta-cognition and back. In B. C. Labate & C. Cavnar (Eds.), *Ayahuasca healing and science*. Cham: Springer.

Viol, A., Palhano-Fontes, F., Onias, H., Araujo, D. B., & Viswanathan, G. M. (2017). Shannon entropy of brain functional complex networks under the influence of the psychedelic ayahuasca. *Scientific Reports, 7*(1), 1–13. https://doi.org/10.1038/s41598-017-06854-0.

Vollenweider, F. X., & Kometer, M. (2010). The neurobiology of psychedelic drugs: Implications for the treatment of mood disorders. *Nature Reviews Neuroscience, 11*(9), 642–651. https://doi.org/10.1038/nrn2884.

Watts, R., Day, C., Krzanowski, J., Nutt, D., & Carhart-Harris, R. (2017). Patients' accounts of increased "connectedness" and "acceptance" after psilocybin for treatment-resistant depression. *Journal of Humanistic Psychology, 57*(5), 520–564. https://doi.org/10.1177/0022167817709585.

Chapter 4
Ayahuasca and Psychotherapy: Beyond Integration

Mauricio Diament, Bruno Ramos Gomes, and Luis Fernando Tófoli

Introduction

Ayahuasca is a psychedelic brew that is traditionally used in a wide range of contexts in South America. Several indigenous and mestizo populations in the Amazon consume it, and it is used for many different purposes: healing, divination, hunting, and celebrations, among many others (Labate et al. 2008). Despite all of this diversity, since the Western cultures began learning about ayahuasca, it has been used mainly for healing purposes and self-development. Not only has the use of ayahuasca as a substance been expanded throughout the world, but the various rituals using it have as well. There are many diverse practices using ayahuasca with various understandings about it and a multiplicity of goals. Each tradition has different views regarding what is cured and how.

We are two psychiatrists and a psychologist working in a Brazilian context, where ayahuasca has been used in urban contexts since the end of the last century. We have been participating in ayahuasca rituals in different contexts and have also been attending to patients who use ayahuasca. Many questions could be formulated at this moment about the possible dialogue between ayahuasca and contemporary urban cultures and their views. How could this brew, with such a variety of effects and purposes, be therapeutic in modern cultures? How does it relate to Western forms of care, such as psychotherapy? More than that, can ayahuasca be integrated into regular Western healing practices?

M. Diament, MD (✉) · L. F. Tófoli, MD, PhD
Interdisciplinary Cooperation for Ayahuasca Research and Outreach (ICARO), School of Medical Sciences, University of Campinas (Unicamp), Campinas, São Paulo, Brazil

B. R. Gomes
Interdisciplinary Cooperation for Ayahuasca Research and Outreach (ICARO), School of Medical Sciences, University of Campinas (Unicamp), Campinas, São Paulo, Brazil

Department of Collective Health, School of Medical Sciences, University of Campinas (Unicamp), Campinas, São Paulo, Brazil

© Springer Nature Switzerland AG 2021
B. C. Labate, C. Cavnar (eds.), *Ayahuasca Healing and Science*,
https://doi.org/10.1007/978-3-030-55688-4_4

There is a wide scope of possibilities. In fact, we already have different practices being developed around the world. The aim of this chapter is to help health professionals deal with the use of ayahuasca in their context, focusing on the contemporary use of ayahuasca by Western populations, and how ayahuasca and psychotherapeutic processes can be integrated. In South America, there are different traditions that use ayahuasca to address a wide range of problems. These ailments are usually understood as physical, mental, and spiritual illnesses simultaneously: these three aspects of life are usually intertwined in traditional healing processes (Moure 2005). There are rituals that are connected with spiritual entities, religious contexts, witchcraft, and so on. In most groups, ayahuasca is used as a teacher, a living plant that communicates with individuals through its effects.

On the other side, in Western cultures, the tradition of using substances for treatment treads a very different path. There are huge differences in the practices: Western medicine sets a different role for the patient's experience and has a stronger focus on symptoms.

Even with remarkable differences, the Western practices that are closest to ayahuasca healing processes are those developed with psychedelics in the last century. In a short and prolific period between the 1950s and 1970s, there was an intense development of psychotherapy practices with psychedelic substances, such as LSD or psilocybin (Grinspoon and Bakalar 1997). During this period, substances with effects that are in some ways similar to ayahuasca were used to help patients deal with different problems, such as depression, anxiety, alcoholism, and psychological distress in general. There were two main strategies developed for treatment during this period: psycholytic therapy and psychedelic therapy. The psycholytic method uses small doses to lower defense mechanisms and facilitate access to memories and unconscious material. It was used in Europe much more and is closer to psychoanalytic contexts. Psychedelic therapy is probably better known, since it became part of the counterculture movement in the 1960s. It was more common in North America and used larger doses in an appropriate setting to provide peak and meaningful experiences (Majić et al. 2015).

Psychedelics were studied by the medical sciences and were part of the treatments proposed to patients by psychiatrists. However, social and political pressure banned all types of use in the 1970s. Unfortunately, the studies performed while psychedelics were legal were not rigorously controlled, and results were rarely evaluated using the standard tools accepted today. Nevertheless, a recent meta-analysis of studies on the treatment of alcoholism with LSD from that period, using sound methodology, presented positive results (Krebs and Johansen 2012). Even today, there are psychotherapists who used psychedelics as therapy in the 1960s and still work with different plants and substances, including ayahuasca. In spite of this, with prohibition, a once-growing culture striving to develop the use of those substances for psychotherapeutic purposes came to a sudden end or went underground. It was only after the 1990s that a renewed interest in this subject emerged, with important research centers like Johns Hopkins, New York University, and the Imperial College developing new studies, focusing on LSD, psilocybin, and MDMA as treatments for different mental disorders (Langlitz 2012).

The use of psychedelics in psychotherapy has some characteristics in common with the use of ayahuasca, but, at the same time, there are profound differences when considering the settings. Psychedelics and ayahuasca share a common focus on the important role of the subjective experience and the emphasis on drug-induced mystical states. They all lead to deep alterations in the perception of oneself and the world. On the other hand, while the setting for the therapeutic use of psychedelics is usually a comfortable room where the patient feels safe and is cared by expert therapists, ayahuasca is used in a wide variation of contexts: it can be used in the rainforest or urban areas, in dark environments or bright temples, in huge groups or intimate circles. Each characteristic of the setting may change the experience, influencing what is thought, felt, and lived; and also, all those characteristics are affected by the beliefs, values, and worldviews of each participant in the group.

In fact, some substances that are nowadays known as psychedelics, such as magic mushrooms, have traditional uses in healing rituals embedded in spiritual meaning that, in some ways, are similar to those of ayahuasca. Magic mushrooms were introduced to mainstream Western societies in the 1950s, and after their use spread around the world through the counterculture movement, their traditional characteristics were rapidly lost. Ayahuasca has come to the attention of Western societies more recently. It started its expansion in the 1980s and is spreading through a different pattern. There are various forms of ayahuasca use circulating around the world, so we can say that it is not only a substance that is spreading throughout the world but also that there are diverse forms of brews, traditions, and practices.

Ayahuasca use in the West has also been influenced by the psychedelic practices that started during the twentieth century. Should ayahuasca experiences be part of psychedelic programs, with a preparation session before the ritual and integration sessions after? Or would it be better to acknowledge that there is no need for a single protocol for the brew when it is used for psychotherapeutic processes? Ayahuasca experiences may have a wide variety of impacts on patients and can be managed with different practices. Despite the growing interest in the brew, there is still much to understand about how it functions and its possible therapeutic effects.

Most of the studies investigating the brew have focused on showing its safety for human use and trying to detect its impact in regular users (Grob et al. 1996; Barbosa et al. 2005; Santos et al. 2007; Fabregas et al. 2010; Labate and Cavnar 2014). Due to the frequent reports of people overcoming disorders such as depression and drug dependence with ayahuasca, there are also studies exploring these aspects (Thomas et al. 2013; Halpern et al. 2008; Osorio et al. 2015; Sanches et al. 2016). These studies show a positive impact on drug users life quality and a reduction or cessation in drug consumption. Trials for depression presented promising results, reducing depressive symptoms from 7 to 14 days after the experience with the brew (Palhano-Fontes et al. 2017). In addition, it is important to mention that these studies explore only part of several potential therapeutic applications.

Along with the scientific explorations of ayahuasca effects, there are groups and institutions that are already using ayahuasca to treat drug dependence. These centers exist mainly in Latin America and are also connected or inspired by different traditions. There are mostly anthropological studies about these contexts (Mercante

2017; Gomes 2013; Moure 2005; Loizaga-Velder and Verres 2014; Labate and Cavnar 2014). There are centers in Peru, Chile, Brazil, and Mexico, using it in inpatient contexts, jungle retreats, or even research settings. It seems that there is a scattering of the use of ayahuasca in many different practices. How can ayahuasca also be integrated into psychotherapy, focused not only on patients with addictions, but on the general population?

In the following pages, we will discuss aspects of the many uses of ayahuasca that may be related or relevant to psychotherapy both in theoretical and practical terms. We have included the role of the ayahuasca experience in this chapter, the importance of ritual practices, the ayahuasca visions, challenging experiences with ayahuasca, and mediation with a plant teacher. We do not intend for this list to be exhaustive, but to bring an initial standpoint for a preliminary and general discussion about the ayahuasca-psychotherapy relationship.

The Role of Experience

Ayahuasca use differs from traditional Western medicine because of importance of the role of experience. Several elements in the ayahuasca experience may cause a positive impact on a psychotherapeutic process. First of all, it's important to say that not all ayahuasca experiences are therapeutic, and sometimes they may even bring distress to daily life. How can this type of experience impact a person? The experiential aspect of ayahuasca use is a key aspect to understand how it can influence the psychotherapeutic process. Elements from Donald Winnicott's theory on human development and the important role of meaningful experiences are key to a better understanding of this aspect.

Winnicott was an important English psychoanalyst from the mid-twentieth century. He believed that, through a therapeutic relationship, patients would be able to experience things they had never lived before and, through this, develop aspects of their personality that they couldn't establish earlier in their lives (Naffah Neto 2005). For a meaningful experience, individuals must feel supported and protected by the environment, so that they can evolve together with what is experienced. In a similar manner, we can say that the setting of an ayahuasca experience has to be supportive and provide a sense of protection to the participant. This is the key role of the ritual and its responsible guidance. A supportive and protected setting helps the experience to fully develop. Within this protective environment, individuals may allow themselves to experience transformative psychological processes.

The emphasis on the experience has another consequence: A symbolic interpretation of what is experienced is not absolutely fundamental. From the meaningful experience, symbolizations may arise, but just living the experience itself can also be therapeutic, and may lead to changes in the daily life and how the patient faces challenges in life. So, it would not be absolutely necessary to understand everything that happens during the ritual. For example, if someone has the experience of dying and is reborn in a ritual, this experience itself may have an impact on the person. An interpretation of how this experience relates to one's life, what that means for one's psychic functioning, and so on may not be necessary (Moure 2005).

There is a wide variety of possibilities for transformative experience with ayahuasca. Loizaga-Velder and Verres (2014) present a few observations on how participating in ayahuasca rituals can impact a psychotherapy process. According to the author's study, ayahuasca can:

- Catalyze therapeutic processes
- Help individuals have a better understanding of the underlying causes
- Help to overcome psychological distress
- Reduce cravings after participating in ayahuasca rituals
- Increase a sense of meaning and purpose in life, through transcendental experiences or contact with divine beings
- Lower psychological defense mechanisms
- Promote therapeutically relevant insights
- Support interpersonal awareness

The experience with ayahuasca could, therefore, influence ongoing therapeutic processes. As the authors say, "Ayahuasca should not be understood as a primarily pharmacological intervention. Rather, it should be conceptualized as a catalyst with a therapeutic value that can unfold when the identified variables of set, substance, setting, and integration are appropriately managed" (Loizaga-Velder and Verres 2014, p. 69).

Ritual Use

The ritual is one of the main aspects of the setting. It condenses a whole collection of features that determine the atmosphere where the experience takes place. As previously mentioned, there is a wide variety of rituals with different formulas of lighting, decoration, the placement and positioning of the participants, and other elements. Rituals can be conducted by a shaman or healer, a group, or an entire community. All these differences will determine divergent experiences. The setting and the ritual will also provide a symbolic repertoire for the experience. As described by a healer from São Paulo, Brazil, "it's a theatrical play, but it's completely real at the same time" (Gomes 2016). It's real in the sense that it has a direct effect on the participant's "trip"; it guides the experience, although it does not determine it. The ritual also sets the experience apart from daily life, creating a difference between what's lived inside the ritual and under the effect of ayahuasca, and an everyday routine. This seems to be important: The ritual will be connected with daily life through the meaning of its effects and on how it changes the person's worldview, but it happens in another dimension of time and space.

At the same time, the ritual will be the "supporting actor" of this "play." The main actor here is the brew, acting on the person's experience. If we take the effects of the classical drug-set-setting theory about the use of psychoactive substances, we can consider that the setting creates the conditions for ayahuasca to act on the relationship with the patient, since the brew is perceived as an alterity, a living being. The plant presents its subjectivity and agency through the effects that the patient feels during the ritual. Some aspects of this relationship include suffering and cleansing, visions, and the notion that the plant communicates with the person.

Ayahuasca Visions

Ayahuasca-induced visual effects can play a significant role within a psychotherapeutic approach, even if they are not present in every ayahuasca session. Neural mechanisms that are related to these phenomena have already been described in an fMRI study that showed the activation of several primary visual structures of the brain at a magnitude comparable to natural images with open eyes (Araújo et al. 2012).

The Israeli psychologist Benny Shanon (2002, 2003) extensively described and catalogued the visions reported by himself and 178 other individuals, including indigenous people, urban South American and non-South American ayahuasca drinkers, the visions reported by a Santo Daime Brazilian leader, by a well-known ayahuasca visionary artist, and those available in the anthropological literature at the time. This sample accounted for diverse traditions and settings, from indigenous rituals to independent urban ceremonies, including the main Brazilian ayahuasca syncretic denominations, Santo Daime, União do Vegetal (UDV), and Barquinha. He proposed a categorization for the contents of ayahuasca visions: supernatural, architecture, natural animals, human beings, plants, objects, and personal biography. In general, personal biographical elements appear to be less prevalent and are more common among those interviewed after their first experience with the brew. He also briefly discusses the occurrence and reoccurrence of some contents in light of what he calls the Freudian personal unconscious and the Jungian collective archetypal unconscious—mainly by disagreeing with their utilization in this field— and proposes a cognitivist approach to them, even though this approach is not clearly practical or clinical.

Ayahuasca visions have been named in French, *vision-mouvoir* (literally, "movement-vision") by psychologist Clara Novaes (2011). Based on the reports of urban ayahuasca users (who are probably more prone to undergo psychotherapy), the term refers to the visual experience of ayahuasca as something that makes one move through daily life. This would enable a perception of the very nature of otherwise banal things, details that are not perceived without an altered state of consciousness. It is as if ayahuasca could intensify life—including ordinary life outside of the ritual and its effects—by intensifying the senses during the experience.

The visions produced by ayahuasca, especially those with strong biographical aspects, offer extensive material for psychotherapeutic work. These complex visions usually include the reenactment of past experiences, and dreams and projections for the future. This kind of phenomena is more or less prevalent depending on factors, such as experience with the substance, religious affiliation, and ethnic background (Shanon 2003). The therapist should, therefore, keep in mind that familiarity with the brew also plays an important role.

At any rate, an open and nonjudgmental reception for the patient's reports is vital. Often, visionary states are experienced as real (yet nonpsychotic), such as in encounters with divine religious characters and other beings, or with the realization of the energy net that unites all things. These experiences should be appraised as reported and integrated within the patient's narrative. The professional's theoretical choices and orientation will provide specific interpretations and alternatives to the exegesis and elaboration of ayahuasca visions.

Challenging Experiences with Ayahuasca

Another aspect of the ayahuasca experience that sets it apart from other psychedelics is its intense effect on the body. Ayahuasca may make you vomit, weep, tremble, feel dizzy, cold, or hot. It can make you feel bad, cause diarrhea, and induce terrifying visions. However, all this variety of suffering is not perceived by regular users as a side effect of the substance but as part of an ongoing relationship with the brew. Since the plant is perceived as a living being, it is communicating through the distress one may experience. This part of the experience is also permeated with meaning and may be therapeutic, despite the physical and psychological pain itself.

The purgative effects of the brew are interpreted through the values and conceptions of the social group. For the Peruvian mestizo vegetalistas, ayahuasca is also known as *la purga* ("the purge"). In Santo Daime and the UDV, a challenging experience with the brew is known as *peia* (a regional term for "a beating") (Okamoto da Silva 2004). Considered a fundamental aspect of the experience for many groups, a ritual with more *peia* and *purga* represents a stronger and, sometimes, more effective situation. The suffering can be experienced in many forms and is usually connected with notions, such as punishment, discipline, or cleansing. Sometimes the person may feel "beaten" by the plant, by a spirit invoked by the ritual, or it can be a suffering process that cleanses the person, without a necessary punishment intention.

The act of purging generally makes the person feel better, lighter and many times (not always) makes the individual clearly aware of what their suffering is related to. One important aspect here is that this awakening of conscience is characterized as evidence: it appears for the person as a strong truth, differing from normal thoughts. The act of suffering and purging makes the person feels better later, but it also may lead to changes in daily life. In a sense, it is a catharsis (a psychoanalytical term that literally means "purification") that can be followed by insights on moral faults, mistakes, and patterns that need to change.

Though the term "purge" would suggest that these situations happen only with physical purging, such as vomiting or diarrhea, it is important to keep in mind that experiences that are exclusively mentally or spiritually challenging are also common. Though they are somewhat similar to what is described as a "bad trip" or a challenging experience with other psychedelics (Barrett et al. 2016), once again, the setting and the ritual component fundamentally shape the meanings of this experience for the individual drinking the brew. Just as in other aspects of the ayahuasca experience, it is important that the suffering, punishment, discipline, and catharsis elements be analyzed by the therapist in a nonjudgmental and culturally sensitive way.

The Mediation of a "Plant Third Party"

Another feature of ayahuasca inebriation that is relevant to psychotherapy—and that may be, so to speak, the closest ayahuasca can get to the Socratic dialogue that may occur with a therapist—is the interchange that individuals may experience, within their thoughts, with an element that is partially recognized as nonself. This aspect has been previously described in the ayahuasca literature and has been given different native names. In Santo Daime or the UDV, it may be referred to as *voz do Mestre* (the Master's voice) or be identified as counsel that is offered by spiritual entities related to the brew (Gomes 2013; Melo 2015). This aspect is understood as the intercession of the spirit of ayahuasca in the *vegetalista* traditions (Moure 2005) and is clearly experienced by Westerners as well (Doyle 2012).

The interaction with this "plant third party" may vary. It may come as advice, commands, or comments regarding the person's actions, thoughts, and visions. This description may sound similar to a pseudohallucination, meaning the experience of hearing a voice "inside the head," or "hallucinating" a thought, a symptom that is a characteristic in psychotic disorders. However, ayahuasca drinkers—regardless of how much they rely on or believe in what is said—clearly know that this effect is caused by the brew, and this insight makes the experience differ from that of an actual psychotic phenomena. In Western contexts, there is a feeling that this "voice" or "thought" is somewhat internal, meaning it is a part of the individual's thoughts but, at the same time, has external agency that communicates with the person in a rather unpredictable way.

The communication with this perceived intelligence has evident implications in a psychotherapeutic setting. Issues brought by the patient may have at least three perspectives: that of the patient who is an ayahuasca drinker; the therapist's point of view, manifested in different ways depending on their preferred school of thought; and the mediation of the "plant spirit." This scenario seems to be a peculiar situation when compared to the regular patient-therapist relationship. In a psychotherapeutic environment where ayahuasca is respected and acknowledged as relevant, sometimes there is a plant-patient-therapist relationship. While not necessarily present at the therapeutic setting—unless therapists drink ayahuasca with their patients, which is rare but does occur—this potential mediation between the plant's counsel and the therapist's interventions happens through the patient. This opens new perspectives that, to our knowledge, have not been sufficiently explored by the diverse schools of thought in psychology.

This raises the question regarding how ayahuasca drinking can interact with a psychotherapeutic setting, whether in potion-mediated therapy or in the case of brew drinkers who are also psychotherapy patients. We are going to discuss these potential interactions in the second half of this chapter. But, before proceeding into that, let's visit a quick discussion about the impact of the drug-set-setting triad on the ayahuasca experience.

Ayahuasca: Set and Setting

In practical terms, if we try to set the basis for a psychotherapeutic approach to the ayahuasca experience as a whole, maybe approaches such as the use of statistics and taxonomies, such as those proposed by Shanon, are not as useful as a primary tool. Zinberg's (1984) "drug, set and setting" model could be proposed as a better approach to a patient's report. The "drug" component, which is herein understood as the subject's awareness of the physiological effects of the substance, can help to determine the non-pathological characteristic of these experiences. By knowing where and how ayahuasca interacts with the central nervous system as well as applying basic psychopathology concepts, the ayahuasca visions, for instance, can be understood as non-hallucinatory, since in almost all ayahuasca experiences the individual comprehends that the visual phenomena are a result of a substance-induced state. Hence, individuals can keep up with the understanding that they are the agents of their experience (Dominguez-Clavé et al. 2016), even in more intense episodes when there's an interaction with visionary contents. Exceptions are reported when admixtures are used in the preparation of the brew, like in the case of ayahuasca containing *Solanaceae* plants such as *Datura* and *Brugmansia* (Shanon 2002). Also, persons with a predisposition to psychosis are in greater risk for presenting a true psychotic episode.

The "set," meaning the mindset—personal beliefs and psychological processes previous to the ayahuasca intake—is, as experienced users know, essential for the whole experience and, thus, also to its visionary or emotional component. Even though the patterns can be repetitive to some extent, as noted by Shanon, their interpretation by the individual can vary according to the set and setting.

The setting—or the context of use—also plays an important role in experiences with ayahuasca. For example, one could expect more personal biography elements coming up during an experimental trial that investigates the effects of ayahuasca on depression inside a hospital office than in a neo-shamanic ritual that celebrates the ancient spirits of the Earth. Unfortunately, from the perspective of biomedicine and health sciences, systematic studies of the impact of the setting on the content of ayahuasca experiences have not been undertaken.

Ayahuasca and Psychotherapy

In such a diversity of effects and forms of use, there are also many ways of approaching ayahuasca in the psychotherapeutic processes. Here, we outline only a few basic aspects that may help a therapist deal with their patients' experience with ayahuasca.

The ayahuasca experience may resemble other experiences in some aspects, mostly mystical experiences, but also dreaming. How can a therapist deal with these kinds of experiences? This will depend on their psychotherapeutic approach, but regardless of the therapist's orientation, it is very important to validate the experience. The therapist doesn't need to believe in the reality of the things that the patient reports, such as seeing Buddha or God or being swallowed by a gigantic anaconda. Regardless of the therapist's own beliefs, the ayahuasca experience appears as real to patients as dreams do and, therefore, should be included and discussed in the therapeutic process, if the patient brings it into the session. How was the experience for the patient? How did this impact their understanding of life? Are there other echoes from the experience that the therapist may perceive? Those are a few questions that may be occupying the therapist's mind when working with a patient participating in ayahuasca sessions. Below, we present a few possibilities for bringing together ayahuasca and psychotherapy.

Integration

Integration is not a new word in psychedelic therapy. Since the early days of this practice, providing a moment to help people who had undergone a peak psychedelic experience "reenter" everyday experience has been common (Pahnke et al. 1970; Grof 1980; Majić et al. 2015). Today, however, when it comes to ayahuasca, this "integration" is seen as almost mandatory in some circles, as if the experience with the brew would only be complete after an integration session. What this means is not exactly clear. There may be a more "psychological" perspective; for instance, the idea of integrating unconscious material that was unearthed by the experience into consciousness (Cohen 2017). On the other hand, integration may also have a broader meaning, including the integration of psychedelics within a community, usually a tribal one (Mabit 2007; Goldsmith 2010), or even the nurture of a support group aimed at fostering the assimilation of the use of psychedelics into a meaningful lifestyle (Norris and Megler 2017). In the world of ayahuasca use, there is quite an extensive offering of "professional" integration and other similar postexperience processes.

The use of the term integration, as we understand it, refers to the process of trying to bring together two elements that do not seem to be naturally compatible. There are communities where ayahuasca has a spiritual, social, and political role that is pivotal, such as in certain indigenous groups and some urban or rural communities organized around the consumption of ayahuasca (notably, Céu do Mapiá in the Brazilian State of Amazonas, considered the spiritual capital of the ICEFLU branch of Santo Daime). In these settings, everyday life and the ayahuasca experience are a continuum. People that are part of such groups are less prone to experience a crisis when trying to conceive of their previous ordinary lives in light of the new or astonishing experience that may have shaken their core beliefs, values, or positions in life and that may invite a new approach toward their work, relationships, health, habits, etc. What happens in the rituals is naturally applicable to daily life, which, in turn, is molded to some extent by the entheogenic experience and the cultural aspects related to it.

With the spread of ayahuasca and its entrance into urban centers and Western cultures, a conflict seems to have emerged between a worldview that is individualist, consumerist, and performance-oriented and what is experienced during a few ayahuasca sessions in structured, serious, and appropriate contexts. People often find it difficult to bring this new way of looking at the world and the new insights about their own lives and choices, into their daily life, especially when the experience takes place far from home and in an alien landscape, for instance, at a jungle retreat. They may not always have someone to talk about what they have just experienced, and the intensity of the urge to make life-changing decisions may be a source of anguish and distress to some individuals. In these cases, some kind of help could be useful (Aixalà 2017). So, what kind of support may be needed? Is it mandatory to undergo a psychotherapeutic process? Is there clear evidence that some activities considered to be integration processes, such as painting, stretching, yoga, and collective sharing, will minimize these sensations?

Two styles of integration practices coexist nowadays: one that serves to maximize and optimize and, maybe, to synthesize and symbolize what occurred during the experience; and another is focused on giving some kind of support to people who find themselves having undesirable effects following an experience.

Integration processes that are meant to be a part of the whole experience are usually held by people that have had their own experiences with ayahuasca, not necessarily psychotherapists. This means individuals who empirically learned how to use artistic tools, physical techniques, and sharing circles to maximize the whole experience. Even if it's not mandatory to consider that every ayahuasca experience must be meaningful, integration circles can help some people deal with the numerous unanticipated experiences that may have just transpired and to feel safer among people who have faced similar challenges. When we talk about the ineffability of the entheogenic experience, as well as the intimate and delicate material that may have come to surface, it's important to stress that not everyone feels comfortable sharing their experiences in a circle or at the time the integration process is set. It's also interesting that several people may need no support at all after their experiences, being able to integrate it themselves, sometimes together with friends or by doing things that please them.

Integration processes guided to support people in distress after an ayahuasca experience are more similar to a brief psychotherapy process. However, when it comes to helping people in more serious conditions, it's important to have psychotherapy training and to be able to identify, treat, or address someone who may present with acute anxiety, transient psychosis, paranoia, etc. Usually, this kind of support is not offered as part of retreats or scheduled ceremonies, and the person who is having a bad experience, or someone close, has to seek advice. It's desirable that the psychotherapy professional supporting this person has some experience in the field of ayahuasca or psychedelics, but it's even more important to be open minded and nonjudgmental. Misclassifying a bad ayahuasca experience as pathological may pose even more problems than the experience itself.

Ayahuasca as an Element of Psychotherapy

Unlike the integration perspective, in which the intake of ayahuasca seems to play a central role in the whole process, whether or not therapeutically, ayahuasca can be incorporated by psychotherapists using experiences that take place outside psychotherapy. This means that individuals who are willing to start, or are already in, therapy may benefit from bringing their ayahuasca experiences to the clinician's office and by offering the therapist more elements to work with. Here, as in the integration process for someone in distress, it's not vital that the therapist be an *ayahuasquero* or a psychonaut, although personal experience in psychotherapy is a basic requirement. Likewise, the therapist's personal experimentation with ayahuasca or other psychedelics is desirable, but not at all indispensable. On the other hand, having an open and nonjudgmental mindset is vital.

Also, unlike traditional Western medicine, where pills are prescribed by a doctor, the contemporary use of ayahuasca depends on each person's interest in it. People seek ayahuasca experiences for many different purposes and will establish different relationships with the plant. For a good combination of ayahuasca and psychotherapy, it is important to understand what the patient is expecting and experiencing with ayahuasca. In this composition, the psychotherapist may help the patient understand this relationship, since sometimes this may not be sufficiently clear. For some individuals, just a few rituals at the right moment in their personal process will have a positive impact on their life, helping them to deal with problems and disorders, and that will be enough. For others, ayahuasca use is periodic. Some patients use it fortnightly, monthly, or bimonthly for months or years. Others may use it more frequently, sometimes with ayahuasca occupying a central role in their lives.

When patients are using it regularly, some report experiencing two processes in parallel: one in psychotherapy and another in the sequences of ayahuasca rituals. Sometimes these processes intersect, dialoguing and influencing each other. The process with ayahuasca may bring new contents, or lower defense mechanisms, and help access aspects that are usually not possible in regular therapy sessions.

Another composition is possible: the ayahuasca experience can create conditions for the psychotherapeutic process to continue. There are patients who are addicted to alcohol and drugs who feel a great reduction in their cravings during the weeks after drinking ayahuasca, helping to bring stability to their daily life and enabling a psychotherapeutic process to develop and gradually address the drug addiction issue.

Authors have also mentioned that Westerners are, by principle, not part of the cosmology and cosmopolitics from ayahuasca's native origins (Trichter 2009). Therefore, psychotherapy could be a form of preparation for the experience, as well as increasing its benefits and minimizing some risks:

> The integration of an ayahuasca ceremony into a framework of ongoing psychotherapy would create a greater sense of safety for the client when participating in an ayahuasca ritual. This would allow the participant to explore a situation that often brings up fear, feeling fully prepared. It would also allow the client to share his or her experiences with the therapist post-ceremony. Lastly, this rapport would create an opportunity for the therapist to make interpretations more freely while the ceremony experience is more temporally and affectively grounded within the client. After participating in the ceremony, the affective experiences and the insights that may have been obtained through the ayahuasca ritual could be interpreted, worked through, and integrated into the ongoing therapy. (Trichter 2009, p. 144)

There are plenty of psychotherapy methods. Much has been written about what defines a psychotherapy process, who can be a therapist, and who should not be. This is a discussion that will probably last as long as human beings are interested in their and others' inner worlds. People choose the approach that better suits their worldviews, their cosmologies, their beliefs. Our experience as mental health professionals with diverse orientations—and as individuals who have experienced different methods of psychotherapy and who have had our own personal ayahuasca experiences—shows us that ayahuasca can be well placed as an adjunct element in a psychotherapeutic process. We believe that a well-trained psychotherapist, with enough experience—personal, professional, practical, and theoretical—to handle massive amounts of non-ordinary experiences, sometimes overwhelming ones, can incorporate the experience of their patients who drink ayahuasca, not as a central therapeutic tool but as another element of their lives. How the material resulting from the ayahuasca experience will be incorporated into the clinical narrative depends on the professional's theoretical orientation.

Ayahuasca and the Schools of Psychological Thought

One example of how ayahuasca experiences can be interpreted through diverse theoretical standpoints is Ido Cohen's work (2017). In his doctoral thesis, he used Jungian psychology concepts, such as consciousness (the total sum of mental, physical, and emotional processes available to the person), personal unconsciousness (the shadow), collective unconsciousness (the realm that can be symbolized by different archetypes present in our history, arts, mythology, etc.), and the Self (the entity from which all other instances emerge and that, in a process called individuation, demands to be recognized, integrated, and realized in order to bring the person back to wholeness). Cohen explains that, for Jung, societal and civilizing processes are responsible for a tension between the conscious and the unconscious, from which the human being is disconnected. This tension and disconnection, in turn, are responsible for different suffering experiences. One solution for this suffering would be to bridge this separation between consciousness and unconsciousness and recognize the latter as complementary and necessary for the former; this would bring light to the secret influence of the unconscious on the conscious. This would be accomplished by an intellectual, affective, and artistic approach to the tension between them, liberating the person to connect with their deeper Self and ancestral and collective identities. The author analyzed the experience of English-speaking Westerners participating in an ayahuasca retreat with the objective of evaluating how they integrated what was experienced into their lives a year or less later. Through the concepts explained above and others, Cohen found that, from a Jungian perspective, ayahuasca can be seen as "an entity that allows connection to the Self, clearing the blockages of the ego–Self axis [a constant dialogue between the conscious and unconscious] to allow the individuation process to become more conscious, and for the unconscious material to come forth with greater clarity" (Cohen 2017, p. 115).

Following Freudian and English-speaking post-Freudian psychoanalytic concepts, Gastelumendi (2013) developed a theoretical approach for the Shipibo-Conibo ayahuasca experience. In this perspective, ayahuasca and psychoanalysis are seen as a combination of two universal traditions: medicine and self-knowledge. He presents several important concepts of psychoanalysis and links them to some aspects of the phenomenology of ayahuasca. For instance, the concept of treatment resistance is compared to the fear of losing control and sanity during ayahuasca experiences. The anguishing feeling of disintegration, loss of Self-unity, and death that are not rare during this kind of ritual, and ayahuasca experiences in general, are approached using Winnicott's concept of "fear of breakdown." In a psychotherapeutic process, fear of breakdown may be handled with support for the patient; in a similar fashion, shamans have tools to help people struggle through these moments during the ritual. The fear of death during the experiences is compared to the concepts of fear of castration or the death drive that would be challenged by the process of becoming conscious. Exploring several psychoanalytical concepts and establishing connections with ayahuasca experiences, Gastelumendi sets theoretical grounds for classical psychoanalysis and ayahuasca to converge.

Far from exhausting all possibilities, these two studies can be seen as two instances of how plastic ayahuasca can be when incorporated by two of the main theoretical approaches in psychotherapy. Psychotherapists willing to work with ayahuasca users should try the exercise of approaching it through their current knowledge and practice. This is, in our opinion, a path that remains open for many other contributions.

Closing Remarks

This is a preliminary text on this subject. Many of the ideas expressed here need further confirmation from more studies with different approaches, using input ranging from anthropology to neuroscience. There is a special need for new theoretical approaches from the psychological schools. Ayahuasca still lacks, for instance, a thorough discussion within the field of Lacanian psychoanalysis, or from the cognitive-behavioral therapy perspective.

To conclude, we would like to stress some key points that we believe are very important. First, it is clear that there is not a school of psychotherapetic thought, at the moment, that could be designated as "ayahuasca therapy." More important than defining which technique is the best to be used with the brew is the fact that the settings where ayahuasca is consumed are so diverse and that proclaiming a "right" way to perform psychotherapy with ayahuasca would probably be misleading.

However, this does not mean that specific intricacies of the ayahuasca experience could not be used to develop better integration procedures and innovative ways of incorporating the peculiar features of the ayahuasca trance (such as visions, challenging experiences, and plant teachings) into psychotherapy. On the other hand, an ethical approach from any psychotherapeutic school of thought would be equally desirable, without the necessity of developing a subspecialty within each theoretical framework for ayahuasca use. Again, cultural sensitivity and nonjudgmental approaches are desirable tools for any therapist but are particularly suited for those who have ayahuasca drinkers as patients.

Ayahuasca should not be regarded as a panacea; nevertheless, it may be a catalyst for personal transformation. As such, it should not be ignored by psychotherapy. On the contrary, we believe that a serious agenda of research connecting ayahuasca and psychotherapy could greatly benefit the world of ayahuasca drinkers in ritual, religious, or therapeutic settings and enrich the field of psychotherapy.

Acknowledgments The authors thank Elizabeth Connolly and Ana Florence for reviewing and suggestions.

References

Aixalà, M. (2017). *Developing integration of visionary experiences: A future without integration. Chacruna.net* Retrieved from http://chacruna.net/developing-integration-visionary

Araújo, D. B., Ribeiro, S., Cecchi, G. A., Carvalho, F. M., Sanchez, T. A., Pinto, J. P., et al. (2012). Seeing with the eyes shut: Neural basis of enhanced imagery following ayahuasca ingestion. *Human Brain Mapping, 33*(11), 2550–2560. https://doi.org/10.1002/hbm.21381.

Barbosa, P. C. R., Giglio, J. S., & Dalgalarrondo, P. (2005). Altered states of consciousness and short-term psychological after-effects induced by the first time ritual use of ayahuasca in an urban context in Brazil. *Journal of Psychoactive Drugs, 37*(2), 193–201. https://doi.org/10.108 0/02791072.2005.10399801.

Barrett, F. S., Bradstreet, M. P., Leoutsakos, J.-M. S., Johnson, M. W., & Griffiths, R. R. (2016). The challenging experience questionnaire: Characterization of challenging experiences with psilocybin mushrooms. *Journal of Psychopharmacology, 30*(12), 1279–1295. https://doi. org/10.1177/0269881116678781.

Cohen, I. (2017). *Re-turning to wholeness: The psycho-spiritual integration process of ayahuasca ceremonies in Western participants from a Jungian perspective* (Doctoral dissertation). San Francisco: California Institute of Integral Studies.

Dominguez-Clavé, E., Soler, J., Elices, M., Pascual, J. C., Álvarez, E., de la Fuente Revenga, M., et al. (2016). Ayahuasca: Pharmacology, neuroscience and therapeutic potentials. *Brain Research Bulletin, 126*(1), 89–101. https://doi.org/10.1016/j.brainresbull.2016.03.002.

Doyle, R. (2012). Healing with plant intelligence: A report from ayahuasca. *Anthropology of Consciousness, 23*(1), 28–43. https://doi.org/10.1111/j.1556-3537.2012.01055.x.

Fábregas, J. M., González, D., Fondevila, S., Cutchet, M., Fernández, X., Barbosa, P. C. R., et al. (2010). Assessment of addiction severity among ritual users of ayahuasca. *Drug and Alcohol Dependence, 111*(3), 257–261. https://doi.org/10.1016/j.drugalcdep.2010.03.024.

Gastelumendi, E. (2013). Una Mirada Psicoanalítica a la Experiencia con Ayahuasca [A psycho-analytic look at the experience with ayahuasca]. *Revista Psicoanálisis, 12*, 92–110. Retrieved from: http://www.spp.com.pe/uploads/biblioteca/BiViPsiL/Revista_SPP/Gastelumendi_12.pdf

Goldsmith, N. M. (2010). *Psychedelic healing: The promise of entheogens for psychotherapy and spiritual development.* New York City: Simon & Schuster.

Gomes, B. (2013). Ayahuasca e Recuperação de Pessoas em Situação de Rua [Ayahuasca and recovery of homeless people]. *Saúde e Transformação Social, 4*(2), 91–98.

Gomes, B. (2016). *O Uso Ritual da Ayahuasca na Atenção à População em Situação de Rua* [The ritual use of ayahuasca for the care of homeless people]. Salvador: EDUFBA.

Grinspoon, L., & Bakalar, J. B. (1997). *Psychedelic drugs reconsidered.* New York City: The Lindesmith Center.

Grob, C. S., McKenna, D. J., Callaway, J. C., Brito, G., Neves, E., Oberlaender, G., et al. (1996). Human psychopharmacology of hoasca, a plant hallucinogen used in ritual context in Brazil. *The Journal of Nervous & Mental Disease, 184*(2), 86–94.

Grof, S. (1980). *LSD psychotherapy.* Alameda: Hunter House.

Halpern, J. H., Sherwood, A. R., Passie, T., Blackwell, K. C., & Ruttenber, J. (2008). Evidence of health and safety in American members of a religion who use a hallucinogenic sacrament. *Medical Science Monitor, 14*(8), 15–22.

Krebs, T. S., & Johansen, P.-O. (2012). Lysergic acid diethylamide (LSD) for alcoholism: Meta-analysis of randomized controlled trials. *Journal of Psychopharmacology, 26*(7), 994–1002. https://doi.org/10.1177/0269881112439253.

Labate, B. C., & Cavnar, C. (Eds.). (2014). *The therapeutic use of ayahuasca.* New York City: Springer.

Labate, B. C., Rose, I. S., & Santos, R. G. (2008). *Religiões Ayahuasqueiras: um Balanço Bibliográfico* [Ayahuasca religions: A bibliographic review]. Campinas: Mercado de Letras.

Langlitz, N. (2012). *Neuropsychedelia: The revival of hallucinogen research since the decade of the brain.* Berkeley: University of California Press.

Loizaga-Velder, A., & Verres, R. (2014). Therapeutic effects of ritual ayahuasca use in the treatment of substance dependence – Qualitative results. *Journal of Psychoactive Drugs, 46*(1), 63–72. https://doi.org/10.1080/02791072.2013.873157.

Mabit, J. (2007). Ayahuasca in the treatment of addictions. In T. B. Robert & M. J. Winkelman (Eds.), *Psychedelic medicine: New evidence for hallucinogenic substances as treatments* (Vol. 2, pp. 87–103). Westport: Praeger.

Majić, T., Schmidt, T. T., & Gallinat, J. (2015). Peak experiences and the afterglow phenomenon: When and how do therapeutic effects of hallucinogens depend on psychedelic experiences? *Journal of Psychopharmacology, 29*(3), 241–253.

Melo, R. V. (2015). *Sobre Ayahuasca e Transformação na União do Vegetal* [On ayahuasca and transformation in the União do Vegetal]. Retrieved from http://neip.info/novo/wp-content/uploads/2015/04/rosa_udv_2011.pdf

Mercante, M. S. (2017). Imaginação, Linguagem, Espíritos e Agência: Ayahuasca e o Tratamento da Dependência Química [Imagination, language, spirits and agency: Ayahuasca and the treatment of addiction]. *NAU Revista de Antropologia, 60*(2), 562–587. https://doi.org/10.11606/2179-0892.ra.2017.137322.

Moure, W. G. (2005). *Saudades da Cura: Estudo Exploratório de Terapêuticas de Tradição Indígena na Amazônia Peruana* [longing for the cure: An exploratory study of therapies from the indigenous tradition in the Peruvian Amazon] (Doctoral dissertation). São Paulo: São Paulo University.

Naffah Neto, A. (2005). Winnicott: Uma Psicanálise da Experiência Humana em seu Devir Próprio [Winnicott: A psychoanalysis of human experience in its own becoming]. *Natureza Humana, 7*(2), 433–454.

Norris, L., & Megler, J. D. (2017). *Why a culture of integration is critical for the modern psychedelic movement. Chacruna.net.* Retrieved from: http://chacruna.net/culture-integration-critical-modern-psychedelic-movement-2/

Novaes, C. (2011). *L'Experience Urbaine de l'Ayahuasca: Paysages des Subjectivités Contemporaines* [The urban experience of ayahuasca: Landscapes of contemporary subjectivities] (Doctoral dissertation). Sorbonne: Université Paris Descartes

Okamoto da Silva, L. (2004). *Marachimbé Veio Foi para Apurar. Estudo Sobre o Castigo Simbólico, ou Peia, no Culto do Santo Daime* [Marachimbé has come to investigate study on the symbolic punishment, or *peia*, in the Santo Daime cult] (Doctoral dissertation). São Paulo: PUC-SP.

Osório, F. L., Sanches, R. F., Macedo, L. R., Santos, R. G., Maia-de-Oliveira, J., Wichert-Ana, L., et al. (2015). Antidepressant effects of a single dose of ayahuasca in patients with recurrent depression: A preliminary report. *Revista Brasileira de Psiquiatria, 37*(1), 13–20. https://doi.org/10.1097/JCP.0000000000000436.

Pahnke, W. N., Kurland, A. A., Unger, S., Savage, C., & Grof, S. (1970). The experimental use of psychedelic (LSD) psychotherapy. *JAMA, 212*(11), 1856–1863. https://doi.org/10.1001/jama.1970.03170240060010.

Palhano-Fontes, F., Barreto, D., Onias, H., Andrade, K. C., Novaes, M., Pessoa, J., et al. (2017). Rapid antidepressant effects of the psychedelic ayahuasca in treatment-resistant depression: A randomised placebo-controlled trial. *bioRxiv*, 103531. https://doi.org/10.1101/103531.

Sanches, R. F., Osório, F. L., Santos, R. G., Macedo, L. R. H., Maia-de-Oliveira, J. P., Wichert-Ana, L., et al. (2016). Antidepressant effects of a single dose of ayahuasca in patients with recurrent depression – A SPECT study. *Journal of Clinical Psychopharmacology, 36*(1), 1–5. https://doi.org/10.1097/JCP.0000000000000436.

Santos, R. G., Landeira-Fernandez, J., Strassman, R. J., Motta, V., & Cruz, A. P. M. (2007). Effects of ayahuasca on psychometric measures of anxiety, panic-like and hopelessness in Santo Daime members. *Journal of Ethnopharmacology, 112*(3), 507–513. https://doi.org/10.1016/j.jep.2007.04.012.

Shanon, B. (2002). *The antipodes of the mind: Charting the phenomenology of the ayahuasca experience.* Oxford: Oxford University Press.

Shanon, B. (2003). Os Conteúdos das Visões da Ayahuasca [The contents of the visions of ayahuasca]. *MANA, 9*(2), 109–152.

Thomas, G., Lucas, P., Capler, N. R., Tupper, K. W., & Martin, G. (2013). Ayahuasca-assisted therapy for addiction: Results from a preliminary observational study in Canada. *Current Drug Abuse Reviews, 6*(1), 1–13.

Trichter, S. (2009). Ayahuasca beyond the Amazon: The benefits and risks of a spreading tradition. *The Journal of Transpersonal Psychology, 42*(2), 131–148.

Zinberg, N. (1984). *Drug, set and setting: The basis for controlled intoxicant use.* New Haven: Yale University Press.

Chapter 5
A Qualitative Assessment of Risks and Benefits of Ayahuasca for Trauma Survivors

Jessica L. Nielson, Julie D. Megler, and Clancy Cavnar

Introduction

With the renaissance of psychedelic research and culture, increasing attention has been paid toward their therapeutic potential to treat a variety of disorders. Ayahuasca is a psychedelic tea that has shown promise as a therapeutic option to treat symptoms of depression (Osório et al. 2015; Palhano-Fontes et al. 2017; Sanches et al. 2016) and help with substance abuse (Lawn et al. 2017; Loizaga-Velder and Velder 2014; Thomas et al. 2013). Emerging anecdotal and proposed evidence have also indicated its potential to treat post-traumatic stress disorder (PTSD) (Galvão et al. 2018; Inserra 2018; Nielson and Megler 2014; Tafur 2017). Recent years have seen a boom in ayahuasca tourism, where an increasing number of people from Western countries are seeking out ayahuasca Amazonian ceremonies for psycho-spiritual healing related to trauma. In order to maximize potential benefit and reduce risk of harm, it's important to develop a therapeutic container for administering ayahuasca that is appropriate for mental health disorders such as PTSD. Ayahuasca is taken in various contexts, including traditional indigenous or mestizo ceremonies; South American religious ceremonies, such as Santo Daime and UDV (Labate and MacRae 2010; Labate et al. 2009); spiritual retreat centers catering to "gringos"; as well as ceremonies in underground communities in the United States and other parts of the world where it is illegal (e.g., Europe and Australia) (Labate et al. 2017). Therefore, it is possible to get a sense of what aspects of these different settings

J. L. Nielson (✉)
Department of Psychiatry, Institute for Health Informatics, University of Minnesota, Minneapolis, MN, USA
e-mail: jnielson@umn.edu

J. D. Megler
Entheogenic Research, Integration, and Education, San Francisco, CA, USA

C. Cavnar
Chacruna Institute for Psychedelic Plant Medicines, San Francisco, CA, USA

© Springer Nature Switzerland AG 2021
B. C. Labate, C. Cavnar (eds.), *Ayahuasca Healing and Science*,
https://doi.org/10.1007/978-3-030-55688-4_5

have the potential for contributing to healing traumatizing experiences, especially in the context of treating PTSD.

The combination of various settings and the mental state of the individuals participating in ceremonies relates to a popular notion in the psychedelic research community known as "set and setting" (Hartogsohn 2016; Leary et al. 1969; McElrath and McEvoy 2002). The context under which someone has a psychedelic experience can dictate whether the experience is helpful or harmful to their healing process. Previous risk factors and contraindications have been reported previously that suggest potential harm for people with pre-existing conditions like cardiovascular disease and major psychotic disorders, as well as pharmacological interactions between serotonergic drugs and alkaloids extracted from the different plant admixtures used to make ayahuasca (dos Santos 2013; Gable 2007). The current chapter discusses facets of set and setting for the use of ayahuasca in the context of treating PTSD and potential recommendations for appropriate preparation of the set and setting for modern ayahuasca trauma treatment.

Methods

Data were collected from a retrospective, cross-sectional online survey (UCSF IRB #16-19906) between June 27, 2016, until queried for the current analysis on December 29, 2017. The survey was shared on social media by the Multidisciplinary Association for Psychedelic Studies (MAPS) to recruit participants to the study.

We applied qualitative analysis on survey data from ayahuasca users regarding the perception of their experiences and whether they found them to be dangerous and/or helpful. Respondents' answers were broken up into condition groups based on self-reported history or current diagnosis of PTSD. Open-ended responses in text format were blocked similarly into four conditions for no PTSD (A), declined to answer (B), current PTSD (C), and past PTSD (D) (Fig. 5.1). Text were analyzed by three separate raters, blinded* to condition to identify common and unique themes related to dangerous and helpful experiences reported by the survey respondents. *We note that rater 1 created the blinding code to condition and therefore was not completely blinded during her coding of the responses. Each document was split into whether they responded to the experience being dangerous or helpful for each condition. Raters 1 and 2 coded all four conditions, and rater 3 coded only the PTSD groups (Fig. 5.1).

Qualitative analysis was performed using grounded theory, which aims to carefully examine the text of the data from each respondent and identify themes that emerge from the data (Charmaz 2006; Strauss and Corbin 1998). We applied this within the context of different groups self-reporting various degrees of PTSD severity to determine if those seeking out ayahuasca to treat PTSD need special care beyond what people with no history of PTSD find helpful during ayahuasca ceremonies.

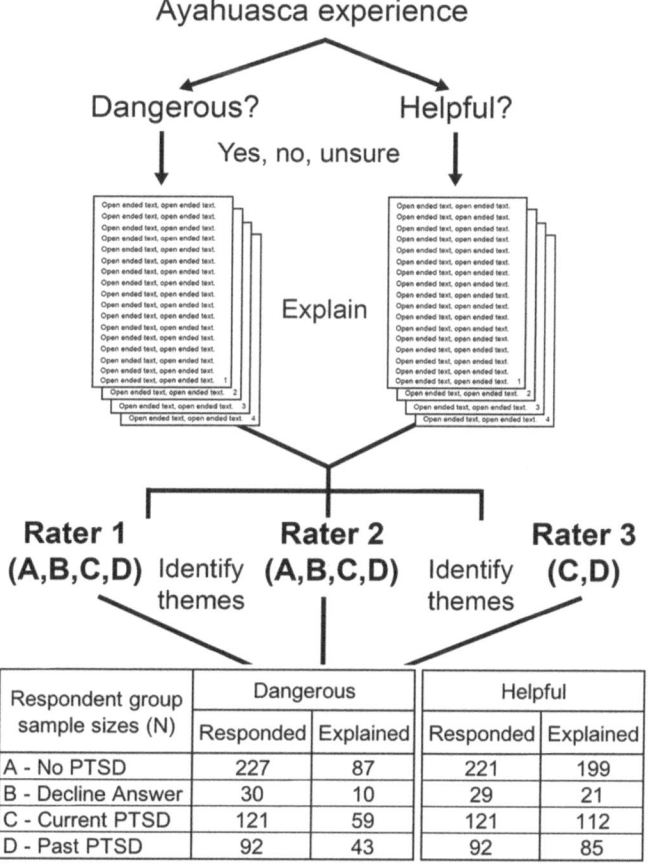

Fig. 5.1 Workflow for identifying themes of dangerous and helpful experiences of ayahuasca use

Rater 1 read through condition A (no history of PTSD) first to get a starter set of themes for potential dangerous and helpful experiences, which were used to help quantify their occurrence in the remainder of the text from the other PTSD conditions. Repetitions of similar ideas or experiences were tallied with matched identified themes, and new themes or unique experiences were recorded as they emerged in the readings. Comments described not only what the participants experienced themselves but anything they had witnessed or observed during their experience in others. These distinctions are not separated for simplicity in the current analyses.

Rater 2 was blinded to condition and coded all four PTSD categories for both helpful and dangerous experiences. First, instances of judgment were noted; many respondents had several mixed replies, and these were parsed out so that individual aspects of experiences could be noted. As particular descriptors added up, categories emerged. Some answers appeared to be speculative, rather than from direct experience, or from someone else's experience, but all were treated as mentions, including responses that indicated the individual had not yet drunk ayahuasca. Despite several outliers, clear patterns emerged from the bulk of the responses, many reflecting complaints heard among ayahuasca user's groups online and elsewhere.

Rater 3 was blinded to condition and began coding condition C (current PTSD) and condition D (past PTSD), by first reading through open-ended responses about dangerous experiences encountered or observed during their ayahuasca ceremony, followed by analyzing the helpful experience open-ended responses. The coding process for this rater began with writing in a spreadsheet phrases that have potential to be themes. As rater 3 read through the reported experiences, she noted new potential themes as they appeared, counting the occurrence of repeats in already identified themes. At the completion of tallying all sections, rater 3 read back through the spreadsheet of identified themes, combined themes and their tallies that were similar in meaning, and then organized themes into categories based on commonalities and overlap.

Rater 1 then combined all three rater responses to identify overlapping and distinct themes between raters. The sum of counts within each theme was converted to a proportion of the total counts each rater reported within each PTSD condition. This was necessary to normalize the variability in total counts each rater submitted. For example, raters 1 and 2 identified roughly the same number of unique phrases related to dangerous experiences; however, their tallies for how frequently they were found within the text was substantially different. These proportions were calculated and graphed together for all raters to visualize how consistent each theme was and whether they changed between PTSD groups, for both helpful and dangerous experiences. Graphs were made using GraphPad Prism® v7.

Results

The results from the retrospective cross-sectional online survey reported here are from respondents who completed the full survey, were over the age of 18, and did not opt out of consent at the end of the survey. Responses from a total of 520 people revealed approximately half the respondents had a self-reported past ($N = 99$) or current ($N = 147$) diagnosis of PTSD, with the rest of the respondents either declining to answer about their PTSD history ($N = 33$) or reporting no previous history or diagnosis of PTSD ($N = 241$). The authors provide the following disclaimer regarding the perceived "dangerous" experiences of ayahuasca. Many of the themes described below are considered by regular users to be normal aspects of the ayahuasca experience. However, the tone with which the answers were provided and the context under whether they agreed them to be "dangerous" or not provided these categorizations for themes as either "helpful" or "dangerous" (see Fig. 5.1).

While most respondents provided answers of either "yes," "no," or "unsure" for questions of whether they experienced either helpful or dangerous aspects of their ayahuasca experience, not everyone was willing to provide details in the open-ended responses. Data for dangerous experiences had substantially less open-ended responses associated with them across all groups (33–49% of answers), compared to helpful experiences which respondents were more willing to provide details about (72–93% of answers). We may speculate as to why this is the case, but this wasn't specifically asked of the survey participants. The authors propose that participants are more inclined to describe the helpful aspects of their experience because they aren't as triggering as the perceived dangerous experiences. Another interpretation may be that the survey was self-selecting for people who are already enthusiastic about ayahuasca and therefore may have an implicit bias toward describing these qualities in more detail than the ones that may be seen as adding controversy to the use of ayahuasca. For the no PTSD group, 94% answered the yes/no/unsure question about dangerous encounters; however, only 38% of those provided further descriptions in open-ended responses as to the details of those dangerous experiences. Similarly, 90% of the respondents that did not disclose their PTSD status answered the simple question, but only 33% provided detailed responses on the dangers they encountered. Interestingly, a higher proportion of the respondents with a past or current diagnosis of PTSD provided open-ended responses about dangers. While only 82% of the current PTSD group answered the simple yes/no/unsure question, 49% of those responses provided additional details about what those dangers were. Compare this to the past PTSD group, where 93% of them answered the simple question, with 47% of them providing more details in the open-ended responses.

Within each PTSD group, approximately 16–20% of the respondents reported that they experienced and/or observed something dangerous about their experience, while 68–76% claimed they did not observe anything dangerous about their experience. However, evaluating only the proportion of respondents who provided open-ended answers to describe specifics about what was helpful or harmful has slightly different results. While there was no difference between PTSD groups on whether they experienced helpful aspects of their experience (92–99%), those who chose to disclose more about their dangerous experiences (39–60%) reported "yes" to encountering some dangerous aspects of ayahuasca, with only 20–44% reporting "no" as to whether they observed these dangers.

From the open-ended responses, we describe below the common themes that arose that each of the raters identified while reading through the open-ended responses for the participants willing to share what they perceived to be dangerous and/or helpful. These themes are summarized in Tables 5.1 and 5.2 across all PTSD groups for simplicity.

Table 5.1 Common themes related to dangerous aspects of an ayahuasca experience

Dangerous themes	Details	Raters
Complications with other participants in the ceremony	Overcrowding, lack of screening for unstable people	1, 2, 3
Psychological complications	Internal fears, panic, intense emotions, feeling out of control, presence of "entities," psychosis	1, 2, 3
Unsafe setting, no supervision	No medical supervision, lack of one-on-one support, fires, language barriers, dangers of the jungle, intoxicated facilitators	1, 2, 3
Physical complications	Increased heart rate and blood pressure, problems breathing, exhaustion, dehydration, unsteady gait	1, 2, 3
Access to purging facilities	Difficult or no access to toilet/bucket, walking to bathroom without help/falling	1, 2, 3
Not being prepared	Not following dieta, contraindicated medications, pre-existing conditions, too soon after trauma, facilitators not educating participants about risks	1, 2, 3
Brew ingredients	Presence of *Datura* or tobacco, non-traditional admixtures	1, 2
Unethical facilitators	Sexual abuse, financial exploitation, legal concerns, "witch doctors"	1, 2, 3
Lack of follow-up care	No process for integration, risks of self-harm	1, 2, 3
Dogmatic practices	Strict religious practices, gender issues	1, 3
Inexperienced facilitators	Not having a shaman, lack of trust due to inexperience, unable to handle psycho-spiritual complications	2, 3
Use of cannabis	Smoking marijuana during ceremony	3

Dangerous Themes

Complications with other participants in the ceremony was a theme that popped up for all three raters. Respondents describe the behaviors and the energy of the other participants to feel dangerous. Some attributed this to poor screening of participants by the center beforehand. Others found it was dangerous to be in ceremony with other participants who are not also taking ayahuasca. There was mention of feeling that other participants may be "energy vampires" that felt unsafe to them, as well as sitting with dark energies brought by others. The feeling that they didn't know the other people in ceremony, leading to feelings of not knowing if they could trust them while all in a vulnerable state, also was mentioned. A very common issue that was discussed was that ceremonies were simply overcrowded, with too many participants crammed into the ceremony space. Some reported that they felt unsafe around "entities" that were brought in by other people, and that their purging was too distracting for their own process to unfold, which became frustrating as well as feeling unsafe. One participant noted that there were teenagers in the ceremony, which they felt was unsafe for the teenager, but not necessarily for the ceremony space or them. Some respondents noticed that there was sexual activity and sexual energy being brought into the space by others that made them feel unsafe, while it

Table 5.2 Common themes related to helpful aspects of an ayahuasca experience

Helpful themes	Details	Raters
Soothing sounds	Shaman singing icaros, playing instruments, sounds of the jungle, familiar music, being able to sit in silence	1, 2, 3
Experienced facilitators and helpers	Having a shaman, having support from helpers, one-on-one attention	1, 2, 3
Supportive, like-minded group	Doing ceremony with others in their community, people they are comfortable with	1, 2, 3
Safe/comfortable space	Having a space to lay down, pillows, blankets, spaces to be alone, being in nature, darkness/limited light	1, 2, 3
Trust	Trusting the facilitators, trusting ayahuasca, being able to let go/surrender	1, 2
Self-support	Meditation, journaling, yoga, prayer, connection to spirits/guides, self-love	1, 2, 3
Ritual components	Smudging/cleansing of space, sacred items (talismans, an altar), mapacho smoke	1, 2, 3
After-care	Post-session meetings with shaman, with group	1, 2
Preparation	Pre-session meetings, setting intentions, following dieta, experience with other psychedelics	1, 2, 3
Comforts and supplements	Places to lay down, pillows/blankets, Agua de Florida, ginger tea for nausea, citrus, flower baths, scented oils, pets	1, 2, 3
Help with purging	Helpers to provide and clear purge buckets, toilet paper/tissue, easy access/help to bathroom, being comfortable purging in front of others	1, 2, 3
Symptom alleviation	Claims of being healed by ayahuasca from symptoms of trauma (e.g., PTSD, depression)	3[a]

[a]Rater 3 categorized this as a theme, whereas raters 1 and 2 took note of it, however didn't create a theme for it since symptom alleviation is an outcome rather than a predictor of potential treatment response

seems one participant noted that others in ceremony were using an audio recorder that they kept hidden from the group, which felt deceptive and dangerous.

Psychological complications experienced by the respondents was also a major theme that felt somewhat dangerous to many of those willing to share their experiences. The most common description was related to feeling out of control, which brought on feelings of fear, panic, and anxiety. More specific psychological complications were also noted that may be unique to smaller groups of respondents but warrant description of potential psychological complications that could arise. Some respondents claimed to have lasting complications beyond their experience, including the development of panic attacks during the months after their ceremony and psychosis for several days after, and one participant claims to have been haunted by an "imp" for a long duration after their ceremony that lived in their home, whose spouse, who did not take ayahuasca, claimed to have seen as well. Other respondents noted feeling threatened by unfriendly "entities" that made them feel scared. Other psychological complications included paranoia, intense emotional reactions, being catatonic, feeling as though they were dying, fear of being persecuted, inner conflict, questioning their identity, fear that they were being subjected to sorcery, terrifying visions, and undergoing spiritual battles, and some felt the experience was itself re-traumatizing.

Unsafe settings and a lack of supervision was another major theme that arose among the respondents. Elements of the setting that felt unsafe included having bonfires or other types of fires/flames in the ceremony space that respondents felt was a potential danger. Some claimed that there was access to weapons that felt very unsafe. Having stairs nearby where participants could trip and fall was noted, as well as unsafe walkways for trying to navigate around that were often too dark too see, in addition to them trying to navigate around while under the influence of aya-huasca. Some respondents felt unsafe being in the Amazon jungle where they felt they were in danger from creatures and animals such as snakes, spiders, and jaguars. There was also mention of being nervous about the language barrier and being around strangers. Others mentioned feeling unsafe due to a lack of supervision from those facilitating or supporting the participants during their experience. Some respondents felt it was unsafe for the facilitator/shaman to also be drinking aya-huasca with them and feeling uncomfortable that no one was "sober" during their experience and therefore did not feel supported. Many people claimed that there wasn't enough, if any, supervision during their experience. This was described as either having no one present or not enough people present to administer one-on-one care to each participant as needed. Respondents also felt it was dangerous that there were no medical and/or psychological professional on-site to help if/when serious complications arose during their experience. This lack of support also extended to making sure people were ready to leave when the experience was over and often feeling they were forced to leave while still under the influence and having to drive home.

Physical complications were another prominent theme, which is to be expected during an ayahuasca experience, especially considering that most people feel nau-sea and vomit/purge during their experience. However, additional complications that went beyond these typical physical reactions included risks of choking while purging and feeling as though they had overdosed or were given too much aya-huasca. Cardiovascular complications were described by many respondents, includ-ing both hypertension and hypotension, tachycardia, chest pain, and trouble breathing. Some respondents felt extreme dehydration due to excessive sweating, convulsions, fainting and/or blackouts, temporary blindness, physical pain, and extreme exhaustion by the end of the experience.

Access to purging facilities was also described as a problem that could intro-duce potential dangers where participants could be harmed. Some people describe more of a discomfort or unwillingness to purge in front of others in the group settings. However, some respondents claimed there wasn't safe access to purging facilities or bathrooms, where they had to get up and walk to a place to purge while intoxicated, where some people would fall and hurt themselves or felt addi-tional stress about where they could purge. Some also claimed that the purging facilities were unsanitary, which could pose risks for infection or other gastrointestinal complications.

Lack of sufficient preparation was another theme that emerged from the respondents, spanning from a lack of information from the retreat center or facilitators to their own personal lack of physical and psychological preparation before the experience. Within this theme, many claimed to experience "preflight" anxiety beforehand, where they felt scared and anxious about what was going to happen because they didn't know what to expect from the experience. Others mentioned dangers associated with not following the pre-prescribed dieta/diet, including weaning themselves off contraindicated medications and avoiding certain foods. Some felt not having any prior experience with other psychedelics was potentially dangerous, while those seeking treatment for PTSD felt they may have taken it too soon after their trauma and were therefore unprepared. The system that responds to fearful memories threats may be hyperactivated early after a trauma. These are the same systems that are activated during an ayahuasca experience that may act like exposure therapy (Nielson and Megler 2014). However, if done too soon after the trauma, the impact of such an intervention may be more harmful to recovery.

Unethical facilitators were mentioned as dangerous themes, which seems to circle around the emerging topic of "ayahuasca tourism." Respondents mentioned feeling as though they were being taken advantage of, either through forms of sexual abuse by the shamans/facilitators or through the theft of their personal belongings. Others mentioned that their facilitators demanded more money from them during their ceremony, threatening to leave them alone if they didn't pay them more. Others felt unsafe from a legal standpoint and worried that they would be arrested for doing ayahuasca. Along that same vein, another theme was related more to the lack of experience of the facilitators. While not mentioned as being unethical, respondents felt unsafe with "neo-shamans" who clearly did not have the experience or expertise necessary to safely facilitate the ayahuasca ceremony. Some felt this was very inappropriate and opened them up to more psycho-spiritual damage because the person facilitating was not prepared to handle such challenging experiences. Respondents mentioned feeling the facilitators or support staffs choice of words to try and sooth them actually made them feel worse.

Lack of follow-up care was another theme that was mentioned. Respondents who felt unsupported afterward mentioned risks of self-harm that could follow without sufficient after-care and discussions with the facilitators to discuss what came up for them and how best to process and integrate that once the ceremony was over.

Dogmatic practices were noted, more specifically related to the ayahuasca religions, including Santo Daime. Some respondents felt uncomfortable and unsafe in these religious settings, claiming to feel as though they were being attacked, with references to Catholic practices and themes resembling bondage, discipline, dominance, submission, sadism, and masochism (BDSM).

Helpful Themes

Soothing sounds and music were one of the most prominent themes mentioned across all participants. Music of some kind, whether it was through the icaros being sung by the shaman, musical instruments and songs being played by the facilitators or others in the group, as well as the sounds of nature, was found to be profoundly helpful for participants. Musical instruments described included drums, flutes, didgeridoos, and chimes, whereas natural sounds that were found to be soothing included the various sounds of jungle life, including birds chirping and insects buzzing, as well as the sound of the rain falling outside. While most respondents found the icaros and singing to be one of the most helpful aspects of the experience, some found the ability to be able to sit in complete silence to be very helpful for them as well.

Experienced facilitators and helpers were another very common theme that respondents reported. Having an experienced shaman present was one of the more prominent topics mentioned within this theme. However, not everyone felt a shaman was necessary, but at least having someone guiding the experience that had sufficient training and experience in the traditional uses of ayahuasca. Other skills respondents reported to be very helpful from their facilitators and support staff included having trained psychotherapists present, and having plenty of one-on-one attention from the facilitators, as well as having someone sober there to take care of them. Some claimed that a healing touch was very helpful from the shaman/facilitator or helpers, while some even felt being able to touch the facilitator or hold/be held was helpful for them. Some mentioned having a facilitator that spoke the same language was helpful, as long as there was sufficient consent from both participant and facilitator regarding the extent of that soothing touch or hug.

Sitting in ceremony with a supportive and like-minded group was also found to be very helpful for respondents during their ayahuasca experience. Being in a group where they felt comfortable with the other people in the ceremony, through either getting to know them beforehand or sitting in a group with friends and family or other shared community, was very important to them. Some mentioned enjoying having a large group, while others prefer either a small group of people or just being alone to be helpful.

Doing ceremony in a safe and comfortable space was another prominent theme that was helpful for a productive and safe ayahuasca experience. Many respondents described being able to be in nature and in a setting that had low light and/or darkness was very helpful for them, meaning being in the jungle and doing ceremony at night was a common setting that people found to be helpful. Being in a familiar place was found to be helpful for some while having sufficient space to move around and express themselves as they needed.

A sense of trust with either the medicine or the space/facilitators was also a very important theme that helped respondents have positive ayahuasca experiences. Many felt that the ability to surrender and "let go" to the experience was crucial for it to be a healing experience. This emerged as a result of the other themes, where if they trusted that ayahuasca itself was safe, that their environment was safe, and that the shaman/facilitator was experienced, then they were able to fully surrender to the experience. Respondents also described a need to trust themselves to alleviate any fears that may arise during the experience and be reassured that the effects are only temporary, whereas some felt it was helpful to trust in a higher power to help them through the experience.

Self-support was discussed where respondents felt that being able to rely on their own tools was profoundly helpful during their ayahuasca experience. Various tools of self-support were mentioned, including meditation, yoga, prayer, breathing techniques, and positive self-talk and self-love. Respondents felt it was important to have their own spiritual practice and connection to guides and spirits beforehand to help guide them through the experience.

Ritual aspects of the ayahuasca experience were also found to be very helpful for respondents. Having an altar in the ceremony space decorated with personal talismans from the participants was noted. Others describe the cleansing ritual from the shaman/facilitator to be profoundly helpful, including feathers, mapacho smoke, incense, sage, fire, and other religious symbols and cleansing rituals to protect the space from dark energies.

Providing participants with access to sufficient after-care and integration techniques was found to be helpful for many respondents. This took the form of post-ceremony meetings with the group and facilitators, as well as check-ins from facilitators after returning home to make sure participants were feeling stable and transitioning back into their daily routine without complications.

Preparation prior to the ayahuasca ceremony was a very important theme that respondents found helpful. Preparations included following the recommended diet and detoxing from any contraindicated medications. Knowing beforehand that they would be returning home to a supportive community was found to be helpful. Respondents with prior experience taking other psychedelics was mentioned as being a good way to prepare for taking ayahuasca. Doing research about the retreat center, the facilitators, and ayahuasca itself was described as being helpful methods of preparation. Having clear intentions and no expectations of outcome was discussed, as well as allowing themselves adequate time to prepare and making sure they were able to be away from home and work to have adequate time to prepare, experience, and recover from their ayahuasca ceremony.

Comforts and supplements during the ayahuasca ceremony were found to help-ful. These comforts appeared to be tailored to individuals, with a few common recommendations including Agua de Florida, essential oils, flower baths, and pillows and blankets for comforts. Additional comforts and supplements included having pets present to interact with, having a journal to write in, ginger tea to help with nausea, and hammocks for laying in, and anti-anxiety medications in some cases were helpful.

Help with purging was another theme that emerged, where respondents described feeling much more comfortable and supported when they didn't have to worry about how, when, or where they could purge. Ensuring that someone could help them to the bathroom or having their own bucket next to them that was consistently cleaned out by helpers was important. Some felt that being able to feel comfortable purging around others was important, especially in group settings where people have purge while others are watching. Access to tissue and toilet paper was also mentioned as helpful so people could clean up after purging.

Symptom alleviation was also described. While not directly related to details of set and setting during the ayahuasca experience, it's worth noting that within the open-ended responses about what was helpful, participants do describe an alleviation of their symptoms because of their ayahuasca experience. Many described that ayahuasca healed them from their trauma and their PTSD symptoms, as well as from depression. Some described that ayahuasca helped them to identify their PTSD and anxiety triggers, while others described having a mystical/spiritual experience that helped alleviate their symptoms.

Differences Between Normal Respondents and Those with a Reported History of PTSD

Figure 5.2 reports on the identified themes for dangerous (Fig. 5.2a–d) and helpful (Fig. 5.2e–h) aspects of the ayahuasca experience, blocked by the self-reported PTSD group, and separated for each rater based on how frequently they observed these themes within each PTSD group. Only raters 1 and 2 coded for respondents with no history of PTSD (Fig. 5.2a, e) or who declined to report their history (Fig. 5.2b, f), and all three raters coded for past (Fig. 5.2c, g) and current PTSD groups (Fig. 5.2d, h). The most noticeable difference between the PTSD groups for dangerous themes was related to physical complications, which were much more prominent in those with a reported history of PTSD (Fig. 5.2c, d). Additionally, people with a history of PTSD reported more dangers associated with the lack of safety and supervision in the setting. Among the helpful themes reported, no major differences between PTSD groups were observed in the distribution across the themes, where there was a consensus across all respondents and raters that the most helpful aspects of their ayahuasca experience were having a shaman and/or experienced facilitator and support staff present and being able to listen to music, icaros, or other soothing sounds (Fig. 5.2e–h).

Discussion

The term "set and setting" was identified in the late 1960s by Timothy Leary, Ralph Metzner, and Richard Alpert to describe not only the importance of the mindset of the individual undergoing a psychedelic experience (set) but also the setting in which the experience takes place (Leary et al. 1969). In both categories of dangerous and helpful themes related to ayahuasca, topics related to "set and setting"

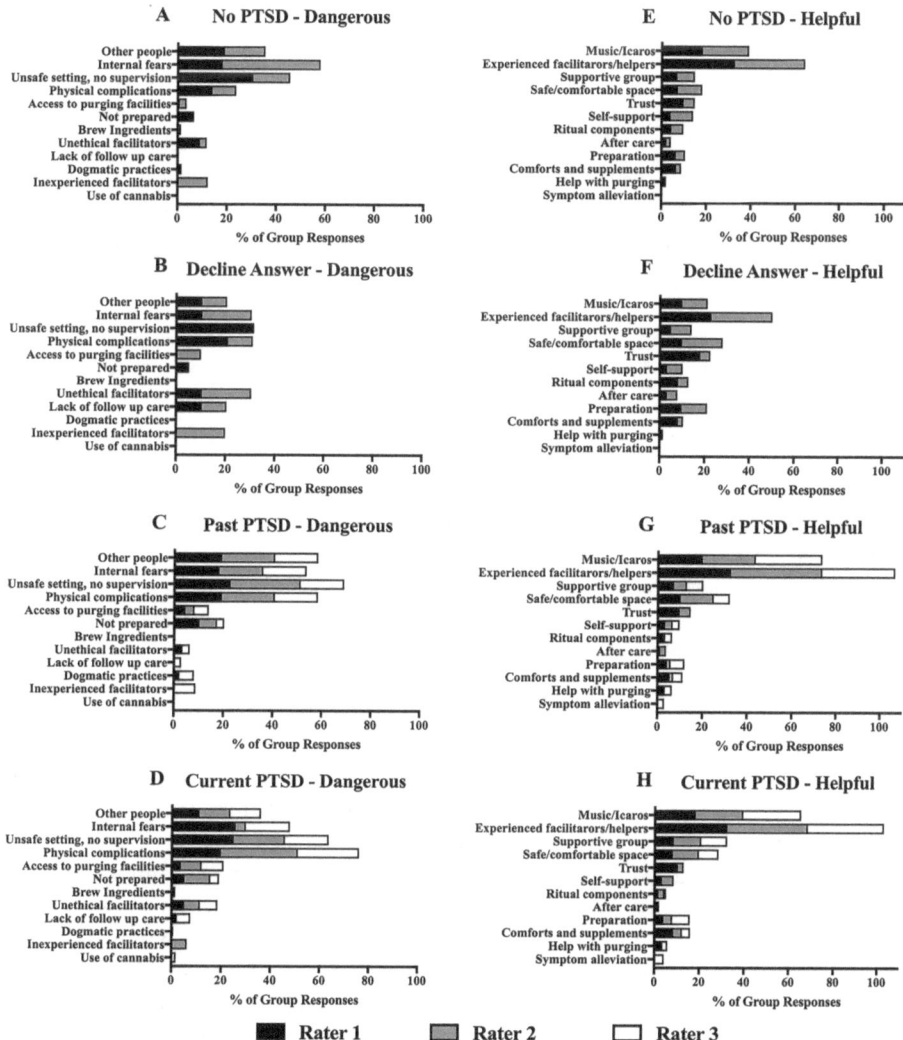

Fig. 5.2 Common themes identified across three independent raters for dangerous (**a–d**) and helpful (**e–h**) aspects of the ayahuasca experience across self-reported PTSD diagnostic categories, including no history of PTSD (**a, e**), respondents that declined to answer their PTSD status (**b, f**), those claiming to be in remission from a past diagnosis of PTSD (**c, g**), and those claiming to have a current diagnosis of PTSD (**d, h**)

appeared to be the most commonly acknowledged across all groups. This indicates that creating a safe and supportive environment and ensuring that participants are prepared and in the right mindset are perhaps the most important factors to achieve beneficial outcomes from a psychoactive experience such as ayahuasca.

In traditional cultures that use plant-based medicines, people grow up being familiarized with such medicines and the ecosystems they come from, allowing participants to enter the psychedelic experience with a knowledge and understanding. Thus, the unpredictable effects of psychedelics are held within the container of their native culture and perspective (Goldsmith 2011). For individuals coming from modern industrial societies, understanding practices from shamanistic paradigms when using psychoactive substances is critical for using psychedelics safely and effectively (Bravo and Grob 1989). Individuals not familiar with the traditional context or container are often left to discover alone a practical application of the transpersonal ayahuasca experience into the modern culture they live in. Bravo and Grob (1989) state:

> In traditional societies, shamans discuss and explain the significance of what transpired during the profoundly altered state of mind that the patient encountered during the session. From the perspective of modern psychedelic psychotherapy, similar emphasis must be placed on utilizing the post psychedelic state to optimize the realization of the therapeutic goals. Comprehensive processing in the hours and days that follow the psychedelic experience is necessary to increase the likelihood that the insight and gains achieved during the drug-facilitated state will be retained. (page 127)

Thus current techniques for "Westerners" to maximize the therapeutic potential of psychedelic experiences have been developed to bring together indigenous practices and the modern context participants live in, as the experiences must be made relevant for the current culture from which the individual will be returning home to (Bravo and Grob 1989).

Considering these results from the standpoint of treatment of PTSD in ayahuasca rituals, it is clear that safety is the overriding concern: the safety of being in a comfortable, secure place; the safety of being with a known group and leader; and the safety of following a known protocol. The dangers of feeling unsafe for participants are that they may manifest anger or paranoia, the feeling that the person is the subject of a plot to harm him or her, and they may react in a defensive style to an imagined offense. The paranoid reaction is one of the more common manifestations of mental delusion found in the world of substance-induced mental disorders and also one of the more common negative mental health problems associated with ayahuasca-induced negative mental states.

Rituals conducted in the dark in a foreign language open the door for doubt and unease in people primed to be suspicious. One way to prevent such moments could be to give familiarity precedence by showing the site of the ritual in the daytime, meeting with the shaman and other participants, and preparing the participants in helping them know how the ritual will proceed. Working only with highly recommended centers and shamans is another way to prevent negative outcomes by ensuring feelings of safety and competence. Adequate staffing will also be a safety factor that can be assessed prior to the ritual. All of these precautions can be understood under the title of "preparation," and, while preparation is important for all ayahuasca drinkers, it is even more so for those who suffer from PTSD.

To reference another chapter from this volume that is related to the current study from the chapter "Ayahuasca and Psychotherapy: Beyond Integration" by Mauricio Diament, Bruno Ramos Gomes, and Luis Fernando Tófoli:

For a meaningful experience, individuals must feel supported and protected by the environment, so that they can evolve together with what is experienced. In a similar manner, we can say that the setting of an ayahuasca experience has to be supportive and provide a sense of protection to the participant. This is the key role of the ritual and its responsible guidance. A supportive and protected setting helps the experience to fully develop. Within this protective environment, individuals may 'let happen' on themselves psychological processes during the experience.

Studies on the use of ayahuasca for the treatment of substance abuse disorders may also be capturing people suffering from PTSD, as substance misuse is often a symptom of unresolved trauma (Jacobsen et al. 2001). "Healing" from addiction is often healing from trauma, as symptoms are faced in sobriety and worked through, rather than buried beneath dopamine dumps provided by substances or behaviors of abuse.

Other symptoms of PTSD that may be exacerbated by ayahuasca rituals include disturbing images regarding the trauma (Nielson and Megler 2014) and emotional lability (Riba et al. 2003, 2006). Sleep disturbance, another symptom of PTSD, does not seem to be a factor beyond the acute intoxication stage of ayahuasca use (Barbosa et al. 2012). In most reported cases, mood is improved after ayahuasca sessions, and ayahuasca has been found to help in cases of treatment-resistant depression (Palhano-Fontes et al. 2017); negative changes in mood are another symptom of PTSD, and mood elevation is one of the more promising effects of ayahuasca on humans (Palhano-Fontes et al. 2014; Sanches et al. 2016). The anxiety and depression common to sufferers of PTSD may well be alleviated by the use of ayahuasca, making progress in therapy more likely.

However, another major factor to account for when considering ayahuasca to treat PTSD is the fact that those with a history of PTSD reported more physical complications associated with cardiovascular function. This is especially important considering that many people living with PTSD will develop some form of cardiovascular disease (Edmondson and Cohen 2013) and may be at greater risk for adverse events related to heart rate and blood pressure regulation during an ayahuasca session. Careful monitoring should be performed by a physician to ensure prospective PTSD patients are safe during any treatment protocols.

Another important factor to consider regarding the appropriate use of ayahuasca to treat people with PTSD is related to whether it should be administered individually, or in a group setting, as is the traditional method. However, those in the past PTSD group reported having more issues with overcrowding or unstable people sharing the ceremony space, compared to the current PTSD group. This is particularly interesting for any sort of group therapy for trauma work. People's nervous systems resonate with each other, so if one individual is agitated, it can cause others to become agitated as well. If someone is agitated and they are in a therapeutic relationship with someone whose nervous system is relaxed, it can create a container whereby the agitated person may start to feel more regulated. This type of resonance with others in the environment is further supported by the results reported in the current study, where a substantial proportion of individuals identified having supportive staff readily available as being helpful during their ayahuasca experience.

Similarly, being around people they felt comfortable with was reported to be helpful and may further support this theory of being in a type of energetic harmony with others in the room.

This concept of people's nervous systems resonating with each other is referred to as "limbic resonance" (Lewis et al. 2001). There has been discussion among researchers involved in ayahuasca and trauma work (anecdotal) about whether or not it is best for ayahuasca ceremony groups aiming to support healing for those with PTSD to be comprised of all individuals with PTSD or a mix of individuals with PTSD and those without. If all the participant in a group have PTSD, it would mean there are not any regulated and calm nervous systems for the individuals with PTSD to resonate with, aside from the person facilitating the ceremony, presumably. Having this type of distribution of only participants with PTSD could potentially lead to an entire room full of nervous systems perceiving some level of threat, which may make the therapy re-traumatizing rather than therapeutic. Therefore, we propose that if PTSD patients are to undergo treatment with ayahuasca in a group setting, the safest care for both individuals with PTSD and those without PTSD should be a small proportion of the total group having PTSD. Another important aspect of having additional able-bodied individuals physically and mentally available is to help participants in the space feel safe. It would be reasonable to assume that the calm presence of support staff allows for ceremony participants to have more regulated nervous systems for similar reasons as described above.

Lastly, we'd like to touch upon the cultural differences inherent in the present study and how those may be perceived as "dangerous" or otherwise abnormal to seekers from outside the traditional culture of ayahuasca use. These are related to issues with encountering "entities" or "spirits" as well as feeling that some of the shamans or facilitators were witch doctors of some sort. While some cases of true psychosis or other major mental illness not compatible with taking ayahuasca may be present in certain individuals, a certain amount of what those consider dangerous may simply be artifacts or misinterpretations of traditional practices that seem dangerous because they are foreign and unfamiliar. What some may consider to be a psychotic experience of seeing "entities" may in fact be a spiritual emergency that often can be navigated by the so-called witch doctor familiar with such encounters. It has been suggested that participants emerging from these types of situations may benefit from integration work back home with a Western-trained psychotherapist familiar with dealing with spiritual crisis (Lewis 2008).

In conclusion, while the current study supports the literature regarding the safety and therapeutic potential of ayahuasca for alleviating symptoms associated with mental health disorders related to trauma, there are some potential risks that clinicians, researchers, and patients should be accounting for if this is to be used to treat PTSD. Therapeutic and ritualistic settings must be equipped with the appropriate monitoring equipment and prophylactic drugs to counteract cardiovascular complications, accompanied by experienced clinicians, a shaman, and trained and supportive staff, and include personal comforts and music to maximize the therapeutic potential of ayahuasca for PTSD. Additionally, for Westerners not familiar with the traditional and indigenous uses of ayahuasca, additional preparation before and integration after may help participants process unfamiliar encounters that may be perceived as dangerous. However, these encounters may be cultural differences that require more time and energy to understand and process, with the appropriate supportive environment upon returning to daily life back home.

Acknowledgments Funding for survey data collection was provided by the Multidisciplinary Association for Psychedelic Studies (MAPS). The authors would like to thank Rick Doblin, Allison Feduccia, and Bia Labate for fruitful discussions during the development of the survey questions.

References

Barbosa, P. C. R., Mizumoto, S., Bogenschutz, M. P., & Strassman, R. J. (2012). Health status of ayahuasca users. *Drug Testing and Analysis, 4*(7–8), 601–609. https://doi.org/10.1002/dta.1383.

Bravo, G., & Grob, C. (1989). Shamans, sacraments, and psychiatrists. *Journal of Psychoactive Drugs, 21*(1), 123–128. https://doi.org/10.1080/02791072.1989.10472149.

Charmaz, K. (2006). *Constructing grounded theory: A practical guide through qualitative analysis.* London: Sage.

de Galvão, A. C. M., de Almeida, R. N., dos Santos Silva, E. A., Freire, F. A. M., Palhano-Fontes, F., Onias, H., et al. (2018). Cortisol modulation by ayahuasca in patients with treatment resistant depression and healthy controls. *Frontiers in Psychiatry, 9*, 185. https://doi.org/10.3389/fpsyt.2018.00185.

dos Santos, R. G. (2013). A critical evaluation of reports associating ayahuasca with life-threatening adverse reactions. *Journal of Psychoactive Drugs, 45*(2), 179–188. https://doi.org/10.1080/02791072.2013.785846.

Edmondson, D., & Cohen, B. E. (2013). Posttraumatic stress disorder and cardiovascular disease. *Progress in Cardiovascular Diseases, 55*(6), 548–556. https://doi.org/10.1016/j.pcad.2013.03.004.

Gable, R. S. (2007). Risk assessment of ritual use of oral dimethyltryptamine (DMT) and harmala alkaloids. *Addiction, 102*(1), 24–34. https://doi.org/10.1111/j.1360-0443.2006.01652.x.

Goldsmith, N. M. (2011). The ten lessons of psychedelic therapy, rediscovered. In *Psychedelic healing: The promise of entheogens for psychotherapy and spiritual development.* Rochester: Healing Arts Press.

Hartogsohn, I. (2016). Set and setting, psychedelics and the placebo response: An extra-pharmacological perspective on psychopharmacology. *Journal of Psychopharmacology, 30*(12), 1259–1267. https://doi.org/10.1177/0269881116677852.

Inserra, A. (2018). Hypothesis: The psychedelic ayahuasca heals traumatic memories via a sigma 1 receptor-mediated epigenetic-mnemonic process. *Frontiers in Pharmacology, 9*, 330. https://doi.org/10.3389/fphar.2018.00330.

Jacobsen, L. K., Southwick, S. M., & Kosten, T. R. (2001). Substance use disorders in patients with posttraumatic stress disorder: A review of the literature. *American Journal of Psychiatry, 158*(8), 1184–1190. https://doi.org/10.1176/appi.ajp.158.8.1184.

Labate, B. C., & MacRae, E. (Eds.). (2010). *Ayahuasca, ritual and religion in Brazil.* London/New York: Routledge.

Labate, B. C., Santana de Rose, I., & dos Santos, R. G. (2009). *Ayahuasca religions: A comprehensive bibliography & critical essays.* Santa Cruz: Multidisciplinary Association for Psychedelic Studies (MAPS.

Labate, B. C., Cavnar, C., & Gearin, A. K. (Eds.). (2017). *The world ayahuasca diaspora: Reinventions and controversies*. New York: Routledge.

Lawn, W., Hallak, J. E., Crippa, J. A., Dos Santos, R., Porffy, L., Barratt, M. J., et al. (2017). Well-being, problematic alcohol consumption and acute subjective drug effects in past-year ayahuasca users: A large, international, self-selecting online survey. *Scientific Reports, 7*(1), 15201. https://doi.org/10.1038/s41598-017-14700-6.

Leary, T., Metzner, R., & Alpert, R. (1969). *The psychedelic experience: A manual based on the Tibetan book of the dead*. London: Academic.

Lewis, S. E. (2008). Ayahuasca and spiritual crisis: Liminality as space for personal growth. *Anthropology of Consciousness, 19*(2), 109–133. https://doi.org/10.1111/j.1556-3537.2008. 00006.x.

Lewis, T., Amini, F., & Lannon, R. (2001). *A general theory of love*. New York: Vintage.

Loizaga-Velder, A., & Verres, R. (2014). Therapeutic effects of ritual ayahuasca use in the treatment of substance dependence--qualitative results. *Journal of Psychoactive Drugs, 46*(1), 63–72. https://doi.org/10.1080/02791072.2013.873157.

McElrath, K., & McEvoy, K. (2002). Negative experiences on ecstasy: The role of drug, set, and setting. *Journal of Psychoactive Drugs, 34*(2), 199–208. https://doi.org/10.1080/02791072.200 2.10399954.

Nielson, J. L., & Megler, J. D. (2014). Ayahuasca as a candidate therapy for PTSD. In *The therapeutic use of ayahuasca* (pp. 41–58). Berlin/Heidelberg: Springer. https://doi. org/10.1007/978-3-642-40426-9_3.

Osório, F. d. L., Sanches, R. F., Macedo, L. R., dos Santos, R. G., Maia-de-Oliveira, J. P., Wichert-Ana, L., et al. (2015). Antidepressant effects of a single dose of ayahuasca in patients with recurrent depression: A preliminary report. *Revista Brasileira de Psiquiatria, 37*(1), 13–20. https://doi.org/10.1590/1516-4446-2014-1496.

Palhano-Fontes, F., Alchieri, J. C., Oliveira, J. P. M., Soares, B. L., Hallak, J. E. C., Galvao-Coelho, N., & de Araujo, D. B. (2014). The therapeutic potentials of ayahuasca in the treatment of depression. In *The therapeutic use of Ayahuasca* (pp. 23–39). Berlin/Heidelberg: Springer. https://doi.org/10.1007/978-3-642-40426-9_2.

Palhano-Fontes, F., Barreto, D., Onias, H., Andrade, K. C., Novaes, M., Pessoa, J., et al. (2017). Rapid antidepressant effects of the psychedelic ayahuasca in treatment-resistant depression: a randomised placebo-controlled trial. *BioRxiv*, 103531. https://doi.org/10.1101/103531.

Riba, J., Valle, M., Urbano, G., Yritia, M., Morte, A., & Barbanoj, M. J. (2003). Human pharmacology of ayahuasca: Subjective and cardiovascular effects, monoamine metabolite excretion, and pharmacokinetics. *The Journal of Pharmacology and Experimental Therapeutics, 306*(1), 73–83. https://doi.org/10.1124/jpet.103.049882.

Riba, J., Romero, S., Grasa, E., Mena, E., Carrió, I., & Barbanoj, M. J. (2006). Increased frontal and paralimbic activation following ayahuasca, the pan-Amazonian inebriant. *Psychopharmacology, 186*(1), 93–98. https://doi.org/10.1007/s00213-006-0358-7.

Sanches, R. F., de Lima Osório, F., dos Santos, R. G., Macedo, L. R. H., Maia-de-Oliveira, J. P., Wichert-Ana, L., et al. (2016). Antidepressant effects of a single dose of ayahuasca in patients with recurrent depression. *Journal of Clinical Psychopharmacology, 36*(1), 77–81. https://doi. org/10.1097/JCP.0000000000000436.

Strauss, A. L., & Corbin, J. M. (1998). *Basics of qualitative research: Grounded theory procedures and techniques* (2nd ed.). Thousand Oaks: Sage.

Tafur, J. (2017). In C. Pincus (Ed.), *The fellowship of the river: A medical doctor's exploration into traditional Amazonian plant medicine*. Phoenix: Espiritu Books.

Thomas, G., Lucas, P., Capler, N. R., Tupper, K. W., & Martin, G. (2013). Ayahuasca-assisted therapy for addiction: Results from a preliminary observational study in Canada. *Current Drug Abuse Reviews, 6*(1), 30–42. Retrieved from http://www.ncbi.nlm.nih.gov/pubmed/23627784

Chapter 6
Ayahuasca and Childhood Trauma: Potential Therapeutic Applications

Daniel Perkins and Jerome Sarris

Childhood Trauma and Adverse Experiences

The last 20 years have seen major developments in the understanding of how childhood trauma (negative events that cause distress and overwhelm a person's ability to cope) can have long-term effects on the health and well-being of adults who have experienced this. Child sexual abuse was first included in global burden of disease and disability estimates in 2004, and there has been a steady accumulation of research and evidence identifying the public health issues and costs associated with various traumatic childhood experiences.

Much of this research has used the framework of adverse childhood experiences or ACEs, which encompass emotional, physical, and sexual abuse, as well as various other adverse events, including growing up in a household in which there is domestic violence, alcohol or drug abuse, or a member with mental illness; criminal behavior or incarceration of a family member; caregiver separation or divorce; and neglect, both physical and emotional (Anda et al. 2010; Felitti et al. 1998). Such experiences have been found to be associated with higher rates of physical and mental illness, disability, and premature death in adulthood (Anda et al. 2010).

It is parents who are responsible for the large majority of these traumas, around 80%, and while some exposure is not uncommon, it is multiple traumatic experiences that result in a compounding of negative effects and the greatest damage to

D. Perkins (✉)
School of Social and Political Science, University of Melbourne, Melbourne, VIC, Australia
e-mail: d.perkins@unimelb.edu.au

J. Sarris
NICM, School of Science and Health, Western Sydney University,
Campbelltown, NSW, Australia

ARCADIA Research Group, The University of Melbourne, Department of Psychiatry,
The Melbourne Clinic, Richmond, VIC, Australia
e-mail: J.Sarris@westernsydney.edu.au

© Springer Nature Switzerland AG 2021
B. C. Labate, C. Cavnar (eds.), *Ayahuasca Healing and Science*,
https://doi.org/10.1007/978-3-030-55688-4_6

future life outcomes (van der Kolk 2005). International data suggest that between 50 and 70% of adults have experienced at least one form of childhood adversity and 30–40% two or more forms (Herringa 2017; Ronald et al. 2006; Rosenman and Rodgers 2004). Such experiences also tend to occur in clusters, with the presence of one adverse experience significantly increasing the likelihood of additional such events (Dong et al. 2004).

The individual and societal costs of adverse childhood experiences are substantial and include increased childhood and adult medical costs, productivity losses, additional child welfare and criminal justice costs, and costs associated with an increased need for special education. In 2010 in the United States, individual lifetime costs related to childhood maltreatment were estimated at $210,000, with the total US economic burden assessed at up to $585 billion (Fang et al. 2012).

This chapter focuses on the use of the Amazonian psychoactive plant brew ayahuasca (typically made from *Banisteriopsis caapi* and *Psychotria viridis*) as a potential therapeutic tool to support the treatment of individuals that have experienced multiple or chronic adverse childhood experiences. Ayahuasca contains dimethyltryptamine (DMT), along with harmala alkaloids, and is currently being investigated for therapeutic efficacy in the treatment of mood, anxiety, and addictive disorders, with further research exploring a range of other medical applications (Frecska et al. 2016). Features of ayahuasca consumption that may be particularly useful with this group include profound psychological insights (often relating to early life events); potential therapeutic efficacy in the treatment of depression, anxiety, and addictions; possible therapeutic neurobiological effects; and enhanced self-awareness. Here we review the nature and impacts of childhood trauma and outline a rationale for considering the therapeutic use of ayahuasca with this group, drawing on existing evidence and early responses to a global survey of ayahuasca consumption.

Effects of Early Trauma on Health and Well-being

The ACE Study (Adverse Childhood Experiences Study), undertaken in the United States by the Centers for Disease Control and Prevention and Kaiser Permanente, was a pivotal study in this area that investigated adverse childhood experiences among 17,337 people aged 50 or older (Felitti et al. 1998). The study found that adverse experiences were far more common than previously recognized and that these had a powerful impact on adult health many decades later, which was cumulative in nature. Those who experienced four or more exposures had a 4 to 12 times increased risk of alcoholism, depression, drug abuse, and attempted suicide compared to those with no exposures. The likelihood of smoking, poor self-rated health, and having had a sexually transmitted disease increased by two to four times, and physical inactivity and obesity 1.4–1.6 times among those with four or more exposures. This group also reported rates of physical health problems including skeletal fractures, hepatitis, ischemic heart disease, cancer, stroke, emphysema, and diabetes 1.6–3.9 times higher than among those with no traumatic childhood exposures (Felitti et al. 1998).

The mechanism of these adult effects was suggested to involve impairment to social, emotional, and cognitive development, which, in turn, led to increased adoption of health risk behaviors and, finally, elevated rates of disease, disability, social problems, and premature death (Felitti et al. 1998).

Since the ACE Study, a multitude of research has obtained similar results in a variety of populations around the world (Norman et al. 2012). These studies confirm serious long-term impacts on adult survivors of multiple or chronic childhood trauma across a wide range of domains, with neglect being as damaging as abuse, despite lower public and research attention. As in the ACE Study, exposure has been found to have a compounding effect with the overall level of impact dependent on the number of childhood adverse experiences. Further, childhood trauma exposure displays a powerful impact in increasing symptom complexity in adults, which does not occur after exposure to adult traumas, with a linear relationship between symptom complexity and the number of traumas experienced prior to 18 (Banducci et al. 2014; Briere et al. 2008; Cloitre et al. 2009).

Exposure to significant early trauma has been found to substantially increase the risk of experiencing various mental health conditions, including anxiety, depression, and personality disorders, as well as the likelihood of developing PTSD following adult trauma exposure. Risk is also elevated for antisocial or violent behavior, suicidality, self-harm, somatization, and sexual disorders (Bellis et al. 2017; Giovanelli et al. 2016; Van der Kolk 2014).

A similar picture is evident for a wide range of physical health conditions including coronary heart disease, stroke, and diabetes (Campbell et al. 2016). An important pathway for these effects is the increased participation in various health risk behaviors. Strong associations have been identified between multiple early traumas and smoking, problematic alcohol and drug use, poor diet, obesity, and physical inactivity (Bellis et al. 2017; Giovanelli et al. 2016; Van der Kolk 2014).

Broader impacts on well-being have also been reported, which span almost all aspects of life and have cumulative impacts over the life course. These included a greater likelihood of not graduating from high school, having a low-skilled job, living in a poor household, impaired parenting and work functioning, being a victim of physical assault or abuse, criminal behavior, and incarceration (Gilbert et al. 2015; Giovanelli et al. 2016; Lee and Tolman 2006; Metzler et al. 2017). General well-being is negatively impacted in areas such as reduced life satisfaction, lower optimism about the future, and not feeling close to other people (Bellis et al. 2017; Hughes et al. 2016). Unsurprisingly, adults in this group also record an overutilization of health care, particularly emergency and specialist health services (Chartier et al. 2010).

Additional Pathways

Although the original pathway theorized for the impacts of trauma on later life health outcomes (social, emotional, and cognitive impairments leading to the adoption of health risk behaviors and ultimately increased morbidity and mortality) has been confirmed in subsequent research, more recent studies have proposed additional more direct routes by which childhood trauma can affect health and well-being.

Inflammation There is evidence that the exposure to multiple or chronic childhood trauma itself may contribute to an increased risk of physical and mental illness by the inducement of chronic immune system inflammation and heightened inflammatory reactivity, which is visible in biomarkers such as elevated C-reactive protein among adult survivors (Danese et al. 2008) (Baumeister et al. 2016). Such inflammation has been associated with a range of physical comorbidities, as well as broad negative effects on cognition, emotional processing, and emotional reactivity, as well as increasing the tendency to aggression (Danese and Baldwin 2017; Fanning et al. 2015). It has even been proposed that increased inflammation may contribute to health risk behaviors such as compensatory overeating (Schrepf et al. 2014). The presence of inflammation among childhood trauma survivors may be another factor explaining the complexity of symptom presentation as well as the increased likelihood of treatment failure (found to be higher where chronic inflammation is present) among this group (Danese and Baldwin 2017; Strawbridge et al. 2015).

Neurobiological Development Other recent research has explored the impact of multiple early traumas on brain development and identified that early adverse experiences are associated with long-term changes in brain structure, function, and foundational rhythms. These changes have impacts on multiple brain regions and networks that are associated with various functions including emotional regulation, somatic signal processing, threat detection, memory, pleasure and reward systems, and arousal (Fisher 2014; Teicher et al. 2016) (Bellis et al. 2017). Again, the number of childhood traumatic exposures is closely linked with the cumulative neurological impact (Anda et al. 2010).

Particularly important appear to be the effects on the amygdala and other limbic structures. Evidence suggests that early adverse events lead to hyperactivity of the amygdala and greater activation when exposed to negative or neutral (but not positive) emotional stimuli, with similar effects observed in some connected brain areas (Dannlowski et al. 2013; McLaughlin et al. 2015; Suzuki et al. 2014). Reductions in gray matter volumes have also been identified in the hippocampus, insula, orbitofrontal cortex, anterior cingulate gyrus, and caudate (Dannlowski et al. 2012; Hein and Monk 2017; McLaughlin et al. 2015; Suzuki et al. 2014). Connected with these changes is an upregulation of the hypothalamic–pituitary–adrenal (HPA) axis, which alters the regulation of glucocorticoids, such as the stress hormone cortisol, and dysregulation of the dopamine system (Cross et al. 2017; Gerra et al. 2009).

These childhood trauma-induced changes have been associated with a wide range of impacts including impeded threat processing and emotional regulation, greater propensity for antisocial behavior and violence, difficulties feeling close to other people, diminished reward sensitivity, and cognitive deficits and impaired executive function (Bellis et al. 2017; Cross et al. 2017; Fortenbaugh et al. 2017; Gould et al. 2012; Hanson et al. 2010; Herringa 2017; Marusak et al. 2015). Further, it appears that some effects, such as increasing threat reactivity and weaker emotional regulation, may increase, not diminish, with age (Herringa 2017). These brain-level changes are likely an important additional pathway between childhood experiences and later life psychiatric illnesses.

Self-Awareness Another area that has gained increased attention is the effect of childhood trauma on self-awareness and self-care. Research with people who have experienced significant early trauma has identified brain regions responsible for essential functions connected with self-awareness, such as the sense of physical orientation, sensory integration, coordination of emotion and thinking, connecting the viscera and emotions, and decision-making, to display substantially reduced activity compared to those without a trauma history. Areas affected include the orbital prefrontal cortex, medial prefrontal cortex, anterior cingulate, posterior cingulate, and insula (Bluhm et al. 2009; Van der Kolk 2014).

The reduced activity in these regions is suggested to be an adaptive response to shut off painful sensations, which also has profound impacts in reducing the capacity to be aware of internal states, emotions, and sensations. The effects of this can be seen in reduced proprioceptive awareness, difficulty distinguishing between what is harmful and healthy, impeded regulation of emotional states, reduced ability to be aware of personal needs and preferences, and, commonly, a loss of sense or purpose and direction (Bellis et al. 2017; Cross et al. 2017; Van der Kolk 2014).

An additional frequent effect for this group is dissociation, which occurs as an escape from physical or emotional distress associated with traumatic experiences (Lanius et al. 2010). This can involve compartmentalizing aspects of psychological functioning and detachment, such as depersonalization, derealization, and emotional numbing. For some people that have experienced childhood trauma, dissociation can result in almost complete avoidance of thoughts, sensations, and emotions associated with traumatic memories. Commonly, this can also include feelings of bodily disconnection and substantial impairments to interoception (Cross et al. 2017).

Diagnosis

Although PTSD is the primary trauma diagnosis in the DSM-5, this does not capture many of the broad impacts associated with multiple or chronic adverse childhood experiences, as described by van der Kolk:

PTSD diagnosis does not capture the developmental effects of childhood trauma: the complex disruptions of affect regulation; the disturbed attachment patterns; the rapid behavioral regressions and shifts in emotional states; the loss of autonomous strivings; the aggressive behavior against self and others; the failure to achieve developmental competencies; the loss of bodily regulation in the areas of sleep, food, and self-care; the altered schemas of the world; the anticipatory behavior and traumatic expectations; the multiple somatic problems, from gastrointestinal distress to headaches; the apparent lack of awareness of danger and resulting self-endangering behaviors; the self-hatred and self-blame; and the chronic feelings of ineffectiveness. (van der Kolk 2005, p.406)

Recognizing the utility of evaluating the impact of childhood trauma in its cumulative form, a number of frameworks have been proposed to capture the broad and multifaceted effects (Cloitre et al. 2009). These include complex PTSD, complex trauma, disorders of extreme stress otherwise not specified (DESNOS), and developmental trauma disorder (DTD) (Briere and Spinazzola 2009). These constructs attempt to group together the range of symptoms that can be experienced as a result of such exposures and typically include things such as changes to affect and impulse regulation, cognitive disturbances (such as low self-esteem, dissociation, and shame), somatization (pain or physical symptoms), chronic relationship difficulties, and mood disturbances (Ford 2010; van der Kolk 2005).

In practice, although some children that experience traumatic adverse events will develop diagnosable PTSD, other diagnoses are more common, in particular separation anxiety disorder, oppositional defiant disorder, and phobic disorders (van der Kolk 2005). In adulthood, again, diagnostic labels other than PTSD, such as borderline and antisocial personality disorders, and dissociative disorders, are commonly used (Herringa 2017; van der Kolk 2005). However, some researchers suggest that childhood trauma can actually be the underlying cause of these and other common psychiatric conditions (Kilrain 2017).

Treatment Complexity

The high complexity of symptoms among adults who have experienced multiple childhood traumas creates significant challenges for treatment. Such exposure is associated with a more severe and chronic course of illness for people with depression, greater symptom severity, and a more unfavorable prognosis among those with bipolar disorder, and greater complexity and poorer outcomes for individuals with substance use disorder, with similar outcomes also likely with other psychiatric conditions (Agnew-Blais and Danese 2016; Gawrysiak et al. 2017; Hyman et al. 2008; Liu 2017; Nelson et al. 2017). While pharmacotherapies are sometimes able to lessen symptoms or arousal, they cannot permanently address underlying dysregulation or lack of self-awareness and can have detrimental effects in blocking the dopamine system (which plays an important role in engagement, motivation, and pleasure), as well as potentially causing weight gain, and increasing the risk of diabetes and physical inactivity (Van der Kolk 2014).

Ayahuasca as a Potential Therapeutic Intervention

The consideration of ayahuasca as a therapeutic tool to assist adults that have experienced multiple or chronic adverse childhood experiences relates to the potential for benefit across a number of areas relevant to the difficulties commonly experienced by this group. These include the potential to revisit, reconceptualize, and resolve adverse childhood experiences; possible therapeutic benefits relating to depression, anxiety, and substance use disorders; potential neurobiological and physiological benefits; as well as enhanced self-awareness and broader well-being.

These areas will be discussed further below, along with other published research and some relevant early qualitative data from a multidisciplinary research project examining the use of ayahuasca in different contexts of consumption globally. This project is using an online survey in six languages (English, Portuguese, Spanish, German, Czech, and Italian) to collect detailed cross-sectional data regarding motivations, patterns, and practices of use and reported effects on health and well-being. The study is being undertaken by researchers from Australia, Brazil, Spain, Switzerland, and the Czech Republic, with respondents to date having consumed ayahuasca across traditional, neo-shamanic, and religious contexts.

Revisit, Reconceptualize, and Resolve Adverse Childhood Experiences

A focus on the resolution of childhood trauma has been identified as a common motivation for people to participate in ayahuasca ceremonies, and this is also the case in our survey with over 20% of 600 respondents listing it as a primary motivation. A likely rationale for this is that a core characteristic of the ayahuasca experience is a type of biographical review of important and defining life events. Among these, events in childhood often take a prominent place. These experiences during ayahuasca sessions can include images, recollections, and a revisiting or re-experiencing of early events, sometimes allowing individuals to speak or communicate with the people involved.

A number of researchers have reported these processes to allow ayahuasca drinkers to connect with or uncover previously forgotten traumatic childhood events with a level of distance or perspective that facilitates reconceptualization and new understandings. Although this can be challenging to go through at the time, it can often lead to catharsis, healing, and new levels of empathy and acceptance for themselves and others (Cavnar 2011; Fernández and Fábregas 2014; Gastelumendi 2010; Ventegodt and Kordova 2016). Early data from the survey also confirms these processes with around 50% of respondents reporting "new understanding of childhood events or situations" as one of the insights or learnings they received during their ayahuasca experiences.

For people with childhood trauma, who, as discussed earlier, commonly suffer from high emotional reactivity as well as dissociation related to their early childhood exposures, ayahuasca may provide a rare opportunity to bring such experiences into consciousness. As Echenhofer (2011, p.165) describes, ayahuasca may be effective in these cases by supporting "an unfolding of spontaneous visual and kinesthetic waking transformative imagery narrative that otherwise is very difficult to bring oneself to experience."

These processes were described by numerous survey respondents with responses such as:

> I now understand more about the traumatic events of my childhood, and how they affected my development as a person, both emotionally and physically, but particularly emotionally. I was given clarity about my past that did not exist before, despite many years of psychotherapy. Although from therapy and other spiritual practices, I knew there had been childhood sexual and other trauma, and had made some progress in coming to terms with this, I still did not know exactly what had happened, why it happened, and how it had affected me. In my first ceremony, I was shown exactly what happened, why it happened, and what the consequences were. Consequently, I became more accepting of myself and more able to make positive changes in my life. I had previously been suicidal but the intense love I felt from the plants and the understanding of life that they gave me made it impossible for me to ever seriously consider this again.

One identified characteristic is that, rather than traumatic childhood experiences being seen in isolation, they are often understood as part of a process, which the individual becomes aware and empowered to change (Fotiou 2012). One survey respondent described this as follows, when answering a question about the most important benefits they received from drinking ayahuasca:

> The understanding of self and the fact that I am me and the negativity I had before that led to self-abuse and depression was a direct result of childhood abuse. I now realize that is not who I am. It is only something that happened to me.

It seems that this reprocessing is facilitated during the ayahuasca experience by the activation of neural systems associated with emotional processing and memory, which enable access to deeply stored emotional trauma, while higher cortical areas are stimulated allowing for the processing and reconceptualizing of such events (Nielson and Megler 2014; Riba et al. 2006).

Healing Depression, Anxiety, and Substance Use Disorders

Although not focused specifically on adult survivors of childhood trauma, there is a growing body of evidence supporting the potential therapeutic use of ayahuasca with depression, anxiety, and substance use disorders: an area of high importance for people with childhood trauma histories due to their heightened risk of these conditions and poorer prognosis. A recent review of 21 ayahuasca studies, including experimental research with animals and humans, and psychological assessments of ayahuasca drinkers, reported evidence that consumption was associated with reductions in both depression and anxiety (dos Santos et al. 2016b). In addition, recent open-label and randomized double-blind placebo-controlled trials of ayahuasca administration to hospital in-patients with treatment-resistant major depression have identified evidence of rapid anti-depressant effects after a single dose (de L Osório et al. 2015; Palhano-Fontes et al. 2017).

Looking at effects on drug or alcohol dependence, another systematic review found strong evidence of improvement in biochemical and behavioral parameters related to drug-induced disorders (Nunes et al. 2016). It is suggested that in the appropriate settings, ayahuasca-assisted treatment can catalyze neurobiological and psychological processes that support substance dependence treatment and prevent the likelihood of relapse (Loizaga-Velder and Verres 2014).

Interestingly, for people with a childhood trauma history, changes in alcohol and drug use appear to partly come about from a new awareness of the development of these behaviors as a response to early traumatic events. This was described as follows by one survey respondent:

> (Ayahuasca) helped to revisit childhood traumas that I had already long forgotten, this helped me to understand the harmful patterns I developed during my life, like alcoholism, and an eating disorder. I implemented this knowledge in my everyday life, which resulted in a great improvement in my relationships and my health.

Potential Neurobiological and Physiological Benefits

The adverse effect of early trauma on neurobiological functioning and physical health is another potential area in which ayahuasca may contribute to improved outcomes. Emerging research indicates that the serotonergic psychedelics, including DMT in ayahuasca as well as LSD, psilocybin, and mescaline, may comprise a powerful new class of anti-inflammatory agents with potential efficacy against a range of physical and mental health inflammatory related diseases—which is relevant, given associations between early abuse and chronic inflammation (Kyzar et al. 2017).

Studies examining the role of DMT in regulating the sigma-1 receptor also indicate it may provide powerful immunomodulatory effects that could potentially be harnessed for the treatment of neurogenerative and autoimmune diseases, as well as for chronic inflammation of the central nervous system and more general cellular protection (Frecska et al. 2013; Szabo et al. 2014).

There are reports of individuals using ayahuasca for the treatment of a wide range of physical health issues, and models of action have been advanced for cancer and Parkinson's disease (Djamshidian et al. 2015; Schenberg 2013; Schmid et al. 2010).

Reported neurobiological effects of ayahuasca of interest include beneficial effects relating to trauma and depression via the regulation of the HPA axis, long-term modulation of serotonin and dopamine systems (which play a key role in various psychiatric disorders), and possible neuroplastic changes underpinning positive long-term behavioral modification (Callaway et al. 1994; Galvao et al. 2018; Nielson and Megler 2014). Alkaloids present in *Banisteriopsis caapi* in ayahuasca have also been found to stimulate neurogenesis in vitro, and imaging studies have reported metabolic and connectivity changes associated with increased psychological capacities 24 h after ayahuasca consumption (Morales-García et al. 2017; Sampedro et al. 2017).

Enhanced Self-Awareness and Well-Being

Particularly important in this area are effects that may assist in increasing self-awareness among people with an early trauma history. For such individuals, low self-awareness can be due to changes in brain function or dissociation, which then manifests in poor self-definition, health risk behaviors, difficulties feeling close to people, and impaired interoception.

Enhanced physical awareness, or interoception, is an important therapeutic element allowing greater connection to bodily sensations and subtle messages in a way that can support self-understanding and the integration of ordinary experience. Experimental ayahuasca research has identified significant activation of the neural systems involved with interception and emotional processing, particularly frontal and paralimbic areas, while qualitative studies have reported increased sensory awareness and an ability to be aware of and reduce dissociative behaviors among drinkers with early trauma (Espinoza 2014; Kaufman 2015; Riba et al. 2006). This was also present among our survey respondents:

> After extreme child abuse, which led to severe complex PTSD, I had dissociated to such a degree that I did not even recognize that I did not consider other people to "exist" in the sense I did. Nor did I possess a concrete sense of self. My name had as much significance as the breath of the wind. Ayahuasca led me to the horrifying realization that these things did in fact exist, which, while tremendously appreciated and invaluable, still has not been fully integrated.

The response of another participant, who felt that their ayahuasca consumption had been positive overall, highlights that such rapid removal of dissociative blockages can also involve an element of risk if not well supported:

> First time I experienced her (ayahuasca) in Peru, 5 ceremonies. Here, Aya (ayahuasca) removed all the blocks that I had which were preventing me to feel the childhood traumas. It did a great job removing the blocks and because my environment was perfect at that time, I felt like in heaven. I felt one with the universe and I obtained a great physical health. I was feeling more, seeing more, all my senses were amplified. I was able to see how I can be in life. After returning home, life was not so perfect anymore, and I started to feel all the traumas and now there was no block, nothing to stop me to FEEL the pain caused by traumas. I was completely unprepared for this and had no idea how to deal with any trauma ... so I went almost crazy and destroyed some relationships with people around me. I was able to put some blocks back after some time. Second time I took Aya in Europe and the same story happened: all blocks removed and felt all the pain possible:) Then finally I met someone who taught me how to deal with these traumas and finally reached a good mental, emotional and also spiritual state.

Reports of enhanced self-awareness in other research identify this across a wide range of areas, such as personal health and desire for a healthy lifestyle, including diet, alcohol consumption, and exercise; enhanced feelings of connection with others; strengthened decision-making; and a greater sense of life purpose and direction (Espinoza 2014; Frecska et al. 2016; Kavenská and Simonová 2015). For example, "After taking ayahuasca I feel less inclined to drink alcohol. I have started meditating, eating healthier and taken steps to build a healthier and happier life. I feel a greater sense of connectedness with everything and everyone around me" (Kaufman 2015, p.125).

A number of studies have also reported broader well-being effects, including profound spiritual experiences of drinkers, often associated with reoriented personal values and life directions (Harris and Gurel 2012; Kaufman 2015; Kavenská and Simonová 2015). One of the largest and most detailed of these identified spiritual effects resulting in radical transformations of peoples' lives. Common themes included new understandings of their own personality and life, deeper appreciation of the divine and sacred, and reflections on and commitment to values, particularly personal responsibility, justice, and love (Shanon 2002). This was described as follows by one of our survey respondents:

> It has greatly increased my self-confidence. It has increased the seriousness with which I conduct my life, examine my values, and choose my actions. My journeys showed me spiritual visions. This has added to my sense of who I am.

Reported effects from people with childhood trauma have included developing a new awareness of their treatment of others, enhanced personal values and integrity, and improvement in interpersonal relationships, a highly problematic area for this group (Kaufman 2015). These themes were also present among our survey respondents, for example:

> A healing with the shaman (during a ceremony) allowed me to "step up" and become more fully present in myself. As a result of this healing, many aspects of my life fell into place on my return, including the resolution of financial issues, greater success in my business, the resolution of a feeling of disassociation, which I had experienced for many years, increased happiness, greater feeling of connection to daily life, and an increased sense of belonging in the world. I also experienced the healing of resentment towards my parents and was able to forgive many childhood instances of emotional abuse.

And,

> It has released deep and profound childhood trauma that affected my adult life, by cheating, stealing, being unfaithful with my girlfriend, nervous or unstable with friends, and always wanting to consume soft drugs.

Broader well-being effects are also confirmed in findings of a recent review of 28 human ayahuasca studies that concluded that, in addition to antidepressant and anti-addictive effects, long-term consumption was associated with enhanced mood and cognition, increased spirituality, and reduced impulsivity, with no evidence of increased psychopathology or cognitive deficits (dos Santos et al. 2016a, b).

Safety

Although there is good evidence that the use of ayahuasca in structured settings is relatively safe, some authors have raised questions about risks for certain individuals who may not possess the psychological resilience required to withstand the powerful psychoactive effects, an issue particularly relevant for those who have experienced significant childhood trauma (Bouso et al. 2017; Trichter 2010). The revisiting of past traumas can lead to new traumatic experiences that may be difficult to assimilate, and psychological distress following ayahuasca ceremonies has been reported, with a small number of case reports of severe depression and psychotic states (Warren et al. 2013). As described earlier, for people exposed to childhood trauma, the rapid deconstruction of dissociative patterns protecting them from overwhelming emotions can be a further risk. Adequate psychosocial supports outside the context of consumption appear particularly important; however, identifying therapists with expertise in the treatment of childhood trauma who are also open to and skilled in working with and integrating psychedelic experiences is a likely challenge.

Final Remarks

Adverse childhood experiences can result in enormous lifetime consequences for exposed individuals, influencing their emotional and physical well-being as well as interpersonal relationships and life directions. The high complexity of symptoms among this group presents a major challenge to standard therapeutic approaches, and there is evidence that ayahuasca may be a useful adjunct to such treatments. Particularly beneficial could be its ability to simultaneously address a broad range of emotional, cognitive, and neurobiological impacts. However, the highly vulnerable nature of this group confirms the need for caution. Although offering promise, there is a need for further research investigating the extent to which proposed benefits occur in practice, as well as potential risks and strategies to best minimize these.

References

Agnew-Blais, J., & Danese, A. (2016). Childhood maltreatment and unfavourable clinical outcomes in bipolar disorder: A systematic review and meta-analysis. *Lancet Psychiatry, 3*(4), 342–349. https://doi.org/10.1016/s2215-0366(15)00544-1.
Anda, R. F., Butchart, A., Felitti, V. J., & Brown, D. W. (2010). Building a framework for global surveillance of the public health implications of adverse childhood experiences. *American Journal of Preventive Medicine, 39*(1), 93–98. https://doi.org/10.1016/j.amepre.2010.03.015.
Banducci, A. N., Hoffman, E., Lejuez, C. W., & Koenen, K. C. (2014). The relationship between child abuse and negative outcomes among substance users: Psychopathology, health,

and comorbidities. *Addictive Behaviors, 39*(10), 1522–1527. https://doi.org/10.1016/j. addbeh.2014.05.023.

Baumeister, D., Akhtar, R., Ciufolini, S., Pariante, C. M., & Mondelli, V. (2016). Childhood trauma and adulthood inflammation: A meta-analysis of peripheral C-reactive protein, interleukin-6 and tumour necrosis factor-alpha. *Molecular Psychiatry, 21*(5), 642–649. https://doi. org/10.1038/mp.2015.67.

Bellis, M. A., Hardcastle, K., Ford, K., Hughes, K., Ashton, K., Quigg, Z., & Butler, N. (2017). Does continuous trusted adult support in childhood impart life-course resilience against adverse childhood experiences – A retrospective study on adult health-harming behaviours and mental well-being. *BMC Psychiatry, 17*(1), 110–110. https://doi.org/10.1186/s12888-017-1260-z.

Bluhm, R. L., Williamson, P. C., Osuch, E. A., Frewen, P. A., Stevens, T. K., Boksman, K., et al. (2009). Alterations in default network connectivity in posttraumatic stress disorder related to early-life trauma. *Journal of Psychiatry & Neuroscience, 34*(3), 187–194.

Bouso, J. C., dos Santos, R. G., Grob, C. S., da Silveira, D. X., McKenna, D. J., de Araujo, D. B., …, & Labate, B. C. (2017). *Ayahuasca: Technical report 2017*. Retrieved from: https:// www.iceers.org/new-edition-of-the-ayahuasca-technical-report/

Briere, J., & Spinazzola, J. (2009). Assessment of the sequelae of complex trauma. In C. A. Courtois & J. D. Ford (Eds.), *Treating complex traumatic stress disorders: An evidence-based guide* (p. 2009). New York: Guilford Press.

Briere, J., Kaltman, S., & Green, B. L. (2008). Accumulated childhood trauma and symptom complexity. *Journal of Traumatic Stress, 21*(2), 223–226. https://doi.org/10.1002/jts.20317.

Callaway, J., Airaksinen, M., McKenna, D., Brito, G., & Grob, C. (1994). Platelet serotonin uptake sites increased in drinkers of ayahuasca. *Psychopharmacology, 116*(3), 385–387. https://doi. org/10.1007/bf02245347.

Campbell, J. A., Walker, R. J., & Egede, L. E. (2016). Associations between adverse childhood experiences, high-risk behaviors, and morbidity in adulthood. *American Journal of Preventive Medicine, 50*(3), 344–352. https://doi.org/10.1016/j.amepre.2015.07.022.

Cavnar, C. (2011). *The effects of participation in ayahuasca rituals on gays' and lesbians' self perception* (Doctoral dissertation). Pleasant Hill: John F. Kennedy University. ProQuest Dissertations & Theses Global database.

Chartier, M. J., Walker, J. R., & Naimark, B. (2010). Separate and cumulative effects of adverse childhood experiences in predicting adult health and health care utilization. *Child Abuse & Neglect, 34*(6), 454–464. https://doi.org/10.1016/j.chiabu.2009.09.020.

Cloitre, M., Stolbach, B. C., Herman, J. L., Kolk, B. v. d., Pynoos, R., Wang, J., & Petkova, E. (2009). A developmental approach to complex PTSD: Childhood and adult cumulative trauma as predictors of symptom complexity. *Journal of Traumatic Stress, 22*(5), 399–408. https://doi.org/10.1002/jts.20444.

Cross, D., Fani, N., Powers, A., & Bradley, B. (2017). Neurobiological development in the context of childhood trauma. *Clinical Psychology: Science and Practice, 24*(2), 111–124. https://doi. org/10.1111/cpsp.12198.

Danese, A., & Baldwin, J. R. (2017). Hidden wounds? Inflammatory links between childhood trauma and psychopathology. *Annual Review of Psychology, 68*(1), 517–544. https://doi. org/10.1146/annurev-psych-010416-044208.

Danese, A., Moffitt, T. E., Pariante, C. M., Ambler, A., Poulton, R., & Caspi, A. (2008). Elevated inflammation levels in depressed adults with a history of childhood maltreatment. *Archives of General Psychiatry, 65*(4), 409–415.

Dannlowski, U., Stuhrmann, A., Beutelmann, V., Zwanzger, P., Lenzen, T., Grotegerd, D., et al. (2012). Limbic scars: Long-term consequences of childhood maltreatment revealed by functional and structural magnetic resonance imaging. *Biological Psychiatry, 71*(4), 286–293. https://doi.org/10.1016/j.biopsych.2011.10.021.

Dannlowski, U., Kugel, H., Huber, F., Stuhrmann, A., Redlich, R., Grotegerd, D., et al. (2013). Childhood maltreatment is associated with an automatic negative emotion processing bias in the amygdala. *Human Brain Mapping, 34*(11), 2899–2909. https://doi.org/10.1002/hbm.22112.

de L Osório, F., Sanches, R. F., Macedo, L. R., dos Santos, R. G., Maia-de-Oliveira, J. P., Wichert-Ana, L., et al. (2015). Antidepressant effects of a single dose of ayahuasca in patients with recurrent depression: A preliminary report. *Revista Brasileira de Psiquiatria, 37*(1), 13–20. https://doi.org/10.1590/1516-4446-2014-1496.

Djamshidian, A., Bernschneider-Reif, S., Poewe, W., & Lees, A. J. (2015). *Banisteriopsis caapi, a forgotten potential therapy for Parkinson's disease? Movement Disorders Clinical Practice, 3*(1), 19–26. https://doi.org/10.1002/mdc3.12242.

Dong, M., Anda, R. F., Felitti, V. J., Dube, S. R., Williamson, D. F., Thompson, T. J., et al. (2004). The interrelatedness of multiple forms of childhood abuse, neglect, and household dysfunction. *Child Abuse & Neglect, 28*(7), 771–784. https://doi.org/10.1016/j.chiabu.2004.01.008.

dos Santos, R. G., Hallak, J. E. C., Bouso, J. C., & Balthazar, F. M. (2016a). The current state of research on ayahuasca: A systematic review of human studies assessing psychiatric symptoms, neuropsychological functioning, and neuroimaging. *Journal of Psychopharmacology, 30*(12), 1230–1247. https://doi.org/10.1177/0269881116652578.

dos Santos, R. G., Osório, F. L., Crippa, J. A. S., & Hallak, J. E. C. (2016b). Antidepressive and anxiolytic effects of ayahuasca: A systematic literature review of animal and human studies. *Revista Brasileira de Psiquiatria, 38*(1), 65–72. https://doi.org/10.1590/1516-4446-2015-1701.

Echenhofer, F. (2011). *Ayahuasca shamanic visions: Integrating neuroscience, psychotherapy, and spiritual perspectives.* Chicago: University of Chicago Press.

Espinoza, Y. (2014). Sexual healing with Amazonian plant teachers: A heuristic inquiry of women's spiritual–erotic awakenings. *Sexual & Relationship Therapy, 29*(1), 109–120.

Fang, X., Brown, D. S., Florence, C. S., & Mercy, J. A. (2012). The economic burden of child maltreatment in the United States and implications for prevention. *Child Abuse & Neglect, 36*(2), 156–165. https://doi.org/10.1016/j.chiabu.2011.10.006.

Fanning, J. R., Lee, R., Gozal, D., Coussons-Read, M., & Coccaro, E. F. (2015). Childhood trauma and parental style: Relationship with markers of inflammation, oxidative stress, and aggression in healthy and personality disordered subjects. *Biological Psychology, 112*, 56–65. https://doi.org/10.1016/j.biopsycho.2015.09.003.

Felitti, V. J., Anda, R. F., Nordenberg, D., Williamson, D. F., Spitz, A. M., Edwards, V., et al. (1998). Relationship of childhood abuse and household dysfunction to many of the leading causes of death in adults: The Adverse Childhood Experiences (ACE) study. *American Journal of Preventive Medicine, 14*(4), 245–258. https://doi.org/10.1016/S0749-3797(98)00017-8.

Fernández, X., & Fábregas, J. M. (2014). Experience of treatment with ayahuasca for drug addiction in the Brazilian Amazon. In B. C. Labate & C. Cavnar (Eds.), *The therapeutic use of ayahuasca.* Heidelberg: Springer.

Fisher, S. F. (2014). *Neurofeedback in the treatment of developmental trauma: Calming the fear-driven brain.* New York: W. W. Norton & Company.

Ford, J. D. (2010). Complex adult sequelae of early life exposure to psychological trauma. In R. A. Lanius, E. Vermetten, & C. Pain (Eds.), *The impact of early life trauma on health and disease: The hidden epidemic.* Cambridge: Cambridge University Press.

Fortenbaugh, F. C., Corbo, V., Poole, V., McGlinchey, R., Milberg, W., Salat, D., et al. (2017). Interpersonal early-life trauma alters amygdala connectivity and sustained attention performance. *Brain and Behavior: A Cognitive Neuroscience Perspective, 7*(5), e00684. https://doi.org/10.1002/brb3.684.

Fotiou, E. (2012). Working with "la medicina": Elements of healing in contemporary ayahuasca rituals. *Anthropology of Consciousness, 23*(1), 6–27. https://doi.org/10.1111/j.1556-3537.2012.01054.x.

Frecska, E., Szabo, A., Winkelman, M. J., Luna, L. E., & McKenna, D. J. (2013). A possibly sigma-1 receptor mediated role of dimethyltryptamine in tissue protection, regeneration, and immunity. *Journal of Neural Transmission (Vienna), 120*(9), 1295–1303. https://doi.org/10.1007/s00702-013-1024-y.

Frecska, E., Bokor, P., & Winkelman, M. (2016). The therapeutic potentials of ayahuasca: Possible effects against various diseases of civilization. *Frontiers in Pharmacology, 7*(35), 1–17. https://doi.org/10.3389/fphar.2016.00035.

Galvao, A. C. M., Almeida, R. N., dos Santos Silva, E. A., de Morais Freire, F. A., Palhano-Fontes, F., Onias, H., …, & Galvao-Coelho, N. L. (2018). A single dose of ayahuasca modulates salivary cortisol in treatment-resistant depression. *bioRxiv*. https://doi.org/10.1101/257238.

Gastelumendi, E. (2010). Ayahuasca: Current interest in an ancient ritual. In K. Miyoshi, Y. Morimura, & K. Maeda (Eds.), *Neuropsychiatric disorders* (pp. 279–286). Tokyo: Springer.

Gawrysiak, M. J., Jagannathan, K., Regier, P., Suh, J. J., Kampman, K., Vickery, T., & Childress, A. R. (2017). Unseen scars: Cocaine patients with prior trauma evidence heightened resting state functional connectivity (RSFC) between the amygdala and limbic-striatal regions. *Drug and Alcohol Dependence, 180*, 363–370. https://doi.org/10.1016/j.drugalcdep.2017.08.035.

Gerra, G., Leonardi, C., Cortese, E., Zaimovic, A., Dell'agnello, G., Manfredini, M., et al. (2009). Childhood neglect and parental care perception in cocaine addicts: Relation with psychiatric symptoms and biological correlates. *Neuroscience and Biobehavioral Reviews, 33*(4), 601–610. https://doi.org/10.1016/j.neubiorev.2007.08.002.

Gilbert, L. K., Breiding, M. J., Merrick, M. T., Thompson, W. W., Ford, D. C., Dhingra, S. S., & Parks, S. E. (2015). Childhood adversity and adult chronic disease: An update from ten states and the District of Columbia, 2010. *American Journal of Preventive Medicine, 48*(3), 345–349. https://doi.org/10.1016/j.amepre.2014.09.006.

Giovanelli, A., Reynolds, A. J., Mondi, C. F., & Ou, S. R. (2016). Adverse childhood experiences and adult Well-being in a low-income, urban cohort. *Pediatrics, 137*(4), e20154016. https://doi.org/10.1542/peds.2015-4016.

Gould, F., Clarke, J., Heim, C., Harvey, P. D., Majer, M., & Nemeroff, C. B. (2012). The effects of child abuse and neglect on cognitive functioning in adulthood. *Journal of Psychiatric Research, 46*(4), 500–506. https://doi.org/10.1016/j.jpsychires.2012.01.005.

Hanson, J. L., Chung, M. K., Avants, B. B., Shirtcliff, E. A., Gee, J. C., Davidson, R. J., & Pollak, S. D. (2010). Early stress is associated with alterations in the orbitofrontal cortex: A tensor-based morphometry investigation of brain structure and behavioral risk. *The Journal of Neuroscience, 30*(22), 7466–7472. https://doi.org/10.1523/JNEUROSCI.0859-10.2010.

Harris, R., & Gurel, L. (2012). A study of ayahuasca use in North America. *Journal of Psychoactive Drugs, 44*(3), 209–215. https://doi.org/10.1080/02791072.2012.703100.

Hein, T. C., & Monk, C. S. (2017). Research review: Neural response to threat in children, adolescents, and adults after child maltreatment – A quantitative meta-analysis. *Journal of Child Psychology and Psychiatry, 58*(3), 222–230. https://doi.org/10.1111/jcpp.12651.

Herringa, R. J. (2017). Trauma, PTSD, and the developing brain. *Current Psychiatry Reports, 19*(10), 69. https://doi.org/10.1007/s11920-017-0825-3.

Hughes, K., Lowey, H., Quigg, Z., & Bellis, M. A. (2016). Relationships between adverse childhood experiences and adult mental well-being: Results from an English national household survey. *BMC Public Health, 16*, 222–233. https://doi.org/10.1186/s12889-016-2906-3.

Hyman, S. M., Paliwal, P., Chaplin, T. M., Mazure, C. M., Rounsaville, B. J., & Sinha, R. (2008). Severity of childhood trauma is predictive of cocaine relapse outcomes in women but not men. *Drug & Alcohol Dependence, 92*(1), 208–216. https://doi.org/10.1016/j.drugalcdep.2007.08.006.

Kaufman, R. (2015). *How might the ayahuasca experience be a potential antidote to Western hegemony: A mixed methods study* (Doctoral dissertation). Santa Barbara: Fielding Graduate University. ProQuest Dissertations & Theses Global database.

Kavenská, V., & Simonová, H. (2015). Ayahuasca tourism: Participants in shamanic rituals and their personality styles, motivation, benefits and risks. *Journal of Psychoactive Drugs, 47*(5), 351–359.

Kilrain, M. V. (2017). DTD: the effects of child abuse and neglect: developmental trauma disorder; not yet officially recognized, results from child maltreatment and has many neurobiologie consequences. *The Clinial Advisor* (5):26–38.

Kyzar, E. J., Nichols, C. D., Gainetdinov, R. R., Nichols, D. E., & Kalueff, A. V. (2017). Review: Psychedelic drugs in biomedicine. *Trends in Pharmacological Sciences, 38*(11), 992–1005. https://doi.org/10.1016/j.tips.2017.08.003.

Lanius, R. A., Vermetten, E., Loewenstein, R. J., Brand, B., Schmahl, C., Bremner, J. D., & Spiegel, D. (2010). Emotion modulation in PTSD: Clinical and neurobiological evidence for a dissociative subtype. *The American Journal of Psychiatry, 167*(6), 640–647. https://doi.org/10.1176/appi.ajp.2009.09081168.

Lee, S. J., & Tolman, R. M. (2006). Childhood sexual abuse and adult work outcomes. *Social Work Research, 30*(2), 83–92.

Liu, R. T. (2017). Childhood adversities and depression in adulthood: Current findings and future directions. *Clinical Psychologist, 24*(2), 140–153. https://doi.org/10.1111/cpsp.12190.

Loizaga-Velder, A., & Verres, R. (2014). Therapeutic effects of ritual ayahuasca use in the treatment of substance dependence–qualitative results. *Journal of Psychoactive Drugs, 46*(1), 63–72.

Marusak, H. A., Martin, K. R., Etkin, A., & Thomason, M. E. (2015). Childhood trauma exposure disrupts the automatic regulation of emotional processing. *Neuropsychopharmacology, 40*(5), 1250–1258. https://doi.org/10.1038/npp.2014.311.

McLaughlin, K. A., Peverill, M., Gold, A. L., Alves, S., & Sheridan, M. A. (2015). Child maltreatment and neural systems underlying emotion regulation. *Journal of the American Academy of Child and Adolescent Psychiatry, 54*(9), 753–762. https://doi.org/10.1016/j.jaac.2015.06.010.

Metzler, M., Merrick, M. T., Klevens, J., Ports, K. A., & Ford, D. C. (2017). Adverse childhood experiences and life opportunities: Shifting the narrative. *Children and Youth Services Review, 72*, 141–149. https://doi.org/10.1016/j.childyouth.2016.10.021.

Morales-García, J. A., de la Fuente Revenga, M., Alonso-Gil, S., Rodríguez-Franco, M. I., Feilding, A., Perez-Castillo, A., & Riba, J. (2017). The alkaloids of *Banisteriopsis caapi*, the plant source of the Amazonian hallucinogen ayahuasca, stimulate adult neurogenesis in vitro. *Scientific Reports, 7*, 5309. https://doi.org/10.1038/s41598-017-05407-9.

Nelson, J., Klumparendt, A., Doebler, P., & Ehring, T. (2017). Childhood maltreatment and characteristics of adult depression: Meta-analysis. *The British Journal of Psychiatry, 210*(2), 96–104. https://doi.org/10.1192/bjp.bp.115.180752.

Nielson, J., & Megler, J. (2014). Ayahuasca as a candidate therapy for PTSD. In B. C. Labate & C. Cavnar (Eds.), *The therapeutic use of ayahuasca* (pp. 41–58). Heidelberg: Springer.

Norman, R. E., Byambaa, M., De, R., Butchart, A., Scott, J., & Vos, T. (2012). The long-term health consequences of child physical abuse, emotional abuse, and neglect: A systematic review and meta-analysis. *PLoS Medicine, 9*(11), e1001349. https://doi.org/10.1371/journal.pmed.1001349.

Nunes, A. A., dos Santos, R. G., Osório, F. L., Sanches, R. F., Crippa, J. A. S., Hallak, J. E. C., et al. (2016). Effects of ayahuasca and its alkaloids on drug dependence: A systematic literature review of quantitative studies in animals and humans. *Journal of Psychoactive Drugs, 48*(3), 195–205. https://doi.org/10.1080/02791072.2016.1188225.

Palhano-Fontes, F., Barreto, D., Onias, H., Andrade, K. C., Novaes, M., Pessoa, J., et al. (2017). Rapid antidepressant effects of the psychedelic ayahuasca in treatment-resistant depression: A randomised placebo-controlled trial. *Psychological Medicine, 49*(4), 655–663.

Riba, J., Romero, S., Grasa, E., Mena, E., Carrió, I., & Barbanoj, M. (2006). Increased frontal and paralimbic activation following ayahuasca, the pan-Amazonian inebriant. *Psychopharmacology, 186*(1), 93–98. https://doi.org/10.1007/s00213-006-0358-7.

Ronald, A. C., Brian, L. H., Robert, H. P., Jeanne, M., Laura, S., Lawrence, S., et al. (2006). Early life stress and adult emotional experience: An international perspective. *The International Journal of Psychiatry in Medicine, 36*(1), 35–52. https://doi.org/10.2190/5R62-9PQY-0NEL-TLPA.

Rosenman, S., & Rodgers, B. (2004). Childhood adversity in an Australian population. *Social Psychiatry and Psychiatric Epidemiology, 39*(9), 695–702. https://doi.org/10.1007/s00127-004-0802-0.

Sampedro, F., de la Fuente Revenga, M., Valle, M., Roberto, N., Domínguez-Clavé, E., Elices, M., et al. (2017). Assessing the psychedelic "after-glow" in ayahuasca users: Post-acute neurometabolic and functional connectivity changes are associated with enhanced mindfulness capacities. *The International Journal of Neuropsychopharmacology, 20*(9), 698–711. https://doi.org/10.1093/ijnp/pyx036.

Schenberg, E. E. (2013). Ayahuasca and cancer treatment. *SAGE Open Medicine, 1.* https://doi.org/10.1177/2050312113508389.

Schmid, J. T., Jungaberle, H., & Verres, R. (2010). Subjective theories about (self-)treatment with ayahuasca. *Anthropology of Consciousness, 21*(2), 188–204. https://doi.org/10.1111/j.1556-3537.2010.01028.x.

Schrepf, A., Markon, K., & Lutgendorf, S. K. (2014). From childhood trauma to elevated C-reactive protein in adulthood: The role of anxiety and emotional eating. *Psychosomatic Medicine, 76*(5), 327–336. https://doi.org/10.1097/PSY.0000000000000072.

Shanon, B. (2002). *The antipodes of the mind: Charting the phenomenology of the ayahuasca experience.* Oxford: Oxford University Press.

Strawbridge, R., Arnone, D., Danese, A., Papadopoulos, A., Vives, A. H., & Cleare, A. J. (2015). Inflammation and clinical response to treatment in depression: A meta-analysis. *European Neuropsychopharmacology, 25*(10), 1532–1543. https://doi.org/10.1016/j.euroneuro.2015.06.007.

Suzuki, H., Luby, J. L., Botteron, K. N., Dietrich, R., McAvoy, M. P., & Barch, D. M. (2014). Early life stress and trauma and enhanced limbic activation to emotionally valenced faces in depressed and healthy children. *Journal of the American Academy of Child and Adolescent Psychiatry, 53*(7), 800–813. e810. https://doi.org/10.1016/j.jaac.2014.04.013.

Szabo, A., Kovacs, A., Frecska, E., & Rajnavolgyi, E. (2014). Psychedelic N,N-dimethyltryptamine and 5-methoxy-N,N-dimethyltryptamine modulate innate and adaptive inflammatory responses through the sigma-1 receptor of human monocyte-derived dendritic cells. *PLoS One, 9*(8), e106533. https://doi.org/10.1371/journal.pone.0106533.

Teicher, M. H., Samson, J. A., Anderson, C. M., & Ohashi, K. (2016). The effects of childhood maltreatment on brain structure, function and connectivity. *Nature Reviews. Neuroscience, 17*(10), 652–666. https://doi.org/10.1038/nrn.2016.111.

Trichter, S. (2010). Ayahuasca beyond the Amazon: The benefits and risks of a spreading tradition. *Journal of Transpersonal Psychology, 42*(2), 131–148.

van der Kolk, B. A. (2005). Developmental trauma disorder: Toward a rational diagnosis for children with complex trauma histories. *Psychiatric Annals, 35*(5), 401–408.

Van der Kolk, B. A. (2014). *The body keeps the score: Mind, brain and body in the transformation of trauma.* New York: Penguin Books.

Ventegodt, S., & Kordova, P. (2016). Contemporary strategies in Peru for medical use of the hallucinogenic tea ayahuasca containing DMT: In search of the optimal strategy for the use of medical hallucinogens. *Journal of Alternative Medicine Research, 8*(4), 455–470.

Warren, J. M., Dham-Nayyar, P., & Alexander, J. (2013). Recreational use of naturally occurring dimethyltryptamine—Contributing to psychosis? *Australian and New Zealand Journal of Psychiatry, 47*(4), 398–399. https://doi.org/10.1177/0004867412462749.

Chapter 7
Acute and Long-Term Effects of Ayahuasca on (Higher-Order) Cognitive Processes

Natasha L. Mason and Kim P. C. Kuypers

Introduction

Ayahuasca is a South American psychotropic plant brew, generally consisting of the boiled stems of the *Banisteriopsis caapi* vine, and the leaves of the *Psychotria viridis* bush. *B. Caapi* contains the β-carboline alkaloids harmine, tetrahydroharmine, and harmaline, while *P. Viridis* contains the tryptamine N,N dimethyltryptamine (DMT), a hallucinogen that is structurally similar to serotonin (5-HT). When taken orally, DMT is rendered inactive via monoamine oxidase activity in the gastrointestinal tract (McKenna 2004; McKenna et al. 1984). However, when combined with *B. caapi*, the β-carboline alkaloids inhibit monoamine oxidase activity, allowing DMT to reach the systemic circulation and cross the blood-brain barrier, where it activates $5\text{-}HT_{1A}$, $5\text{-}HT_{2A}$, and $5\text{-}HT_{2C}$ receptors (Dos Santos et al. 2016a, b; Fantegrossi et al. 2008; Riba et al. 2003; Smith et al. 1998).

Similar to other serotonergic hallucinogens, the $5\text{-}HT_{2A}$ receptor activation is the suggested mechanism for the acute subjective effects of ayahuasca, which include perceptual modifications, increased rates of thinking when eyes are closed, and increased emotional lability (Bickel et al. 1976; Riba et al. 2001; Riba et al. 2003). These acute effects usually start between 45 and 60 min post-administration, peak between 60 and 120 min, end after 4 h, and follow a dose-response pattern (Riba et al. 2001).

Ayahuasca has been traditionally used for centuries by indigenous and mestizo populations throughout the Amazon Basin for magical, ritual, and medicinal purposes (Luna 2011; Schultes and Hofmann 1979). However, in the last few decades, there has been an increase in the availability of the brew to non-Amazonian populations (Tupper 2009). Subsequently, there has been an increase of anecdotal reports

N. L. Mason (✉) · K. P. C. Kuypers (✉)
Department of Neuropsychology and Psychopharmacology, Faculty of Psychology and Neuroscience, Maastricht University, Maastricht, The Netherlands
e-mail: natasha.mason@maastrichtuniveristy.nl; k.kuypers@maastrichtuniversity.nl

© Springer Nature Switzerland AG 2021
B. C. Labate, C. Cavnar (eds.), *Ayahuasca Healing and Science*,
https://doi.org/10.1007/978-3-030-55688-4_7

from ayahuasca users regarding the acute and long-term effects of the substance, with many claiming that the substance has positive and therapeutic potential for psychosocial, emotional, and substance-related problems (Barbosa et al. 2005; Kjellgren et al. 2009; Winkelman 2005).

Over the past few years, there has been a renewed interest in psychedelic drugs as potential tools in therapy (Dominguez-Clave et al. 2016; Vollenweider and Kometer 2010). Double-blind experimental studies on the acute and long-term effects of psilocybin, a classic hallucinogen whose psychoactive mechanism is also mediated through 5-HT_{2A} receptor activation, have suggested that a psychedelic session can produce life-altering experiences and long-term improvements in well-being and behavior in healthy volunteers (Griffiths et al. 2006; Hasler et al. 2004).

Case-control studies and surveys have suggested that psychedelic users suffer from less psychopathology (Da Silveira et al. 2005; Grob et al. 1996; Hendricks et al. 2015) and display lower patterns of alcohol use (Doering-Silveira et al. 2005). Importantly, clinical trials have suggested that, when taken in an appropriate psychotherapeutic setting, psychedelics such as ayahuasca elicit anxiolytic, antidepressive, and anti-addictive effects (Dos Santos et al. 2016a, b; Palhano-Fontes et al., this volume, Chap. 2), even for patients with treatment-resistant psychopathologies (Carhart-Harris et al. 2017; Oehen et al. 2013; Palhano-Fontes et al. 2019), with symptom reduction persisting weeks after ingestion (Osorio et al. 2015).

The main outcome of interest for the abovementioned studies in both healthy participants and patients has been subjective effects and symptom alleviation, with little attention being paid to the (higher-order) cognitive processes that may be enhanced or that may play a role in subjective experience or symptom alleviation of the disorders. For example, changes in behavior and lifestyle may result from a psychedelic's capacity to stimulate processes, such as flexible (creative) thinking, empathy, and emotion regulation, which are crucial for everyday interactions, and have been found to be decreased in certain disorders like depression, anxiety, and post-traumatic stress disorder (PTSD) (Aldao et al. 2010; Baas et al. 2008; Davis 2009; Nietlisbach et al. 2010; Parlar et al. 2014; Todd et al. 2015; Tull et al. 2007). For instance, individuals with these pathologies display repetitive, rigid, and pathological patterns of negative and compulsive thoughts (Chamberlain et al. 2006; Dos Santos et al. 2016a, b). Similarly, empathic changes are particularly evident in mood disorders, like depression, and have even been associated with symptom severity. Donges et al. (2005), for example, demonstrated that inpatients with major depressive disorder (MDD) displayed reduced awareness of other's emotions; this decrease was associated with elevated symptoms of depression. Similarly, Cusi et al. (2011) found that MDD outpatients reported significantly reduced levels of empathy compared to matched controls, with greater reductions in emotion recognition related to a greater number of past depressive episodes. Finally, emotion regulation is considered a primary feature of mental health (Gross and Muñoz 1995), with maladaptive patterns of emotion regulation impairing daily life functioning and supporting symptoms of psychopathology (Cole et al. 1994) in disorders like depression and PTSD (Joormann and Gotlib 2010; Tull et al. 2007).

The aim of the present chapter was therefore to review the acute and long-term effects of ayahuasca on (higher-order) cognitive processes, such as flexible (creative) thinking, empathy, and emotion regulation, and look for the link between these cognitive effects and subjective mood state and well-being. To that end, the literature has been searched for anecdotal reports and studies in which ayahuasca was administered either in a controlled experimental setting or a quasi-experimental setting, where participants took the substance at an ayahuasca retreat or ceremony. For a summary of included studies, refer to Table 7.1.

Flexible, Creative Thinking

Cognitive flexibility is the readiness with which one can selectively switch between mental processes to generate appropriate behavioral responses and is an important skill that allows individuals to accurately respond to their changing environment (Dajani and Uddin 2015). Within cognitive flexibility is creative thinking, a multi-component construct that includes convergent (CT) and divergent thinking (DT). CT is considered a process of generating a single optimal solution to a particular problem, emphasizing speed, accuracy, and logic. In contrast, DT is a process used to generate many new ideas in a context where more than one solution is correct, like in a brainstorming session, where generating as many innovative ideas or solutions on a particular issue as possible is the ultimate goal. Although both CT and DT are important in creative activities, DT is a more useful estimate of the potential for creative thought in daily life (Runco and Acar 2012).

In order to assess DT and creativity, researchers employ two categories of tests, namely, objective, psychometric tests, and more subjective, expert opinions (Sessa 2008). Psychometric tests typically assess DT, focusing on three different parameters: fluency, defined as the number of ideas an individual gives; originality, defined as the statistical infrequency or uniqueness of ideas; and flexibility, which represents the number of different conceptual categories the ideas cover (Beketayev and Runco 2016). Examples of these tests include Guilford's Alternative Uses Test (1967) Torrance Tests of Creative Thinking, the Pattern/Line Meanings Task, and the Picture Concept Test, composed of stimuli from the Wechsler Preschool and Primary Scale of Intelligence and the Wechsler Intelligence Scale for Children. Conversely, the expert opinion method employs a group of judges assessing persons' creative output. The judges make their assessments individually and then collate their views to establish an overall rating or measure. An example of this is the Consensual Assessment Technique (Amabile 1982).

There is a notable amount of anecdotal reports of increased creative capacity after psychedelic use (Frecska et al. 2012; Harman et al. 1966). Numerous writers and artists, including the likes of Aldous Huxley and Ken Kesey, have acknowledged the influence psychedelic drugs had on their work (Nutt 2012). Accordingly, a survey of 91 artists with at least one psychedelic experience reported that 70% had claimed that the psychedelic experience had affected the content of their work, with

Table 7.1 Summary of studies assessing effects of ayahuasca on cognitive processes

Type of study	Design (BS/WS)	Participants (#)	Reference measure/group	User (type)	Effect (type)	Cognitive/emotional tests and effect	Mood/well-being measures and effect	References
QE	WS	26	Baseline	E	A	PCT: CT ↓↓; DT ↑↑ PLMT: -	Mood and well-being ↑↑	Kuypers et al. (2016)
QE	WS	57 at baseline; 31 at 4 weeks	Baseline	E and N	S;L	PCT: DT -; CT ↑↑	Depression, anxiety, and stress measures↓↓; well-being measures ↑↑	Uthaug et al. (2018)
QE	BS	Ayahuasca users (40); control (21)	Control group	E	L	Torrance tests of creative thinking: originality ↑↑	No	Frecska et al. (2012)
QE	WS	45	Baseline	E and N	S	DERS: emotional nonacceptance, emotional interference, lack of control subscales ↓↓	Mindfulness measures: ↑↑ in observing, acting with awareness, non-judging, non-reacting, and decentering	Domínguez-Clavé et al. (2019)
QE	WS	12	Baseline	N	L	DERS ↓↓ (ns)	No	Thomas et al. (2013)
QE	BS	Ayahuasca users (127); controls (115)	Control group	E	L	The frontal systems behavior scales: users ↓↓ disinhibition/emotional dysregulation scale	Well-being test ↑↑; psychopathology measures ↓↓	Bouso et al. (2012)
E	WS and BS	24 (12 received ayahuasca; 12 received placebo)	Baseline and control group	N	S	fMRI; emotion regulation task Δ in activity in subgenual anterior cingulate cortex, amygdala, insula, and ventrolateral prefrontal cortex after ayahuasca intake	Depression ratings ↓↓	Palhano-Fontes et al. (2017)

E experimental, *QE* quasi-experimental, *A* acute, *S* subacute, *L* long term, *O* observational, *E* experienced, *N* naïve, *PCT* Picture Concept Test, *PLMT* Pattern/Line Meaning Test, *DERS* Difficulties in Emotion Regulation Scale, *VAS* Visual Analog Scale

54% claiming there had been a noticeable improvement in their artistic technique (Krippner 1968). Similarly, a recent review of past experimental studies suggests that, albeit with limited evidence, there is a positive association between creativity and psychedelic use (Iszáj et al. 2017). However, due to different objectives and varying methodological details, like timing of the measurements relative to dosing (e.g., immediately after intake or longer-term effects), the nature of this relationship was not clarified. More recent observational studies provide stronger objective evidence that psychedelics like psilocybin and 5-MeO-DMT have acute (Prochazkova et al. 2018) and persisting effects (Mason et al. 2019; Uthaug et al. 2019) on creativity.

Anecdotal reports suggest that ayahuasca also elicits similar effects as other psychedelic substances (Narby and Huxley 2004; Shanon 2000), with ayahuasca users scoring higher on creativity-related variables versus nonusers (Franquesa et al. 2018). However, to date, only three studies have objectively assessed the effect of ayahuasca on creativity. In order to address this gap in knowledge, a recent study by our group (Kuypers et al. 2016) employed a quasi-experimental design to assess the acute effect of ayahuasca on creative thinking in participants of ayahuasca sessions. Twenty-six experienced ayahuasca users from two spiritual ayahuasca workshops were invited to complete the Patterns/Line Meaning Task (PLMT), assessing CT, and the Picture Concept Test (PCT), assessing both CT and DT, both before ayahuasca intake and during acute inebriation, around 2 h after ayahuasca intake. While no statistically significant effects for the PLMT were found, ayahuasca intake significantly modified DT and CT as measured by the PCT. Specifically, CT was decreased after intake, whereas DT increased. Additionally, participants showed significant increases in visual analog scales assessing subjective effects in mood, like happiness, euphoria, fear, and well-being. Subsequently, another study by our group assessed the subacute and longer-term effects of ayahuasca on creative thinking by utilizing a similar approach (Uthaug et al. 2018). In this study, 57 participants from ayahuasca ceremonies in the Netherlands and Colombia were invited to complete the PCT before ayahuasca intake, the day after ayahuasca intake, and 4 weeks after ayahuasca intake. While no significant effects on DT were found, CT was increased the morning after and 4 weeks after use. The finding that ayahuasca did not have lasting effects on DT is in contrast to a previous study by Frecska (2012), in which 40 participants from an ayahuasca ritual in Brazil completed a creativity test before, and 2 days after, a 2-week-long ayahuasca ceremony. Frecska et al. (2012) found that ingestion of ayahuasca significantly increased the number of original solutions participants gave on the Torrance Tests of Creative thinking, a test assessing DT. An important difference between the two studies is the amount of ayahuasca consumed, with the latter study measuring DT after repeated ingestion of ayahuasca. Thus, taken together, these studies provide evidence that ayahuasca acutely enhances DT, with potential dose-dependent persisting effects.

In accordance with the aforementioned behavioral evidence, a neuroimaging study showed that ayahuasca causes an acute decrement in the functional connectivity in parts of the default mode network, a cluster of brain areas that is active when an individual is awake but not focused on the external environment (Palhano-Fontes et al. 2015). It was suggested that this decrease in distinction between brain networks allows the brain to operate with greater interconnectedness, resulting in a more cognitively flexible state that could therapeutically interrupt the maladaptive patterns seen in cognitively inflexible disorders (Carhart-Harris et al. 2012, 2014a, b; Dos Santos et al. 2016a, b; Tagliazucchi et al. 2014).

Empathy

Empathy, a multifaceted construct consisting of emotional empathy and cognitive empathy—respectively, the ability to feel and understand what another person is experiencing—is thought to play a key role in motivating moral and prosocial behavior (Decety 2011). Importantly, it has been suggested that cognitive flexibility is a prerequisite for empathy and, thus, impaired cognitive flexibility could mediate impaired empathy (Cusi et al. 2011; Grattan and Eslinger 1989).

Commonly used assessments of empathy include questionnaires and paradigms using nonverbal static stimuli. Questionnaires like the Interpersonal Reactivity Index (Davis 1983) provide a trait measure of both cognitive and emotional empathy. Paradigms like the Reading the Mind in the Eyes Test (Baron-Cohen et al. 2001) or the Facial Emotion Recognition Test (Kemmis et al. 2007) provide a state measure of cognitive empathy. Due to the limitations of these paradigms, a further photo-based paradigm, the Multifaceted Empathy Test (MET), was designed to assess state measures of both cognitive and emotional empathy (Dziobek et al. 2008). The MET has since been shown to be sensitive to the effects of psychedelic substances (Dolder et al. 2016; Hysek et al. 2014; Kuypers et al. 2014; Pokorny et al. 2017; Preller et al. 2015; Schmid et al. 2014).

Recent studies assessing the acute effects of serotonergic psychedelics indicate that ±3,4-methylenedioxymethamphetamine (MDMA), psilocybin, and LSD similarly alter empathy. Specifically, MDMA and psilocybin have been found to enhance emotional empathy on the MET (Hysek et al. 2014; Kuypers et al. 2014; Kuypers et al. 2017; Preller et al. 2015; Schmid et al. 2014), with MDMA also impairing the identification of negative emotions (fearful, angry, and sad faces) as measured by the Facial Emotion Recognition Task (FERT) (Hysek et al. 2014). Similarly, LSD has been found to enhance emotional empathy on the MET and decrease recognition of fearful and sad faces, as measured by the FERT (Dolder et al. 2016). Interestingly, a recent study also suggests that enhancement of *emotional* empathy may outlast the acute phase, with increases found up to 7 days after use of psilocybin in a naturalistic setting (Mason et al. 2019). The specificity of emotional and cognitive empathy suggests that emotional empathy is dependent on state variables, and cognitive empathy requires a (trait) ability to identify another's emotions (Pokorny et al. 2017).

Although anecdotal reports suggest that ayahuasca increases empathy (Dobkin de Rios and Rumrrill 2009), there have been no objective studies to date that have assessed this. Due to similarities in the mechanisms of action—namely, activation through the 5-HT system—findings from the previous psychedelic studies add strength to the hypothesis that ayahuasca would enhance emotional empathy. Currently, our group is assessing this hypothesis. To do this, we are visiting ayahuasca ceremonies throughout the Netherlands and administering the MET, as well as measures of creativity and subjective well-being, to participants at three different time points: before they take ayahuasca, the morning after they have taken ayahuasca, and 7 days after they have taken ayahuasca. With this, we hope to assess both the short- and longer-term effects of ayahuasca on emotional and cognitive empathy, as well as the correlation with creativity and subjective well-being.

Emotion Regulation

Emotion regulation is defined as the ongoing processes by which individuals influence the emotions they have, when they have them, and how they experience and express them (Gross 1998). According to Gratz and Roemer (2004), emotion regulation can be conceptualized as involving multiple processes, including the ability to be aware of, to understand, and to accept our emotions, to control impulsive behaviors when feeling negative emotions, and to choose suitable emotion regulation strategies in order to meet personal goals and situational demands. The process model of emotion regulation proposes five strategies individuals use during emotion regulation (Gross 1998). These strategies include "situation selection," which involves choosing to physically approach or avoid an emotionally relevant situation; "situation modification," which involves modification of the emotional impact of the situation; "attentional deployment," which involves directing ones attention towards or away from an emotional situation; "cognitive change or reappraisal," which involves choosing what meaning you attach to the emotional situation; and "response modification," which involves influencing your behavioral, physiological, and experiential response tendencies once they have been elicited. Importantly, these strategies can be either explicit, conscious, and voluntary, or implicit, nonconscious, and automatic (Gyurak et al. 2011). Interestingly, empathy deficits have been suggested to serve as a potential trigger of emotion dysregulation (Schipper and Petermann 2013).

Assessments of emotion regulation include self-report questionnaires, psychophysiological measurements, and responses to emotional stimuli. Questionnaires include the Difficulties in Emotion Regulation Scale (DERS) (Gratz and Roemer 2004), developed to assess each of the proposed multiple processes of emotion regulation, as well as the flexible use of strategies to modulate emotional strategies. Psychophysiological assessments include measurements of heart rate variability, which has been proposed as a psychophysiological marker of emotion regulation capacity (Visted et al. 2017; Williams et al. 2015). In addition, neuroimaging studies investigating the underlying circuitry of emotion regulation have employed various tasks consisting of basic emotional stimuli in order to provoke certain emotions in participants (Goldin et al. 2008; McRae et al. 2008).

Anecdotal reports, animal models, and observational and experimental studies suggest that ayahuasca has emotion and mood-altering properties (Dos Santos et al. 2016a, b; Nunes et al. 2016). A within-subjects, double-blind, placebo-controlled study found that, acutely, ayahuasca was associated with lower scores on scales assessing panic and hopelessness-related states in individuals with over 10 years of ayahuasca experience (Santos et al. 2007). Furthermore, a study by Barbosa et al. (2016) found that, relative to non-ayahuasca using controls, ayahuasca users displayed lower depression scores. Similarly, an open-label trial with major depressive patients found that, under the acute influence of ayahuasca, patients displayed improvements in emotional withdrawal and blunted affect and significantly decreased scores in depression-related scales, with depression symptoms remaining decreased up until 21 days after administration (Osorio et al. 2015; Sanches et al. 2016). In addition, the study reported increased blood perfusion in areas of the brain implicated in the regulation of mood and emotional states, a response associated with antidepressive effects (Sanches et al. 2016). Furthermore, providing perhaps the strongest experimental evidence to date, a recent randomized, double-blind, placebo-controlled trial found that a single dose of ayahuasca reduced symptoms of depression (Palhano-Fontes et al. 2019) and feelings of suicidality (Zeifman et al. 2019) for up to 7 days in a treatment-resistant depression population. Another recent interview study with individuals who were diagnosed with eating disorders found that the majority of the interviewees reported improvements in psychological symptoms following ceremonial ayahuasca drinking, which they partially attributed to increased emotion regulation and processing (Lafrance et al. 2017).

Despite evidence for ayahuasca's emotion and mood-altering properties, studies directly assessing the effects of ayahuasca on emotion regulation are sparse. The most recent study utilized an observational design to assess the impact of one ayahuasca session on emotion regulation (Domínguez-Clavé et al. 2019). Forty-five volunteers completed the DERS prior to and 24 h after the session. Findings showed that volunteers scored lower in emotional nonacceptance, emotional interference, and lack of control 24 h after ayahuasca ingestion. Interestingly, a subset of participants displaying borderline personality disorder traits also demonstrated lower scores in emotional interference after ayahuasca ingestion. Accordingly, a previous observational study found similar results in a separate clinical group. Namely, 12 participants in an ayahuasca-assisted addiction treatment retreat were assessed at baseline and at 6-month follow-up on several psychological and behavioral factors, including emotion regulation (Thomas et al. 2013). Findings showed that participants' scores improved (albeit not significantly) on the DERS. Finally, a study by Bouso et al. (2012) assessed the even longer-term impact of repeated ayahuasca use on various factors, including general psychological well-being and mental health. They found that, compared to controls, ayahuasca users scored higher in emotion regulation at both baseline and 1-year follow-up, as shown by lower values on the items in the Frontal Systems Behavior Scale assessing disinhibition and emotional dysregulation. Users also scored higher on subjective psychological well-being and lower on psychopathology measures.

A recent imaging study evaluated the mechanisms by which ayahuasca modulates emotion processing of patients with depression (Palhano-Fontes et al. 2017). Twenty-four patients with treatment-resistant depression completed two fMRI sessions 1 day before and 1 day after ayahuasca or placebo intake. During the fMRI session, participants performed an emotion regulation task (Ochsner et al. 2012) in which they viewed a series of images with different emotional content and were instructed to passively look at neutral and negative images, as well as to positively reappraise negative images (e.g., by imagining a happy ending). Behavioral data suggested that the group who received ayahuasca rated all images (neutral, negative, and reappraised) more positively after treatment, compared to the placebo treatment. Analysis of the imaging data showed that brain areas involved in emotion regulation processes, such as the subgenual anterior cingulate cortex, amygdala, insula, and ventrolateral prefrontal cortex, were differently modulated in the patients treated with ayahuasca when compared to placebo. Interestingly, pre-posttreatment increases in subgenual anterior cingulate cortex activity, in particular, also positively correlated with reductions in depressive symptoms 1 day posttreatment.

Recent evidence also suggests that ayahuasca can enhance mindfulness abilities, which involve nonjudgmental attention to present-moment experience (Farb et al. 2012). In a study by Soler et al. (2016), 25 individuals were assessed on a measure of mindfulness capacity (the Five Facets of Mindfulness Questionnaire; FFMQ) before and 24 h after they attended an ayahuasca session. After ayahuasca intake, individuals displayed a significant reduction in automatic negative judgmental processing of experiences and inner reactivity, two facets of the FFMQ. These results were replicated by Domínguez-Clavé et al. (2019), who found the same decrease in judgmental processing and reactivity 24 h after an ayahuasca session, and Sampedro et al. (2017), who found similar results at both 24 h and 2 months after ayahuasca intake.

Although these studies do not directly assess emotion regulation, evidence suggests that mindfulness and emotion regulation have a strong, possibly overlapping, relationship. Specifically, studies suggest that the practice of mindfulness (cultivating awareness and acceptance of the present moment) is associated with healthy emotion regulation and may play a causal role (Roemer et al. 2015). Biological evidence supporting this relationship also stems from a previous study investigating the neural correlates of increased mindfulness capacities after ayahuasca intake (Sampedro et al. 2017). Findings suggested that enhanced connectivity between the anterior cingulate cortex (previously implicated in ayahuasca-enhanced emotion regulation and depressive symptom reduction in Palhano-Fontes et al. (2019)) is associated with increased nonjudgmental processing 1 day and 2 months after ayahuasca use.

Discussion

The aim of this review was to assess whether ayahuasca has the ability to enhance (higher-order) socio/cognitive processes, like creative, flexible thinking, empathy, and emotion regulation, and to look for the link between these cognitive effects and subjective mood states and well-being. Although objective evidence is limited, previous studies with ayahuasca and similar psychedelics, like psilocybin and LSD, support the notion that ayahuasca can enhance these processes. Importantly, evidence is given to suggest that this enhancement outlasts the acute stage, thus potentially persisting over time. These findings have important implications for the therapeutic utility of ayahuasca.

Previously, it has been shown that creative DT can enhance and strengthen psychological flexibility by allowing individuals to generate new and effective cognitive, emotional, and behavioral strategies on their own, ultimately helping them to adopt adaptive interpretations and coping styles (Forgeard and Elstein 2014). Since ayahuasca promotes flexible cognition, it was suggested that ayahuasca possesses qualities that can facilitate a therapeutic process (Kuypers et al. 2016) and make it suited, for example, to psychedelic-assisted psychotherapy (Bouso et al. 2008). Specifically, the ability of ayahuasca to increase DT acutely (Kuypers et al. 2016) and in the longer term (Frecska et al. 2012) could help patients relive events, recalling various associations without inhibition (Bouso et al. 2008; Frecska et al. 2012, 2016). The subacute effects, e.g., longer-term enhancements in CT (Uthaug et al. 2018), could then be suited in a second "integration" session, in which patients discuss the experiences they had on ayahuasca and find strategies that help them cope with intensive emotions (Kuypers et al. 2016; Mason et al. 2019).

Empathy has been implicated as a factor in symptom severity of disorders like depression. Thus, it could be hypothesized that enhancing an individual's empathic abilities could play a role in symptom alleviation. Furthermore, the addition of a pharmacological agent that can enhance empathy could play an important role in psychotherapy and in the patient-therapist relationship. Namely, the increase in empathy could enhance feelings of openness and trust in the patient that could strengthen the therapeutic alliance between patient and therapist (Oehen et al. 2013). The quality of this relationship has been seen to be particularly important for disorders like PTSD (Charuvastra and Cloitre 2008). Previous research suggests that 5-HT_{2A} agents have the potential to enhance empathy, with enhancements even outlasting the acute phase and positively correlating with increased feelings of well-being (Mason et al. 2019). However, in the case of ayahuasca, this has yet to be shown.

Collectively, being able to monitor and evaluate our emotional experience, accept our emotional reactions, and modulate our emotional responses to meet goals and situational demands is crucial for everyday interactions (Visted et al. 2017). Ayahuasca's ability to reduce symptom pathology of disorders like depression may be due to its ability to alter an individual's engagement with their emotions. For example, enhancing nonjudgmental awareness and reducing reactivity to emotional events may facilitate a healthy engagement with emotions and emotional memories that may lead to an individual experiencing their emotions in a less reactive state instead of avoiding them or responding negatively (Chambers et al. 2009; Hayes and Feldman 2004; Soler et al. 2016). This avoidance, termed "experiential avoidance," is a problem of emotion regulation (Hayes and Feldman 2004), with those displaying higher experiential avoidance also reporting higher emotional reactivity (Sloan 2004).

The reviewed studies are not without their limitations. Most studies employed a quasi-experimental, naturalistic design with experienced ayahuasca users (Bouso et al. 2012; Domínguez-Clavé et al. 2019; Frecska et al. 2012; Kuypers et al. 2016; Uthaug et al. 2018). In such a design, factors like composition of the brew (dose) and the context of the experience differ between studies and need to be considered when looking at the effects of ayahuasca on cognition and subjective experience.

Ayahuasca has previously been shown to display dose-dependent subjective (Riba et al. 2001) and physiological effects (Riba et al. 2003). In accordance with this, a neuroimaging study showed decreased amygdala activity after a low dose of ayahuasca, while another study using a higher dose found increased activity in the same region (Riba et al. 2006). For further information regarding underlying neural correlates, see Dominguez-Clavé et al., this volume, Chap. 1.

Additionally, the context of the experience, or the "set" and the "setting," are two proposed key factors in elucidating the positive and therapeutic potential of psychedelics. "Set" refers to the mental state of the individual prior to and during the experience, like thoughts, mood, and expectations. "Setting" is the physical and social environment that the drug is taken in (Leary et al. 1963). Previous research and anecdotal evidence have shown that both mindset and the social environment play an important role in the outcome of the experience (Shewan et al. 2000). Participants regularly attending ayahuasca ceremonies could inherently have a biased mindset, in that individuals who seek ayahuasca are predisposed to do so for personal growth or to treat an illness. Setting is also of particular importance, as ayahuasca is commonly taken in supportive and ceremonial or religious settings (Lawn et al. 2017).

The aforementioned factors can be controlled for by assessing the effects of ayahuasca in a randomized, double-blind, placebo-controlled experimental design. This design is the gold standard of psychopharmacological research, allowing for control over factors that could affect the study results. To demonstrate this, we will sketch a hypothetical study in which the aim is to assess the effects of ayahuasca on a (higher-order) cognitive process, for example, empathy. As a researcher, you recruit participants, screen them according to your inclusion criteria and, if those are met, you randomly assign them to a drug condition (ayahuasca or placebo), subsequently eliminating selection bias from the researcher and balancing the groups with respect to known (e.g., social differences) and unknown (e.g., genetic differences) factors. On the testing day, the participant comes to the lab, a standardized, "sterile" space identical for each participant (thus controlling for setting), where they are given a drug by the researcher. The drug will be controlled, i.e., it will either be a standardized placebo or a "pharmaceutical" version of ayahuasca, where purity and dose have been established. Importantly, neither the participant nor the researcher knows whether they received ayahuasca or placebo (hence "double blind"), subsequently controlling for expectation effects (e.g., *expecting* that ayahuasca increases empathy). If all of these steps are followed, the researcher has effectively cancelled out the potential influence of confounding variables that can skew study results (e.g., bias, setting, drug composition, and expectation) and thus is able to directly assess the effect of ayahuasca alone. If the data show that participants who took ayahuasca scored higher in empathy compared to the placebo, we can say with more certainty that it is the ayahuasca increasing this cognitive process, and not other external factors.

That being said, a placebo-controlled, double-blind experimental design is currently difficult to implement due to multiple issues related to this scheduled substance. First, approval (e.g., by an institutional review board) is needed to be able to work with the substance, and due to the scheduling of, and stigma surrounding, psychedelic substances, crucial information required to gain such approval is missing. This information includes (but is not limited to) scientific data demonstrating their biological and psychological safety. Furthermore, due to scheduling of these substances in most countries, special licenses are required in order to obtain them for an experimental study, and "pharmaceutical" versions of the drug are required in order to be allowed to administer them to participants. All of these factors together can lead to a vicious cycle: the lack of scientific data results in an inability to study these substances, which means no data can be acquired.

Thus, although less able to assess "cause and effect," the quasi-experimental, naturalistic designs that have been reviewed in this chapter offer a unique opportunity to explore the effects of ayahuasca, overcoming some of the hurdles that scientists face. Namely, via an observational design, scientists can gather necessary data (like safety data) without the need to acquire a special license or a pharmaceutical version of the drug, as the scientist is not the person organizing the ayahuasca ceremony or administering the substance. Additionally, with this design some of the limitations that controlled designs face can be overcome; namely, collecting data in a more externally valid setting, i.e. the cultural or social setting in which people usually consume the substance, versus a "sterile" laboratory. Overall, future studies should employ both types of designs in a complementary approach, with naturalistic studies acquiring externally valid data that researchers can reference when applying for approval for experimental studies that more accurately assess cause and effect.

Future avenues of research include the impact of enhanced cognitive processes on mood and well-being. Although most of the studies assessed participants' mood and well-being in some form, none of them showed the relationship between subjective states and measures of creativity and emotion regulation. Additionally, as studies found a time-related differentiation of effects of ayahuasca on these outcomes—for example, enhancements in emotion regulation were seen both acutely and long term, whereas decrements in CT were seen acutely and enhancement long term—future studies could take both stages into account. Specifically, studies assessing the longevity of the effects are warranted, to see how long ayahuasca's "window of opportunity" lasts.

Conclusion

This review assessed literature to determine whether ayahuasca has the ability to enhance (higher-order) socio/cognitive processes, like creative, flexible thinking, empathy, and emotion regulation and, consequently, also, mood and well-being. While the number of studies is small, findings suggest that ayahuasca enhances socio-cognitive skills acutely and in the longer term. Future clinical research into the therapeutic effects of ayahuasca could assess the relationship between the effect on (higher-order) cognitive and emotional processes and the role both play in the symptom alleviation in the pathological population.

References

Aldao, A., Nolen-Hoeksema, S., & Schweizer, S. (2010). Emotion-regulation strategies across psychopathology: A meta-analytic review. *Clinical Psychology Review, 30*(2), 217–237. https://doi.org/10.1016/j.cpr.2009.11.004.

Amabile, T. M. (1982). Social psychology of creativity: A consensual assessment technique. *Journal of Personality and Social Psychology, 43*(5), 997–1013. https://doi.org/10.1037/0022-3514.43.5.997.

Baas, M., De Dreu, C. K., & Nijstad, B. A. (2008). A meta-analysis of 25 years of mood-creativity research: Hedonic tone, activation, or regulatory focus? *Psychological Bulletin, 134*(6), 779–806. https://doi.org/10.1037/a0012815.

Barbosa, P. C., Giglio, J. S., & Dalgalarrondo, P. (2005). Altered states of consciousness and short-term psychological after-effects induced by the first time ritual use of ayahuasca in an urban context in Brazil. *Journal of Psychoactive Drugs, 37*(2), 193–201. https://doi.org/10.1080/02791072.2005.10399801.

Barbosa, P. C., Strassman, R. J., da Silveira, D. X., Areco, K., Hoy, R., Pommy, J., Thoma, R., & Bogenschutz, M. (2016). Psychological and neuropsychological assessment of regular hoasca users. *Comprehensive Psychiatry, 71*, 95–105. https://doi.org/10.1016/j.comppsych.2016.09.003.

Baron-Cohen, S., Wheelwright, S., Hill, J., Raste, Y., & Plumb, I. (2001). The "Reading the Mind in the Eyes" test revised version: A study with normal adults, and adults with Asperger syndrome or high-functioning autism. *Journal of Child Psychology and Psychiatry, 42*(2), 241–251. https://doi.org/10.1111/1469-7610.00715.

Beketayev, K., & Runco, M. A. (2016). Scoring divergent thinking tests by computer with a semantics-based algorithm. *Europe's Journal of Psychology, 12*(2), 210–220. https://doi.org/10.5964/ejop.v12i2.1127.

Bickel, P., Dittrich, A., & Schoepf, J. (1976). Altered states of consciousness induced by N,N-dimethyltryptamine (DMT). *Pharmakopsychiatr Neuropsychopharmakol, 9*(5), 220–225. https://doi.org/10.1055/s-0028-1094495. (Eine experimentelle Untersuchung zur bewusstseinsverandernden Wirkung von N,N-Dimethyltryptamin (DMT)).

Bouso, J. C., Doblin, R., Farre, M., Alcazar, M. A., & Gomez-Jarabo, G. (2008). MDMA-assisted psychotherapy using low doses in a small sample of women with chronic posttraumatic stress disorder. *Journal of Psychoactive Drugs, 40*(3), 225–236. https://doi.org/10.1080/02791072.2008.10400637.

Bouso, J. C., Gonzalez, D., Fondevila, S., Cutchet, M., Fernandez, X., Ribeiro Barbosa, P. C., Alcazar-Corcoles, M. A., Araujo, W. S., Barbanoj, M. J., Fabregas, J. M., & Riba, J. (2012). Personality, psychopathology, life attitudes and neuropsychological performance among ritual users of ayahuasca: A longitudinal study. *PLoS One, 7*(8), e42421. https://doi.org/10.1371/journal.pone.0042421.

Carhart-Harris, R. L., Erritzoe, D., Williams, T., Stone, J. M., Reed, L. J., Colasanti, A., Tyacke, R. J., Leech, R., Malizia, A. L., Murphy, K., Hobden, P., Evans, J., Feilding, A., Wise, R. G., & Nutt, D. J. (2012). Neural correlates of the psychedelic state as determined by fMRI studies with psilocybin. *Proceedings of the National Academy of Sciences of the United States of America, 109*(6), 2138–2143. https://doi.org/10.1073/pnas.1119598109.

Carhart-Harris, R. L., Leech, R., Hellyer, P. J., Shanahan, M., Feilding, A., Tagliazucchi, E., Chialvo, D. R., & Nutt, D. (2014a). The entropic brain: A theory of conscious states informed by neuroimaging research with psychedelic drugs. *Frontiers in Human Neuroscience, 8*, 20.

Carhart-Harris, R. L., Leech, R., & Tagliazucchi, E. (2014b). How do hallucinogens work on the brain? *Journal of Psychophysiology, 71*(1), 2–8.

Carhart-Harris, R. L., Bolstridge, M., Day, C. M. J., Rucker, J., Watts, R., Erritzoe, D. E., Kaelen, M., Giribaldi, B., Bloomfield, M., Pilling, S., Rickard, J. A., Forbes, B., Feilding, A., Taylor, D., Curran, H. V., & Nutt, D. J. (2017). Psilocybin with psychological support for treatment-resistant depression: Six-month follow-up. *Psychopharmacology*. https://doi.org/10.1007/s00213-017-4771-x.

Chamberlain, S. R., Fineberg, N. A., Blackwell, A. D., Robbins, T. W., & Sahakian, B. J. (2006). Motor inhibition and cognitive flexibility in obsessive-compulsive disorder and trichotillomania. *The American Journal of Psychiatry, 163*(7), 1282–1284. https://doi.org/10.1176/appi.ajp.163.7.1282.

Chambers, R., Gullone, E., & Allen, N. B. (2009). Mindful emotion regulation: An integrative review. *Clinical Psychology Review, 29*(6), 560–572. https://doi.org/10.1016/j.cpr.2009.06.005.

Charuvastra, A., & Cloitre, M. (2008). Social bonds and posttraumatic stress disorder. *Annual Review of Psychology, 59*, 301–328. https://doi.org/10.1146/annurev.psych.58.110405.085650.

Cole, P. M., Michel, M. K., & Teti, L. O. D. (1994). The development of emotion regulation and dysregulation: A clinical perspective. *Monographs of the Society for Research in Child Development, 59*(2–3), 73–102.

Cusi, A. M., Macqueen, G. M., Spreng, R. N., & McKinnon, M. C. (2011). Altered empathic responding in major depressive disorder: Relation to symptom severity, illness burden, and psychosocial outcome. *Psychiatry Research, 188*(2), 231–236. https://doi.org/10.1016/j.psychres.2011.04.013.

Da Silveira, D. X., Grob, C. S., de Rios, M. D., Lopez, E., Alonso, L. K., Tacla, C., & Doering-Silveira, E. (2005). Ayahuasca in adolescence: A preliminary psychiatric assessment. *Journal of Psychoactive Drugs, 37*(2), 129–133. https://doi.org/10.1080/02791072.2005.10399792.

Dajani, D. R., & Uddin, L. Q. (2015). Demystifying cognitive flexibility: Implications for clinical and developmental neuroscience. *Trends in Neurosciences, 38*(9), 571–578. https://doi.org/10.1016/j.tins.2015.07.003.

Davis, M. H. (1983). Measuring individual differences in empathy: Evidence for a multidimensional approach. *Journal of Personality and Social Psychology, 44*(1), 113–126. https://doi.org/10.1037/0022-3514.44.1.113.

Davis, M. A. (2009). Understanding the relationship between mood and creativity: A meta-analysis. *Organizational Behavior and Human Decision Processes, 108*(1), 25–38. https://doi.org/10.1016/j.obhdp.2008.04.001.

Decety, J. (2011). Dissecting the neural mechanisms mediating empathy. *Emotion Review, 3*(1), 92–108.

Dobkin de Rios, M., & Rumrrill, R. (2009). A hallucinogenic tea, laced with controversy. Ayahuasca in the Amazon and the United States. *Journal of Travel Medicine, 16*(4), 298–298. https://doi.org/10.1111/j.1708-8305.2009.00325.x.

Doering-Silveira, E., Grob, C. S., de Rios, M. D., Lopez, E., Alonso, L. K., Tacla, C., & Da Silveira, D. X. (2005). Report on psychoactive drug use among adolescents using ayahuasca within a religious context. *Journal of Psychoactive Drugs, 37*(2), 141–144. https://doi.org/10.1080/02791072.2005.10399794.

Dolder, P. C., Schmid, Y., Müller, F., Borgwardt, S., & Liechti, M. E. (2016). LSD acutely impairs fear recognition and enhances emotional empathy and sociality. *Neuropsychopharmacology, 41*, 2638. https://doi.org/10.1038/npp.2016.82. https://www.nature.com/articles/npp201682#supplementary-information.

Dominguez-Clave, E., Soler, J., Elices, M., Pascual, J. C., Alvarez, E., de la Fuente Revenga, M., Friedlander, P., Feilding, A., & Riba, J. (2016). Ayahuasca: Pharmacology, neuroscience and therapeutic potential. *Brain Research Bulletin, 126*(Pt 1), 89–101. https://doi.org/10.1016/j.brainresbull.2016.03.002.

Domínguez-Clavé, E., Soler, J., Pascual, J. C., Elices, M., Franquesa, A., Valle, M., Alvarez, E., & Riba, J. (2019). Ayahuasca improves emotion dysregulation in a community sample and in individuals with borderline-like traits. *Psychopharmacology, 236*(2), 573–580.

Dominguez-Clavé, E., Elices, M., Morales Garcia, J. A., Pascual, J. C., Soler, J., Pérez-Castillo, A., & Riba, J. (2021). Ayahuasca as an unusually versatile therapeutic agent: From molecules to meta-cognition and back. In B. C. Labate & C. Cavnar (Eds.), *Ayahuasca healing and science. in this volume.*

Donges, U. S., Kersting, A., Dannlowski, U., Lalee-Mentzel, J., Arolt, V., & Suslow, T. (2005). Reduced awareness of others' emotions in unipolar depressed patients. *The Journal of Nervous and Mental Disease, 193*(5), 331–337.

Dos Santos, R. G., Osorio, F. L., Crippa, J. A., & Hallak, J. E. (2016a). Antidepressive and anxiolytic effects of ayahuasca: A systematic literature review of animal and human studies. *Revista Brasileira de Psiquiatria, 38*(1), 65–72. https://doi.org/10.1590/1516-4446-2015-1701.

Dos Santos, R. G., Osório, F. L., Crippa, J. A. S., Riba, J., Zuardi, A. W., & Hallak, J. E. C. (2016b). Antidepressive, anxiolytic, and antiaddictive effects of ayahuasca, psilocybin and lysergic acid diethylamide (LSD): A systematic review of clinical trials published in the last 25 years. *Therapeutic Advances in Psychopharmacology, 6*(3), 193–213. https://doi.org/10.1177/2045125316638008.

Dziobek, I., Rogers, K., Fleck, S., Bahnemann, M., Heekeren, H. R., Wolf, O. T., & Convit, A. (2008). Dissociation of cognitive and emotional empathy in adults with Asperger syndrome using the Multifaceted Empathy Test (MET) [journal article]. *Journal of Autism and Developmental Disorders, 38*(3), 464–473. https://doi.org/10.1007/s10803-007-0486-x.

Fantegrossi, W. E., Murnane, K. S., & Reissig, C. J. (2008). The behavioral pharmacology of hallucinogens. *Biochemical Pharmacology, 75*(1), 17–33. https://doi.org/10.1016/j.bcp.2007.07.018.

Farb, N. A. S., Anderson, A. K., & Segal, Z. V. (2012). The mindful brain and emotion regulation in mood disorders. *Canadian Journal of Psychiatry. Revue Canadienne De Psychiatrie, 57*(2), 70–77. http://www.ncbi.nlm.nih.gov/pmc/articles/PMC3303604/.

Forgeard, M. J. C., & Elstein, J. G. (2014). Advancing the clinical science of creativity. *Frontiers in Psychology, 5*, 613. https://doi.org/10.3389/fpsyg.2014.00613.

Franquesa, A., Sainz-Cort, A., Gandy, S., Soler, J., Alcázar-Córcoles, M. Á., & Bouso, J. C. (2018). Psychological variables implied in the therapeutic effect of ayahuasca: A contextual approach. *Psychiatry Research, 264*, 334–339.

Frecska, E., More, C. E., Vargha, A., & Luna, L. E. (2012). Enhancement of creative expression and entoptic phenomena as after-effects of repeated ayahuasca ceremonies. *Journal of Psychoactive Drugs, 44*(3), 191–199. https://doi.org/10.1080/02791072.2012.703099.

Frecska, E., Bokor, P., & Winkelman, M. (2016). The therapeutic potentials of ayahuasca: Possible effects against various diseases of civilization. *Frontiers in Pharmacology, 7*. https://doi.org/10.3389/fphar.2016.00035.

Goldin, P. R., McRae, K., Ramel, W., & Gross, J. J. (2008). The neural bases of emotion regulation: Reappraisal and suppression of negative emotion. *Biological Psychiatry, 63*(6), 577–586.

Grattan, L. M., & Eslinger, P. J. (1989). Higher cognition and social behavior: Changes in cognitive flexibility and empathy after cerebral lesions. *Neuropsychology, 3*(3), 175–185. https://doi.org/10.1037/h0091764.

Gratz, K. L., & Roemer, L. (2004). Multidimensional assessment of emotion regulation and dysregulation: Development, factor structure, and initial validation of the difficulties in emotion regulation scale. *Journal of Psychopathology and Behavioral Assessment, 26*(1), 41–54.

Griffiths, R. R., Richards, W. A., McCann, U., & Jesse, R. (2006). Psilocybin can occasion mystical-type experiences having substantial and sustained personal meaning and spiritual significance. *Psychopharmacology, 187*(3), 268–283; discussion 284–292. https://doi.org/10.1007/s00213-006-0457-5.

Grob, C. S., McKenna, D. J., Callaway, J. C., Brito, G. S., Neves, E. S., Oberlaender, G., Saide, O. L., Labigalini, E., Tacla, C., Miranda, C. T., Strassman, R. J., & Boone, K. B. (1996). Human psychopharmacology of hoasca, a plant hallucinogen used in ritual context in Brazil. *The Journal of Nervous and Mental Disease, 184*(2), 86–94.

Gross, J. J. (1998). The emerging field of emotion regulation: An integrative review. *Review of General Psychology, 2*(3), 271–299.

Gross, J. J., & Muñoz, R. F. (1995). Emotion regulation and mental health. *Clinical Psychology: Science and Practice, 2*(2), 151–164.

Gyurak, A., Gross, J. J., & Etkin, A. (2011). Explicit and implicit emotion regulation: A dual-process framework. *Cognition & Emotion, 25*(3), 400–412. https://doi.org/10.1080/02699931.2010.544160.

Harman, W. W., McKim, R. H., Mogar, R. E., Fadiman, J., & Stolaroff, M. J. (1966). Psychedelic agents in creative problem-solving: A pilot study. *Psychological Reports, 19*(1), 211–227.

Hasler, F., Grimberg, U., Benz, M. A., Huber, T., & Vollenweider, F. X. (2004). Acute psychological and physiological effects of psilocybin in healthy humans: A double-blind, placebo-controlled dose-effect study. *Psychopharmacology, 172*(2), 145–156. https://doi.org/10.1007/s00213-003-1640-6.

Hayes, A. M., & Feldman, G. (2004). Clarifying the construct of mindfulness in the context of emotion regulation and the process of change in therapy. *Clinical Psychology: Science and Practice, 11*(3), 255–262. https://doi.org/10.1093/clipsy.bph080.

Hendricks, P. S., Thorne, C. B., Clark, C. B., Coombs, D. W., & Johnson, M. W. (2015). Classic psychedelic use is associated with reduced psychological distress and suicidality in the United States adult population. *Journal of Psychopharmacology, 29*(3), 280–288. https://doi.org/10.1177/0269881114565653.

Hysek, C. M., Simmler, L. D., Schillinger, N., Meyer, N., Schmid, Y., Donzelli, M., Grouzmann, E., & Liechti, M. E. (2014). Pharmacokinetic and pharmacodynamic effects of methylphenidate and MDMA administered alone or in combination. *The International Journal of Neuropsychopharmacology, 17*(3), 371–381. https://doi.org/10.1017/s1461145713001132.

Iszáj, F., Griffiths, M. D., & Demetrovics, Z. (2017). Creativity and psychoactive substance use: A systematic review. *International Journal of Mental Health and Addiction, 15*(5), 1135–1149. https://doi.org/10.1007/s11469-016-9709-8.

Joormann, J., & Gotlib, I. H. (2010). Emotion regulation in depression: Relation to cognitive inhibition. *Cognition and Emotion, 24*(2), 281–298.

Kemmis, L., Hall, J. K., Kingston, R., & Morgan, M. J. (2007). Impaired fear recognition in regular recreational cocaine users [journal article]. *Psychopharmacology (Berl), 194*(2), 151–159. https://doi.org/10.1007/s00213-007-0829-5.

Kjellgren, A., Eriksson, A., & Norlander, T. (2009). Experiences of encounters with ayahuasca "the vine of the soul". *Journal of Psychoactive Drugs, 41*(4), 309–315. https://doi.org/10.1080/02791072.2009.10399767.

Krippner, S. (1968). The psychedelic state, the hypnotic trance, and the creative act. *Journal of Humanistic Psychology, 8*(1), 49–67.

Kuypers, K. P. C., de la Torre, R., Farre, M., Yubero-Lahoz, S., Dziobek, I., Van den Bos, W., & Ramaekers, J. G. (2014). No evidence that MDMA-induced enhancement of emotional empathy is related to peripheral oxytocin levels or 5-HT1a receptor activation. *PLoS One, 9*(6), e100719. https://doi.org/10.1371/journal.pone.0100719.

Kuypers, K. P. C., Riba, J., de la Fuente Revenga, M., Barker, S., Theunissen, E. L., & Ramaekers, J. G. (2016). Ayahuasca enhances creative divergent thinking while decreasing conventional convergent thinking. *Psychopharmacology (Berl), 233*(18), 3395–3403. https://doi.org/10.1007/s00213-016-4377-8.

Kuypers, K. P. C., Dolder, P. C., Ramaekers, J. G., & Liechti, M. E. (2017). Multifaceted empathy of healthy volunteers after single doses of MDMA: A pooled sample of placebo-controlled studies. *Journal of Psychopharmacology, 31*(5), 589–598. https://doi.org/10.1177/0269881117699617.

Lafrance, A., Loizaga-Velder, A., Fletcher, J., Renelli, M., Files, N., & Tupper, K. W. (2017). Nourishing the spirit: Exploratory research on ayahuasca experiences along the continuum of recovery from eating disorders. *Journal of Psychoactive Drugs, 49*(5), 427–435.

Lawn, W., Hallak, J. E., Crippa, J. A., Dos Santos, R., Porffy, L., Barratt, M. J., Ferris, J. A., Winstock, A. R., & Morgan, C. J. A. (2017). Well-being, problematic alcohol consumption and acute subjective drug effects in past-year ayahuasca users: A large, international, self-selecting online survey. *Scientific Reports, 7*(1), 15201. https://doi.org/10.1038/s41598-017-14700-6.

Leary, T., Litwin, G. H., & Metzner, R. (1963). Reactions to psilocybin administered in a supportive environment. *The Journal of Nervous and Mental Disease, 137*, 561–573.

Luna, L. E. (2011). Indigenous and mestizo use of ayahuasca. An overview. In R. G. Santos (Ed.), *The ethnopharmacology of ayahuasca* (pp. 1–19). Trivandrum: Transworld Research Network.

Mason, N. L., Mischler, E., Uthaug, M. V., & Kuypers, K. P. (2019). Sub-acute effects of psilocybin on empathy, creative thinking, and subjective well-being. *Journal of Psychoactive Drugs, 51*(2), 123–134.

McKenna, D. J. (2004). Clinical investigations of the therapeutic potential of ayahuasca: Rationale and regulatory challenges. *Pharmacology & Therapeutics, 102*(2), 111–129. https://doi.org/10.1016/j.pharmthera.2004.03.002.

McKenna, D. J., Towers, G. H., & Abbott, F. (1984). Monoamine oxidase inhibitors in South American hallucinogenic plants: Tryptamine and beta-carboline constituents of ayahuasca. *Journal of Ethnopharmacology, 10*(2), 195–223.

McRae, K., Ochsner, K. N., Mauss, I. B., Gabrieli, J. J., & Gross, J. J. (2008). Gender differences in emotion regulation: An fMRI study of cognitive reappraisal. *Group Processes & Intergroup Relations, 11*(2), 143–162.

Narby, J., & Huxley, F. (2004). *Shamans through time*. New York City: Penguin.

Nietlisbach, G., Maercker, A., Rossler, W., & Haker, H. (2010). Are empathic abilities impaired in posttraumatic stress disorder? *Psychological Reports, 106*(3), 832–844. https://doi.org/10.2466/pr0.106.3.832-844.

Nunes, A. A., dos Santos, R. G., Osório, F. L., Sanches, R. F., Crippa, J. A. S., & Hallak, J. E. (2016). Effects of ayahuasca and its alkaloids on drug dependence: A systematic literature review of quantitative studies in animals and humans. *Journal of Psychoactive Drugs, 48*(3), 195–205.

Nutt, D. J. (2012). *Drugs – without the hot air: Minimising the harms of legal and illegal drugs*. Cambridge: UIT Cambridge.

Ochsner, K. N., Silvers, J. A., & Buhle, J. T. (2012). Functional imaging studies of emotion regulation: A synthetic review and evolving model of the cognitive control of emotion. *Annals of the New York Academy of Sciences, 1251*, E1.

Oehen, P., Traber, R., Widmer, V., & Schnyder, U. (2013). A randomized, controlled pilot study of MDMA (+/– 3,4-Methylenedioxymethamphetamine)-assisted psychotherapy for treatment of resistant, chronic post-traumatic stress disorder (PTSD). *Journal of Psychopharmacology, 27*(1), 40–52. https://doi.org/10.1177/0269881112464827.

Osorio, F. L., Sanches, R. F., Macedo, L. R., Santos, R. G., Maia-de-Oliveira, J. P., Wichert-Ana, L., Araujo, D. B., Riba, J., Crippa, J. A., & Hallak, J. E. (2015). Antidepressant effects of a single dose of ayahuasca in patients with recurrent depression: A preliminary report. *Revista Brasileira de Psiquiatria, 37*(1), 13–20. https://doi.org/10.1590/1516-4446-2014-1496.

Palhano-Fontes, F., Andrade, K. C., Tofoli, L. F., Santos, A. C., Crippa, J. A., Hallak, J. E., Ribeiro, S., & de Araujo, D. B. (2015). The psychedelic state induced by ayahuasca modulates the activity and connectivity of the default mode network. *PLoS One, 10*(2), e0118143. https://doi. org/10.1371/journal.pone.0118143.

Palhano-Fontes, F., Onias, H., Novaes, M., Andrade, K. C., Arcoverde, E., Maia-de-Oliveira, J. P., & de Araújo, D. B. (2017). Emotion processing and the antidepressant effects of ayahuasca in treatment-resistant depression: An fMRI study. *Os efeitos antidepressivos da ayahuasca, suas bases neurais e relação com a experiência psicodélica, 94.*

Palhano-Fontes, F., Barreto, D., Onias, H., Andrade, K. C., Novaes, M. M., Pessoa, J. A., Mota-Rolim, S. A., Osorio, F. L., Sanches, R., Dos Santos, R. G., Tofoli, L. F., de Oliveira Silveira, G., Yonamine, M., Riba, J., Santos, F. R., Silva-Junior, A. A., Alchieri, J. C., Galvao-Coelho, N. L., Lobao-Soares, B., Hallak, J. E. C., Arcoverde, E., Maia-de-Oliveira, J. P., & Araujo, D. B. (2019). Rapid antidepressant effects of the psychedelic ayahuasca in treatment-resistant depression: A randomized placebo-controlled trial. *Psychological Medicine, 49*(4), 655–663. https://doi.org/10.1017/s0033291718001356.

Palhano-Fontes, F., Mota-Rolim, S. A., Lobao-Soares, B., Galvao-Coelho, N. L., Maia-de-Oliveira, J. P., & Araujo, D. B. (2021). New evidence of the antidepressant effects of ayahuasca. In B. C. Labate & C. Cavnar (Eds.), *Ayahuasca healing and science*, in this volume.

Parlar, M., Frewen, P., Nazarov, A., Oremus, C., MacQueen, G., Lanius, R., & McKinnon, M. C. (2014). Alterations in empathic responding among women with posttraumatic stress disorder associated with childhood trauma. *Brain and Behavior: A Cognitive Neuroscience Perspective, 4*(3), 381–389. https://doi.org/10.1002/brb3.215.

Pokorny, T., Preller, K. H., Kometer, M., Dziobek, I., & Vollenweider, F. X. (2017). Effect of psilocybin on empathy and moral decision-making. *International Journal of Neuropsychopharmacology, 20*(9), 747–757. https://doi.org/10.1093/ijnp/pyx047.

Preller, K., Pokorny, T., Krähenmann, R., Dziobek, I., Stämpfli, P., & Vollenweider, F. (2015). The effect of 5-HT2A/1a agonist treatment on social cognition, empathy, and social decision-making. *European Psychiatry, 30*, 22.

Prochazkova, L., Lippelt, D. P., Colzato, L. S., Kuchar, M., Sjoerds, Z., & Hommel, B. (2018). Exploring the effect of microdosing psychedelics on creativity in an open-label natural setting. *Psychopharmacology, 235*(12), 3401–3413.

Riba, J., Rodriguez-Fornells, A., Urbano, G., Morte, A., Antonijoan, R., Montero, M., Callaway, J. C., & Barbanoj, M. J. (2001). Subjective effects and tolerability of the south American psychoactive beverage ayahuasca in healthy volunteers. *Psychopharmacology, 154*(1), 85–95.

Riba, J., Valle, M., Urbano, G., Yritia, M., Morte, A., & Barbanoj, M. J. (2003). Human pharmacology of ayahuasca: Subjective and cardiovascular effects, monoamine metabolite excretion, and pharmacokinetics. *The Journal of Pharmacology and Experimental Therapeutics, 306*(1), 73–83. https://doi.org/10.1124/jpet.103.049882.

Riba, J., Romero, S., Grasa, E., Mena, E., Carrio, I., & Barbanoj, M. J. (2006). Increased frontal and paralimbic activation following ayahuasca, the pan-Amazonian inebriant. *Psychopharmacology, 186*(1), 93–98. https://doi.org/10.1007/s00213-006-0358-7.

Roemer, L., Williston, S. K., & Rollins, L. G. (2015). Mindfulness and emotion regulation. *Current Opinion in Psychology, 3*, 52–57.

Runco, M. A., & Acar, S. (2012). Divergent thinking as an indicator of creative potential. *Creativity Research Journal, 24*(1), 66–75. https://doi.org/10.1080/10400419.2012.652929.

Sampedro, F., de la Fuente Revenga, M., Valle, M., Roberto, N., Domínguez-Clavé, E., Elices, M., Luna, L. E., Crippa, J. A. S., Hallak, J. E., & de Araujo, D. B. (2017). Assessing the psychedelic "after-glow" in ayahuasca users: Post-acute neurometabolic and functional connectivity changes are associated with enhanced mindfulness capacities. *International Journal of Neuropsychopharmacology, 20*(9), 698–711. https://doi.org/10.1093/ijnp/pyx036.

Sanches, R. F., de Lima Osorio, F., Dos Santos, R. G., Macedo, L. R., Maia-de-Oliveira, J. P., Wichert-Ana, L., de Araujo, D. B., Riba, J., Crippa, J. A., & Hallak, J. E. (2016). Antidepressant effects of a single dose of ayahuasca in patients with recurrent depression: A SPECT study. *Journal of Clinical Psychopharmacology, 36*(1), 77–81. https://doi.org/10.1097/jcp.0000000000000436.

Santos, R. G., Landeira-Fernandez, J., Strassman, R. J., Motta, V., & Cruz, A. P. (2007). Effects of ayahuasca on psychometric measures of anxiety, panic-like and hopelessness in Santo Daime members. *Journal of Ethnopharmacology, 112*(3), 507–513. https://doi.org/10.1016/j. jep.2007.04.012.

Schipper, M., & Petermann, F. (2013). Relating empathy and emotion regulation: Do deficits in empathy trigger emotion dysregulation? *Social Neuroscience, 8*(1), 101–107. https://doi.org/1 0.1080/17470919.2012.761650.

Schmid, Y., Hysek, C. M., Simmler, L. D., Crockett, M. J., Quednow, B. B., & Liechti, M. E. (2014). Differential effects of MDMA and methylphenidate on social cognition. *Journal of Psychopharmacology, 28*(9), 847–856. https://doi.org/10.1177/0269881114542454.

Schultes, R. G., & Hofmann, A. (1979). *Plants of the gods: Origins of hallucinogenic use.* New York: McGraw-Hill.

Sessa, B. (2008). Is it time to revisit the role of psychedelic drugs in enhancing human creativity? *Journal of Psychopharmacology, 22*(8), 821–827. https://doi.org/10.1177/0269881108091597.

Shanon, B. (2000). Ayahuasca and creativity. *MAPS Newsletter, 10*(3), 18–19. https://maps.org/ news-letters/v10n3/10318sha.pdf.

Shewan, D., Dalgarno, P., & Reith, G. (2000). Perceived risk and risk reduction among ecstasy users: The role of drug, set, and setting. *International Journal of Drug Policy, 10*(6), 431–453.

Sloan, D. M. (2004). Emotion regulation in action: Emotional reactivity in experiential avoidance. *Behaviour Research and Therapy, 42*(11), 1257–1270. https://doi.org/10.1016/j. brat.2003.08.006.

Smith, R. L., Canton, H., Barrett, R. J., & Sanders-Bush, E. (1998). Agonist properties of N,N-dimethyltryptamine at serotonin 5-HT2A and 5-HT2C receptors. *Pharmacology, Biochemistry, and Behavior, 61*(3), 323–330.

Soler, J., Elices, M., Franquesa, A., Barker, S., Friedlander, P., Feilding, A., Pascual, J. C., & Riba, J. (2016). Exploring the therapeutic potential of ayahuasca: Acute intake increases mindfulness-related capacities. *Psychopharmacology (Berl), 233*(5), 823–829. https://doi. org/10.1007/s00213-015-4162-0.

Tagliazucchi, E., Carhart-Harris, R. L., Leech, R., Nutt, D., & Chialvo, D. R. (2014). Enhanced repertoire of brain dynamical states during the psychedelic experience. *Human Brain Mapping, 35*(11), 5442–5456. https://doi.org/10.1002/hbm.22562.

Thomas, G., Lucas, P., Capler, N. R., Tupper, K. W., & Martin, G. (2013). Ayahuasca-assisted therapy for addiction: Results from a preliminary observational study in Canada. *Current Drug Abuse Reviews, 6*(1), 30–42.

Todd, A. R., Forstmann, M., Burgmer, P., Brooks, A. W., & Galinsky, A. D. (2015). Anxious and egocentric: How specific emotions influence perspective taking. *Journal of Experimental Psychology. General, 144*(2), 374–391. https://doi.org/10.1037/xge0000048.

Tull, M. T., Barrett, H. M., McMillan, E. S., & Roemer, L. (2007). A preliminary investigation of the relationship between emotion regulation difficulties and posttraumatic stress symptoms. *Behavior Therapy, 38*(3), 303–313. https://doi.org/10.1016/j.beth.2006.10.001.

Tupper, K. W. (2009). Ayahuasca healing beyond the Amazon: The globalization of a traditional indigenous entheogenic practice. *Global Networks, 9*(1), 117–136. https://doi. org/10.1111/j.1471-0374.2009.00245.x.

Uthaug, M. V., van Oorsouw, K., Kuypers, K. P. C., van Boxtel, M., Broers, N. J., Mason, N. L., Toennes, S. W., Riba, J., & Ramaekers, J. G. (2018). Sub-acute and long-term effects of ayahuasca on affect and cognitive thinking style and their association with ego dissolution. *Psychopharmacology, 235*(10): 2979–2989.

Uthaug, M. V., Lancelotta, R., van Oorsouw, K., Kuypers, K., Mason, N., Rak, J., Šuláková, A., Jurok, R., Maryška, M., & Kuchař, M. (2019). A single inhalation of vapor from dried toad secretion containing 5-methoxy-N, N-dimethyltryptamine (5-MeO-DMT) in a naturalistic setting is related to sustained enhancement of satisfaction with life, mindfulness-related capacities, and a decrement of psychopathological symptoms. *Psychopharmacology (Berl), 236*(9):2653–2666.

Visted, E., Sørensen, L., Osnes, B., Svendsen, J. L., Binder, P.-E., & Schanche, E. (2017). The association between self-reported difficulties in emotion regulation and heart rate variability: The salient role of not accepting negative emotions. *Frontiers in Psychology, 8*, 328. https://doi.org/10.3389/fpsyg.2017.00328.

Vollenweider, F. X., & Kometer, M. (2010). The neurobiology of psychedelic drugs: Implications for the treatment of mood disorders. *Nature Reviews. Neuroscience, 11*(9), 642–651. https://doi.org/10.1038/nrn2884.

Williams, D. P., Cash, C., Rankin, C., Bernardi, A., Koenig, J., & Thayer, J. F. (2015). Resting heart rate variability predicts self-reported difficulties in emotion regulation: A focus on different facets of emotion regulation. *Frontiers in Psychology, 6*, 261. https://doi.org/10.3389/fpsyg.2015.00261.

Winkelman, M. (2005). Drug tourism or spiritual healing? Ayahuasca seekers in Amazonia. *Journal of Psychoactive Drugs, 37*(2), 209–218. https://doi.org/10.1080/02791072.2005.10399803.

Zeifman, R., Palhano-Fontes, F., Hallak, J., Nunes, E. A., Maia-de-Oliveira, J. P., & de Araujo, D. B. (2019). The impact of ayahuasca on suicidality: Results from a randomized controlled trial. *Frontiers in Pharmacology, 10*, 1325.

Chapter 8
Healing with the Brew: Ayahuasca's Reconfiguration of "Addiction"

Piera Talin

Introduction

In this chapter, I provide an ethnographically grounded, qualitative analysis of the experiences of people recovering from substance dependence with the ritual use of ayahuasca. A range of clinical, biomedical, and ethnographic studies have suggested that the psychoactive Amazonian brew ayahuasca plays a positive role in overcoming a range of mental health issues, such as substance dependence, anxiety, and depression. A growing corpus of biomedical studies—many of them reported in this book—examining the pharmacological effects of ayahuasca seems to give credence to anthropological research that has long recorded its therapeutic uses in traditional settings.

Ayahuasca is a shamanic herbal brew, most commonly composed of the admixture of the native Amazonian *Banisteriopsis caapi* vine and the leaves of the *Psychotria viridis* shrub. Over the course of the twentieth century, ayahuasca was integrated into a range of syncretic Christian religious practices, giving rise, in Brazil, to three religions: Santo Daime, União do Vegetal, and Barquinha (Labate and MacRae 2010). These spread alongside neoshamanic uses to Latin American urban centers and were subsequently "globalized" (Tupper 2008). In this process, ayahuasca was translated, so to speak, to fulfill new and distinct objectives, such as therapeutic ones (Labate and Cavnar 2014). New ceremonialized (Barbira Freedman 2014) ritual ayahuasca forms emerged, which seek to harness and manage the powerfully evocative, visionary, and emetic experience to a range of therapeutic ends.

The existing literature on ayahuasca and addiction shows that physiological and psychological mechanisms are deeply enmeshed (Bouso and Riba 2014; Fernández and Fábregas 2014; Talin and Sanabria 2017; Thomas et al. 2013). I argue that psychedelic substances such as ayahuasca are very peculiar pharmacological tools

P. Talin (✉)
CERMES3 (Center de recherche, médecine, sciences, santé, santé mentale, société, UMR 8211), Villejuif, France
e-mail: piera.talin@cnrs.com

© Springer Nature Switzerland AG 2021   137
B. C. Labate, C. Cavnar (eds.), *Ayahuasca Healing and Science*,
https://doi.org/10.1007/978-3-030-55688-4_8

whose value needs to be made sense of within specific ecologies of use and care that are not yet easily comprehensible within the biomedical paradigm. When assessing the use of ayahuasca treatments for addiction, the risk is that addiction is taken for granted as primarily a biological problem and that emphasis is placed mainly on the pharmacological efficacy of ayahuasca. There is, indeed, a lot to know about the pharmacokinetics of ayahuasca and its possible mechanisms of action. However, I argue that these effects can only be rendered fully meaningful within certain settings. My approach thus echoes Mesturini Cappo's (2018), who suggests that we need to view ayahuasca as "entangled," that is, as inextricably caught up in relations that go beyond the human and beyond any narrowly defined understanding of its agency. Mesturini Cappo reminds us of the colonial and postcolonial rationale at work in efforts to reduce ayahuasca's efficacy to its biochemical action. My ethnographic data point to some key extra-pharmacological elements that I believe need to be assessed in equal measure to pharmacological mechanisms. I argue that the success of such treatments cannot be reduced to either the pharmacological effect of ayahuasca or the ritual. Indeed, one without the other is very unlikely to be effective at all. My goal in this chapter is to pave some further ground toward accounting for the potent interfusion of these distinct elements. I adopt the term *curar* (to cure or heal) that my informants use to signify a process that involves more than abstinence. This often involves a rescripting of the past or a transformation of people's understanding and experience of substance use. Addiction itself, as a category, unraveled in the experience of its healing in the context of ceremonial ayahuasca use.

The healing that I present here emerges out of a loose assemblage of empirical, case-by-case assessments within structured networks of support. This loose assemblage gives the ritual space an iterative and reflexive dimension, thereby enabling a form of care that is attentive to the specificity of different situations and contexts. This contrasts, in my informants' perspectives and narratives, with the highly structured approaches that characterize standard addiction treatment programs. The practices of care and networks of support that are provided in and around the regular ritual practices of ayahuasca religious groups I study do not assume to know addiction in some generalizable way. If anything, they bring about an increased awareness of the specificity of each situation and of each person's particular circumstances.

In studying how ayahuasca healing for substance dependence works, I uncovered some of the normativities that people struggling with substance dependence encountered in mainstream addiction treatment centers. This brings me to think in terms of fluctuations and stabilizations of habit rather than in terms of addiction as irreversible rigid essence (Fraser 2016).

Methods: Situating My Study

In this chapter, I focus on the recovery trajectories of persons struggling with substance dependence—ranging from alcohol, tobacco, and antidepressants to crack cocaine and heroin—who find support in overcoming their "addiction" in regular ayahuasca churches or spiritual groups (i.e., outside of ayahuasca centers that have developed specific protocols dedicated to helping people with substance dependence). There is an important literature on addiction recovery with ayahuasca (Labate and Cavnar 2014; Loizaga-Velder and Verres 2014; Mabit and Martin 2007; Mercante 2013; Ramos Gomes 2016), but much remains to be investigated, particularly as claims about ayahuasca's efficacy in addiction recovery become more mainstream and new experimental practices are developed in Latin America and beyond. I draw on observations and interviews with ritual experts and participants who give or receive support to those with substance dependence histories within the context of ayahuasca ritual practice. These are structured ritual spaces whose purpose and function span well beyond healing. While they are not explicitly promoted as healing spaces, participants may be drawn to them because they have heard of the therapeutic benefits attributed to ayahuasca.

The multisited ethnographic research on which this chapter is based was carried out among Italian and Brazilian chapters of the Santo Daime Church and among urban Brazilian spiritual communities that emerged from, but have broken with, the original ayahuasca religions, groups often referred to as "neoayahuasquero" in the literature. In Italy, I conducted ethnographic research among the two largest Santo Daime groups associated with the ICEFLU, with from 30 to 80 ritual participants, and two smaller ones, with about 20 participants, in northern and central Italy. One of these groups is an established community where people live together and have a dedicated ritual space, while the others are networks of participants who live in different places and meet regularly for the rituals and other activities. Ethnographic work was also conducted in the Céu do Mapiá community in the Brazilian state of Amazonas, one of the largest Santo Daime communities, founded in 1983 by Padrinho Sebastião Mota de Melo. This community of roughly 1000 inhabitants stands apart from the others I studied by its size and isolation. It is in the Purus National Park, a conservation zone, and organized around an ideal of spiritual and ecological community life.

Methodologically, the research draws on in-depth narrative interviews, exploring the beliefs and expectations surrounding drug use and practical experimentation, and on ethnographic research using long-term fieldwork and participant observation within ayahuasca communities.

The narratives of people who overcame substance dependence through the ritual use of ayahuasca were collected, recorded, transcribed, and systematically analyzed. In-depth interviews were conducted with ritual experts and with a physician and a psychologist working at a public center for addiction treatment. Finally, in-depth interviews were also conducted with community members responsible for "holding space" during ayahuasca rituals; these individuals spoke about witnessing the recovery processes of people struggling with addiction or about their own recovery.

Based on the rituals I observed, ceremonies involving ayahuasca tend to be performed after dark, and participants are invited to wear white clothing. New participants are introduced to the ritual after preliminary contacts and screening have been carried out with the group's leadership. Ayahuasca doses are adapted to each person and carefully evaluated on a case-by-case basis. Inside the large *salão* (ritual space), chairs are positioned around a central table on which ritual objects have been carefully positioned, including flowers, candles, and icons. To one side, a table covered with a white cloth is set up, from which the ayahuasca brew or Daime sacrament is served. Designated helpers carefully assist each participant to their place and "hold the space," tending to practical matters and helping where needed. The ceremony begins with prayers, songs, or hymns that open a *trabalho* (ritual "work") that will last 4–6 h.

The Material-Semiotic Efficacies of Ayahuasca

Pharmaceuticals, drugs, and herbal remedies have pharmacological, or material, properties and effects as well as social and symbolic ones (Barry 2005; Gomart 2002; Hardon and Sanabria 2017; Sanabria 2016). The two are indissociable. Such substances affect bodies and social identities simultaneously. They are political and pharmacological agents. Anthropological work on drugs has shown that the social setting and cultural and individual expectations that shape the consumption of a substance deeply shape the substance's pharmacological efficacy. It is, in fact, rather nonsensical, from many traditional and indigenous perspectives, to differentiate the two, as Western science attempts to through randomized clinical trials. For this reason, feminist philosopher Wilson (2015) advocates transcending approaches that focus on either pharmacology or meaning and other dichotomies that still plague our understandings of pharmaceuticals and drugs.

Indeed, ayahuasca does not function therapeutically in quite the same way as most of the drugs in the Western pharmacopeia are understood to function (and are evaluated). That is to say, it requires a whole system of support to make it efficacious. As Tupper (2008) has shown, contemporary ayahuasca drinking practices are framed by a global drug policy regime predicated on scientific materialism that explains drug efficacy solely by reference to biochemistry and psychopharmacology. While this captures the important psychopharmacological properties of the brew, it misses an equally interesting variety of phenomena. These concern the unique potential of substances such as ayahuasca to catalyze powerful ritual experiences into lasting, durable change. As Loizaga-Velder and Verres (2014) note, this potential is dependent on such experiences being appropriately managed and integrated through trained guidance that helps direct the psychological dynamics that ayahuasca instigates. In what follows, I review the potential of structured ceremony in providing such support, calling attention to the shared meaning that is constructed through collective ritual participation. I do so by drawing on a range of theoretical approaches in social sciences and anthropologies of drugs to explore the interesting zone where meaning and matter, symbolism, and substance intermingle and reconfigure habits, experiences, and understandings. A better comprehension and qualification of the events that take shape in this zone of intermingling is key, I suggest, to grasping the complex, potent, ambiguous, and promising therapeutic potential of ayahuasca.

Drawing on Wilson's insights, I ask about the serotonergic effects of rituals that are afforded to individuals seeking support in their recovery trajectories. This approach invites us to embark on conceptual and empirical work that does not replicate existing distinctions within social studies of healing between mind and body, pharmacology and culture, etc. that do little to elucidate the complex intertwined trajectories of healing.

Drawing on material-semiotic analyses of how drugs are made to matter and rendered efficacious, I consider in what follows how ayahuasca can catalyze cathartic transformations under certain conditions (Menozzi 2011). The scripting of the experience—the way the effects of ayahuasca can be made to concretize lasting changes under adequate guidance—is a process that challenges existing conceptual vocabularies for the effects of interventions that, as we have seen, tend to ascribe efficacy to either substance or context. The extensive anthropological literature on the performative efficacy of ritual has examined the role of sound, smell, kinesthesia, and the use of ritual objects in making rituals work (Calabrese 2013; Csordas 1988; Sax 2004; Tambiah 1977). This literature points to ways in which bodies and the human sensorium are transformed by ritual participation. The work on psychedelic rituals shows the critical and tangible importance of scripts, such as trip reports (Doyle 2011) or music (Brabec De Mori 2012; Kaelen et al. 2015), in shaping and giving form to the experience. Brazilian anthropologist Mercante (2013) writes of addiction curing with ayahuasca as a long process of social reintegration. Ayahuasca rituals enable a "directing" of the effects of the brew; they circumscribe a field of possibilities, containing the experience and providing tools to render it meaningful. This process blends the somatic, symbolic, and collective dimensions of the experience. The layering of effects is what makes such interventions so powerfully cathartic and catalytic for those who experience them.

I suggest that focusing on the ambiguous, dense, material-semiotic space where conceptual distinctions between matter and signification collapse is a necessary step to complement existing accounts of ayahuasca's efficacy. This allows me to better qualify, alongside social theorists Gomart and Hennion (1999), not "what acts" but rather "what occurs." Pharmacological type explorations—or "what acts" questions—place agency in the thing in a causal sequence. By contrast, the emphasis on "what occurs" enables us to focus on what emerges in the unfolding of an event, such as a ceremony. That is, it brings the focus to the space of potentiality between pure agentic subjectivity and pure chemical determination (Gomart and Hennion 1999). Carlo, one of my informants, reflects on this deep intermingling of the chemical and the ritual in the following interview extract:

> In the days after a trabalho [ritual work], it requires much less effort to reduce the methadone; you can halve the methadone dose in a few days. *The stronger the trabalho is, the easier it is.* … When the work is strong, the ayahuasca has this pharmacological effect and improves your condition.

In this person's experience, the intensity of the ritual directly affects the experience of withdrawal, making it more manageable. The reference to "work," here, suggests that the intensity of the experience is not about the intensity of the dose but rather of the quality of the ritual experience that the person had.

Francesco, who had a long career as a drug user and as a successful musician, described his first experience of ayahuasca. He uses terms that mingle the biological and the symbolic, a description perhaps shaped by his religious background:

> I was asked to try not to take drugs for at least twelve hours before the ceremony. I was in the train with my heart beating, the withdrawal symptoms beginning, emotions, nervousness. This is what the "medicine of the forest" [ayahuasca] found when it came into my body. I was sweating and shaking. They put me in the shower, they gave me a pair of white pants. I sat there with all my prejudices: "Who are these people?" Then I let myself go, I said, "okay, let's try." After fifteen minutes, I see this strange light in the center of the table, after twenty minutes it became a lighthouse and then the rest is something completely indescribable. The incredible result is that since that day, almost five years ago, I have only taken two painkillers for back pain. No other drugs whatsoever. … The strongest feeling I can describe is that I didn't understand anything, but it was like being washed from the inside with Christ's tears. I say it in the religious sense. Like transparent water, or blood, or something that cleanses. I felt sanded from the inside. In fact, in my opinion it reset my body and capacity to feel, so much that it simply cancelled the craving. … The Daime reset all my sensations, it washed and cleaned the receptors from all the substances I abused. … The Daime completely erases the desire of the body and the mind. The body doesn't want it [heroin] anymore. Even when I felt the desire, the idea, then I felt a sensation of nausea, a poisoning. Then, in the following months, there isn't only the physical care or social re-education, there is also a resolution of the deepest suffering, the questions of life, which are the ones that you carry forever.

Many of the people I had the opportunity to speak with about how partaking in ayahuasca ceremonies helped them overcome long struggles with substance dependence relayed similar instances of ineffability in attempting to capture "what occurs": the efficacy of the ritual blends both materiality and potent symbolism.

A further aspect of this material-semiotic dimension of the efficacy of ayahuasca ceremonies concerns the question of purging. The term *limpeza* (purge or cleanse) used in the ayahuasca rituals I studied collapses the material and symbolic domains. Wagner, who has been officiating ayahuasca ceremonies within the Santo Daime tradition for several decades, explains purging as a deeply psychosomatic process: "people are expelling attachments, behaviors, or negative emotions through the catharsis of the body." Marco is 45 years old and began using cocaine and heroin at the age of 22. After several attempts, he realized he could not stop and began using methadone until 2007, when he had his first experience with Daime and stopped using drugs altogether:

> At that time, I used to take methadone in syrup. I was very intoxicated and very ill. This session took place over two days. The first day was a real ordeal, I felt that I had to throw up, but I couldn't and so I was in this limbo. I had no revelations, it was just suffering, and after the first day, I wanted to go home. "If you stay, you might have a chance of opening to something new. If you leave, you know exactly what will happen." someone told me. So, I stayed. The next day, the ayahuasca was much more gentle with me and I was able to purge.
>
> I threw up things that I never saw again, they were like black snakes or worms. I had the feeling I was vomiting something psychic or spiritual. There was this material support, but

it was connected to something else. I don't know exactly what happened, but in the days after, I was deeply relieved. It was as if a black cloak that enveloped me and all my perceptions was suddenly removed, and so many new things could receive light. Suddenly, this cloak had been removed and I could breathe.

This cleansing process can take different forms, including the catharsis brought about by the flow of tears. For example, Antonia explains: "I didn't have visions and I didn't vomit. Instead, I started crying thankful tears. For a person overburdened with guilt, it was an extraordinary event." Francesco explains that ayahuasca helped him gain a clearer experience of his body: "What I can eat, or not. I received from my use of Daime a method of caring for my body. Daime cleans." Roberto, a long-term heroin user who was in and out of addiction treatment centers for many years, reports that, while the congregation was singing, at the height of a ritual, he had a vision of himself journeying at great speed inside a tunnel:

> I had often asked myself how the twenty years of drug use had modified my physical and mental conditions. How would I be if I had not taken so many drugs? While they were singing, I felt a turmoil inside myself. I felt like my fifteen-year-old self again. I entered a tunnel and no longer felt my body. In the tunnel, I began to smell all the scents, to feel all the effects and taste all the drugs I ever used. I tasted everything, all the heroin and all the cocaine. Heroin, as it was when I started, is not the same as today's heroin. In all this time heroin, and its quality, changed many times, it was grey, then beige, then yellowish, then white. I felt the effect of each change. This experience was so strong for me. When the effects of the vision faded, I felt like a young boy again with the joie de vivre of a young boy. I was happy in my heart and I began to smile. I entered the ritual depressed and came out very happy. I felt better and my eyes were shining. Ayahuasca had wiped everything clean and I felt like a new person, just as if I had never used drugs in my soul, in my body, in the wholeness of myself.

A few days later, he had a relapse and took heroin again. Doing so, however, did not lead him back down the same process of addiction. He explains that this relapse experience enabled him to realize that he did not have the same attachment to heroin: "I needed to understand how disgusting heroin is. I experienced the clear difference between dirty and clean, and realized that I didn't want heroin anymore."

Wilson (2015) notes that 95% of the body's serotonin is found in the neural networks that innervate the gut (Wilson 2015, p. 66). Perhaps it is not surprising, then, that the gut, through the central practice of purging evoked above, is so involved in ayahuasca experiences that pharmacological studies consistently associate it with the serotonergic mesolimbic pathways. Her analysis of the somatic entanglement of affects reminds us that these are always heavily modulated by social, suggestive, spatial, placebo, material, cultural, symbolic, and semiotic events. This imbrication of the mental and enterological upends distinctions between the gut and mood, between vomiting and emotion: "The vicissitudes of ingestion and vomiting are complex thinking enacted organically" (Wilson 2015, p. 63). This invites us—in keeping with the narratives of people who undertake the perilous and dedicated trabalho (work) of healing within ayahuasca ceremonial contexts—to deeply question the categories with which we come to understand both disease and healing.

Ayahuasca Reconfigures Addiction, and Our Understandings of It

Much of the literature on ayahuasca treatment for addiction does not critically engage with the notion of addiction itself, taking it relatively for granted. This can be understood as a strategic requirement for ascertaining the validity of this treatment modality within biomedical frameworks and conceptualizations of addiction. Yet, as social theories of addiction have shown, addiction is a complex, enmeshed, and biosocial process that cannot be reduced to a simple medical fact (Garriott and Raikhel 2015). There are many addictions (Campbell and Lovell 2012; Netherland and Hansen 2017). In my ethnography, which took place entirely outside of specialized addiction treatment centers, I seldom encountered the term "addiction." People spoke of healing problems of *dependência* (being dependent), which had less reductionist connotations than terms such as *vicio* (literally, vice) or addiction. Drawing on their narratives, I aim to pave further ground beyond "denaturalizing" addiction as a disease (Raikhel and Garriott 2013). My suggestion is that ayahuasca reconfigures the very meaning and experience of addiction. The people I encountered had for the most part undergone standard addiction treatments, which they often experienced as highly normative and, at times, coercive. They spoke of these standard treatments as providing tools to manage, rather than recover from, their addictive habits or "patterns," as some referred to them.

Because ayahuasca has a potent pharmacological action, there is a risk of overstating its psychopharmaceutical efficacy. Yet, ayahuasca is no magic bullet. In fact, what is interesting about substances such as ayahuasca is that they explicitly challenge the magic-bullet rationale in biomedicine. Magic bullets are targeted therapeutic agents understood to cure the cause of illness by intervening at the point of pathogenesis. However, the restoration of wellness often defies, complexifies, and troubles such causal understandings of health-making.

What is clear in my informants' narratives is that the substance does not bring about healing alone. The ritual in which it is rendered therapeutic is integral to its effect. For example, Francesco had previously used peyote, which is also known to be effective in treating addiction (Calabrese 2013), but doing so had not brought about recovery from his heroin addiction. He explained this was because: "I used it as a drug," that is, without therapeutic intention. What my findings make clear is that it takes more than drinking ayahuasca for durable changes to take place. While ayahuasca may, pharmacologically speaking, facilitate such changes, these are rendered effective in the longer term within certain contexts and collective structures that enable a reconfiguration of one's self-understanding and insights into one's addictive patterns. Marco spoke of such patterns as "loops":

> Methadone was part of the loop. For two to three months, I used to smoke heroin, then take methadone. Psychologically and spiritually the deeper issues were never addressed. The roots of the problem were still there. It was always a matter of time, an opportunity to restart the loop. I have done fifteen or twenty of these loops, each lasting from two to four months, with more or less the same characteristics. I always felt wrong in what I was doing. That it

wasn't right. … With ayahuasca, I had this view from the top. I saw how everything was part of the loop of addiction. … The Daime showed me how these forces of addiction act. I talk about it as if they are conscious, because that's how I perceive them. These forces aren't interested so much in the substance, but to what we bring out when we are addicted to something. If you take away their support, they find other supports, and in the world in which we live now, there can be so many different supports. People who haven't had problems with substances are dependent on a lot of other things, unaware that these things are a support for these forces of addiction to act in them. These destructive effects may not be as striking as with heroin, perhaps. But addiction can prove harmful in many more ways.

I had the opportunity to meet again with Roberto and Carlo, two interviewees, several months after their first extended interview. The collective discussion took place during a Santo Daime event and included a psychiatrist who had earlier treated Roberto in a residential addiction treatment center in Italy. The doctor qualified methadone as "a wonderful medicine that corrects the abstinence from heroin" but recognized that it was also highly addictive, making it, in her words, more "insidious" than heroin. Reflecting on Roberto's experience, and his many relapses, she recognized that the clinic's approach had not worked for him. Given what a difficult case he had been, she stated that she was stunned by how effective ayahuasca had been in his recovery process. Roberto replied:

I tried to decrease the methadone. The clinic's prescription was 140 mg of methadone for several years, and then maybe I could decrease it to 80. For the clinic, there was no problem in taking methadone for the rest of my life. But methadone is as much an enslavement as any heroin addiction. I could not abide by the rules of the residential program because they detracted [sic] me from my will to recover. In the residential program you can't do what you want. They are always controlling you, everywhere, even in the bedrooms. The Santo Daime community where I lived for some time is not a community for addiction treatment, with gates and closed doors, doors with no handles, to stop people getting out at night. There are no prohibitions on meeting with others. The clinic is like a prison. Yes, you choose to go into a residential program, but often it isn't really a choice, because when you arrive, you've hit rock bottom, you are forced to go to such a place. In the Daime community, the rules are self-imposed by the community members. There is a lot of freedom, and I thrived in that freedom. It is a healthy place. It's only 15 km from a big city with one of Italy's largest drug markets, and I never once went there to buy cocaine or heroin. I didn't feel the need nor the desire for the drugs. It was fully my choice to be there. I think that is essential for the quality of recovery.

The people I encountered in my research often had come to ayahuasca rituals after unsuccessful attempts to recover through outpatient clinics or psychiatric residential centers. Most of my interviewees felt out of place in such centers, where they felt stigmatized as socially dysfunctional "junkies." One of my informants joked that in the center through which he had regularly transited, he was told he was an "atypical addict," highlighting the implicit normative construct the institution holds of a "normal" addict. Further, my informants noted that such institutions tend to normalize certain kinds of drugs over others. While they navigated through their recoveries using ayahuasca to titrate off their substitution treatments, for example, they were told that ayahuasca was a "drug" while, at the same time, being prescribed heavy doses of methadone or psychiatric medications that several interviewees highlighted as deeply paradoxical.

Caring for Specificity

For the people I encountered, feeling part of a community is a key aspect of the healing experience. This is consistent with the literature that has shown that regular church attendance has a positive impact on health. For example, studies have found that weekly church attendance can add up to 3 years to one's life (Hall 2006) and that religious observance stimulates immunity and is associated with lower blood pressure (Koenig and Cohen 2002; Woods et al. 1999). These effects may be the product of an increase in supportive social relationships, in exposure to a context favoring healthy behaviors, and in the cultivation of a sense of meaning and coherence. Luhrmann (2013) has argued that the experience of a positive interaction with the supernatural is good for people and may be at the heart of what gives religion health-boosting properties.

This brings us to the final section of this chapter, in which I examine the potential of fluid and adaptable ritual forms as unique forms of caregiving. There is a phenomenal variety of patterns according to which people use drugs and sometimes develop problems (Weinberg 2013). Yet, what transpires from my informants' narratives is that addiction treatment institutions are rarely able to attend to the specificities of people's particular drug use predicaments. In contrast, there are tailored and dynamic care practices in the spaces where ayahuasca is ritually used. I provide a brief ethnographic description of these practices to shed light on their potential therapeutic importance.

During the ritual process itself, there are highly structured forms of caregiving, in which dedicated helpers are responsible for attending to the possible needs of participants. They are specially trained to provide noninvasive supervision of the experience to allow each participant to safely experience his or her cathartic process. Their role is characterized by minimal physical and verbal contact, but they remain at hand to help should it be needed (such as providing a place to rest, some water, a tissue, or a place to purge). Ritual leaders pay great attention to how such care is provided, so that it is neither intrusive nor neglectful. For many of the people I interviewed, the experience of partaking in a collective process of this nature, of receiving this kind of nondirective care, was, in and of itself, deeply transformative. The conducting of the ritual—particularly in emergent practices such as those that break with the highly formalized ayahuasca religions—is iterative, as ritual leaders are highly attuned to what is happening in the congregation and can adapt and respond accordingly. For example, if someone has a difficult experience during a ritual, a specific hymn may be "activated." As one ritual leader explained, hymns have "vibrational influences that connect people with their own powerful mechanism of self-healing." While rituals are highly structured, there is also space to adjust to what is unfolding.

This brings us to reflect on what "good care" could mean in the treatment of addiction. Mol et al. (2010, p. 14) define good care as a "persistent tinkering in a world full of complex ambivalence and shifting tensions." Their point is that care is fundamentally contextual. The theoretical concern with care is itself a method; it requires a sustained detailed ethnographic attention to the subtleties of caring. Such practices are necessarily local, responsive, and heterogeneous, bringing together complexities and frictions that resist universal principles. The practice-based approach advocated by Mol and colleagues emphasizes the reflexive and experimental nature of care that involves "persistent tinkering," making care a perpetual endeavor, one that is never finished and always in the making. Such tinkering blurs the boundaries of the objects configured through care. It is my suggestion that the tinkering that operates in ayahuasca rituals reconfigures (for the people experiencing healing) the notion and contours of what addiction is. I argue that it is precisely by making these normative contours unstable, by caring differently, that such ayahuasca rituals can come to be experienced as effective.

The community of Mapiá has a long experience addressing substance dependence or *vício* as people in this context usually refer to problematic uses of drugs and alcohol. Also, for this reason, in 2002, Spanish psychiatrist Josep Maria Fabregas established an ayahuasca addiction treatment center close to Mapiá, from which many collaborations were established (Fernández and Fábregas 2014). The program of the clinic was based on different forms of psychotherapy and the participation of the patients to the Santo Daime rituals. The clinic has been closed for several years, but Céu do Mapiá still draws people with complex substance dependence trajectories for healing.

In Mapiá, there are different places dedicated to healing, including the Santa Casa Padrinho Manoel Corrente. At the Santa Casa, patients are attended to with a variety of herbal medicines, acupuncture, and physical therapy. The center provides overnight monitoring of patients as well as healing "works" with ayahuasca. It provides different forms of support: a person may be housed in the Santa Casa itself or in a neighboring home through informal arrangements loosely overseen by the ritual leaders, an example of affinities that may arise between different elements. When payment is not an option, a person contributes to the upkeep and development of the home or space with their time and labor, which is understood to be an integral part of recovery. For example, Renato, who was recovering from crack addiction, had been hosted in a little house next to the Santa Casa where he took care of the garden. He developed close bonds with one of the older families, oscillating between living in this family house and returning to the little house next to the Santa Casa: "I live here, I live there, it depends. People here give me freedom to be in peace." What I wish to emphasize with these brief ethnographic descriptions is that there is no a priori spatial or social differentiation between people in recovery, people passing through, and people living in the community permanently.

In the more than 30 years since its foundation, the community of Céu do Mapiá has experienced several important changes. Some of the changes reported by the population of the community include when the first television arrived after more than 10 years, when the founder Padrinho Sebastião Mota de Melo died, when the quality of life significantly increased with the diffusion of the Santo Daime in the rest of the world, and when the growth of the population started to include new inhabitants not following the religion and its intense calendar. Along the years, the presence of addiction and people in search of healing also shaped the nature of

Mapiá. Each specific situation and person that found their way to Mapiá participated in shaping the way the community and the ritual leaders attending to the work
of healing approached the problem and tinkered with the creation of different
responses.

The community itself is undergoing a profound generational transition, as youth
turn away from the doctrine and experiment with alcohol and drugs. In 2009, a new
terreiro was opened with a specific focus on supporting young people who are
"abusing" alcohol, as the terreiro leader referred to it. His narrative blends references to the "spirits," or traces of past conflicts, of the indigenous people who were
killed on the land and of the importance of the structure that is provided by the ritual
calendar and behavioral prescriptions of the Santo Daime.

Elsewhere, I have argued that a sense of belonging is generated through the collective labor of preparing and participating in ritual, and this, together with the sharing of ayahuasca, works to produce a sense of "spiritual kinship" (2013). Among
ayahuasca congregations, the community is often referred to with the kinship term
irmandade (brotherhood), evocative of the intimate nature of the ties and mutual
obligations created in such contexts. In this chapter, I wish to highlight the fact that
"community" is not limited to humans but includes a broader ecology of beings (De
la Cadena 2015). This lively web of beings is not one that has undergone the total-
izing erasures of colonial ecologies, as Myers (2017) eloquently puts it. Plants,
animals, and their spirits, and the spirits that live in the landscape and forest, are part
of this broadly conceived community. Doyle (2011) coined the term "ecodelic" to
refer to a signature aspect of psychedelic experiences, namely, their capacity to
bring about an understanding that we are all part of a "densely interconnected eco-
system" (p. 20).

This experience is particularly strong in Brazil, where the rituals I observed were
on the outskirts of cities, in spaces privileging a relationship to natural features such
as the beach, or vegetation groves. For many people who travel to Mapiá, the immer-
sion in this Amazonian landscape is a key feature of their therapeutic journey.
Within the Santo Daime doctrine, the brew is considered a sentient being in addition
to the trees, flowers, animals, birds, insects, wind, water, and sky. The fact that the
brew comes from the Amazonian forest is not incidental for many of those who
experience it there, and its efficacy derives from the everyday practices and activities
in this ecological web where it is grown, harvested, and consumed. For example,
Alfonso says: "I drink the brew on the edge of the *igarapé* to observe a snake, to
listen to the jaguar in the woods. I become enchanted and am taken to places that do
not exist on this earth, only in ayahuasca, with this song of the jaguar and the mon-
keys, the birds, to travel to the other world. I have a strong connection to the
eagles here."

Feliciano grew up in a Latin American capital and began using alcohol and
cocaine when he was young. He first encountered the Santo Daime in his home city
and found the experience helpful in managing his cocaine addiction. He traveled to
Brazil and made a first visit to Mapiá, but, returning to his home city, he was found
to be re-engaged with his addictive patterns. Eventually, he moved to Mapiá, where
he helps others who are struggling with addiction. In his narrative, life in the city,

especially for the poor and more vulnerable, is violent and alienating, disconnected from nature. He speaks of the community of Mapiá as a refuge, not just from drugs, but from the intensity of urban life that can lead to drugs. This community, founded on an ideal of sustainability, a simpler pace, and a direct contact with the means of existence, gives meaning and support to his life.

Conclusion

Fraser (2016) argues that addiction experts generally do not conceive of addiction as a unitary disease that can be addressed by narrowly conceived medical responses; nevertheless, such ideas endure and continue to garner support in the absence of strategic alternatives. She suggests that, among many addiction service providers, "there is little faith in addiction as a unitary coherent phenomenon that can be readily addressed by dedicated narrowly conceived responses, yet this idea continues to be promulgated because strategic alternatives are absent. *From where might these alternatives emerge?*" (2016, p. 14 emphasis added). My goal has been to highlight an alternative approach to addiction that is based on a radically different understanding of the therapeutic substances' efficacy and which reconfigures understandings and experiences of "addiction."

I pointed to the conceptual challenges biomedical studies of ayahuasca face in making sense of the complex, dynamic, and amplified efficacy of psychedelic-assisted interventions within an epistemology that presumes a radical distinction between the "pharmaceutical" and the "social." Within such an epistemology, it becomes difficult to account for the specific, situated, and contextual efficacy that healers witness and patients experience. In presenting the situated, specific, and iterative dimensions of care practiced in ritual settings, my aim is to show that standardization according to universal or externally determined norms can inhibit the potential of such rituals to respond dynamically to what a situation requires. Supporting such a dynamic of care and making it legible within current evidentiary norms presents a considerable challenge. It invites us to collectively move beyond a discussion about whether psychedelic-assisted interventions are effective in the treatment of addiction and to ask instead for whom and under what conditions they can be.

The ritual uses of ayahuasca to heal addiction that I studied are based on practices of care enacted during the ayahuasca rituals that extend into everyday life via the dense networks that connect people within the communities organized around the ritual use of ayahuasca. Here, I want to suggest that this social component of ayahuasca ritual experience is fundamental to understand its healing efficacy. Hence, to return to Fraser's important question, the alternative I have presented is not simply one based on the pharmacological properties of ayahuasca but on the *ritual* use of ayahuasca, which gives the substance its social, temporal, affective, and existential qualities. In this sense, ayahuasca ritual use for addiction healing is more than the treatment of illness. It consists in the active participation and building of different habits in a person's world. Such healing trajectories are based on dynamic and situated forms of care and space holding, whose aims are, in a sense, much broader than overcoming addictive patterns, seeking instead to support people in giving meaning to their lives. What is noteworthy is the way such groups adapt and respond to the specificities of people's recovery trajectories. Thus, while methadone substitution treatments are not common in Brazil, in the European context, people are experimenting with different means of gradually coming off methadone, with the ritual use of ayahuasca as a further step in their recovery process.

Acknowledgments I am particularly grateful to those who shared their stories with me and to Emilia Sanabria for the inestimable collaboration and support until the final stage of this chapter. Thanks also to Anita Hardon, Bia Labate, Mariana Rios, and Emily Yates-Doerr for their precious feedback on earlier versions of this chapter. This research was made possible, thanks to generous funding from the European Research Council Starting Grant n°757589 "Healing Encounters: Reinventing an indigenous medicine in the clinic and beyond" based at CERMES3 (Université de Paris, EHESS, CNRS).

References

Barbira Freedman, F. (2014). Shamans' networks in Western Amazonia: The Iquitos-Nauta Road. In B. C. Labate & C. Cavnar (Eds.), *Ayahuasca shamanism in the Amazon and beyond* (pp. 130–158). Oxford: Oxford University Press.

Barry, A. (2005). Pharmaceutical matters: The invention of informed materials. *Theory, Culture & Society, 22*(1), 51–69.

Bouso, J. C., & Riba, J. (2014). Ayahuasca and the treatment of drug addiction. In B. C. Labate & C. Cavnar (Eds.), *The therapeutic use of ayahuasca* (pp. 95–110). Berlin: Springer.

Brabec De Mori, B. (2012). About magical singing, sonic perspectives, ambient multinatures, and the conscious experience. *Indiana, 29,* 73–101.

Calabrese, J. (2013). *A different medicine: Postcolonial healing in the Native American Church.* Oxford: Oxford University Press.

Campbell, N. D., & Lovell, A. M. (2012). The history of the development of buprenorphine as an addiction therapeutic. *Annals of the New York Academy of Sciences, 1248*(1), 124–139.

Csordas, T. J. (1988). Elements of charismatic persuasion and healing. *Medical Anthropology Quarterly, 2*(2), 121–142.

De la Cadena, M. (2015). *Earth beings: Ecologies of practice across Andean worlds.* Durham: Duke University Press.

Doyle, R. (2011). *Darwin's pharmacy: Sex, plants, and the evolution of the noösphere.* Seattle: University of Washington Press.

Fernández, X., & Fábregas, J. M. (2014). Experience of treatment with ayahuasca for drug addiction in the Brazilian Amazon. In B. C. Labate & C. Cavnar (Eds.), *The therapeutic use of ayahuasca* (pp. 161–182). Berlin: Springer.

Fraser, S. (2016). Articulating addiction in alcohol and other drug policy: A multiverse of habits. *International Journal of Drug Policy, 31,* 6–14.

Garriott, W., & Raikhel, E. (2015). Addiction in the making. *Annual Review of Anthropology, 44*(1), 477–491.

Gomart, E. (2002). Methadone: Six effects in search of a substance. *Social Studies of Science, 32*(1), 93–135.

Gomart, E., & Hennion, A. (1999). A sociology of attachment: Music amateurs, drug users. *The Sociological Review, 47*(S1), 220–247.

Hall, D. (2006). Religious attendance: More cost-effective than Lipitor? *Journal of the American Board of Family Medicine, 19*(2), 103–109.

Hardon, A., & Sanabria, E. (2017). Fluid drugs: Revisiting the anthropology of pharmaceuticals. *Annual Review of Anthropology, 46*, 117–132.

Kaelen, M., Barrett, F. S., Roseman, L., Lorenz, R., Family, N., Bolstridge, M., et al. (2015). LSD enhances the emotional response to music. *Psychopharmacology, 232*(19), 3607–3614.

Koenig, H., & Cohen, H. (2002). *The link between religion and health: Psychoneuroimmunology and the faith factor*. Oxford: Oxford University Press.

Labate, B. C., & Cavnar, C. (2014). *The therapeutic use of ayahuasca*. Berlin: Springer.

Labate, B. C., & MacRae, E. (2010). *Ayahuasca, ritual and religion in Brazil*. New York City: Routledge.

Loizaga-Velder, A., & Verres, R. (2014). Therapeutic effects of ritual ayahuasca use in the treatment of substance dependence--qualitative results. *Journal of Psychoactive Drugs, 46*(1), 63–72.

Luhrmann, T. M. (2013). Making God real and making God good: Some mechanisms through which prayer may contribute to healing. *Transcultural Psychiatry, 50*(5), 707–725.

Mabit, J., & Martin, S. (2007). Ayahuasca in the treatment of addictions: Evidence for hallucinogenic substances as treatments. In M. Winkelman & T. Roberts (Eds.), *Psychedelic medicine: New evidence for hallucinogenic substances as treatments* (pp. 87–103). Westport: Praeger.

Menozzi, W. (2011). The Santo Daime legal case in Italy. In B. C. Labate & H. Jungaberle (Eds.), *The internationalization of ayahuasca* (pp. 379–400). Berlin: LIT Verlag.

Mercante, M. S. (2013). A ayahuasca e o tratamento da dependência [Ayahuasca and the treatment of dependence]. *Mana, 19*(3), 529–558. https://doi.org/10.1590/S0104-93132013000300005.

Mesturini Cappo, S. (2018). What ayahuasca wants: Notes for the study and preservation of an entangled ayahuasca. In B. C. Labate & C. Cavnar (Eds.), *The expanding world ayahuasca diaspora* (pp. 157–176). London: Routledge.

Mol, A., Moser, I., & Pols, J. (2010). Care: putting practice into theory. *Care in Practice. On Tinkering in Clinics, Homes and Farms.* 7–25.

Myers, N. (2017). Ungrid-able ecologies: Decolonizing the ecological sensorium in a 10,000-year-old naturalcultural happening. *Catalyst: Feminism, Theory, Technoscience, 3*(2), 1–24.

Netherland, J., & Hansen, H. (2017). White opioids: Pharmaceutical race and the war on drugs that wasn't. *BioSocieties, 12*(2), 217–238.

Raikhel, E., & Garriott, W. (2013). *Addiction trajectories*. Durham: Duke University Press.

Ramos Gomes, B. (2016). *O uso ritual da Ayahuasca na atenção à população em situação de rua* [The ritual of ayahuasca with attention to the homeless]. Salvador: Edufba.

Sanabria, E. (2016). *Plastic bodies: Sex hormones and menstrual suppression in Brazil*. Durham: Duke University Press.

Sax, W. S. (2004). Healing rituals: A critical performative approach. *Anthropology & Medicine, 11*(3), 293–306.

Talin, P. (2013). Santo Daime between the Amazon and the West: Community life at the base of new social bonding. *Quaderni di Thule*, Documents of the 35th International Americanist Conference, May 3–10, 2013, Perugia, 13, (pp. 1133–1143). http://neip.info/texto/santo-daime-tra-amazzonia-e-occidente-vita-comunitaria-e-spiritualita-alla-base-di-nuovi-legami-familiari-e-sociali/

Talin, P., & Sanabria, E. (2017). Ayahuasca's entwined efficacy: An ethnographic study of ritual healing from 'addiction. *International Journal of Drug Policy, 44*, 23–30.

Tambiah, S. J. (1977). The cosmological and performative significance of a Thai cult of healing through meditation. *Culture, Medicine and Psychiatry, 1*(1), 97–132.

Thomas, G., Lucas, P., Capler, N. R., Tupper, K. W., & Martin, G. (2013). Ayahuasca-assisted therapy for addiction: Results from a preliminary observational study in Canada. *Current Drug Abuse Reviews, 6*(1), 30–42.

Tupper, K. W. (2008). The globalization of ayahuasca: Harm reduction or benefit maximization? *International Journal of Drug Policy, 19*(4), 297–303.

Weinberg, D. (2013). Post-humanism, addiction and the loss of self-control: Reflections on the missing core in addiction science. *The International Journal on Drug Policy, 24*(3), 173–181.

Wilson, E. (2015). *Gut feminism*. Durham: Duke University Press.

Woods, T., Antoni, M., Ironson, G., & Kling, D. (1999). Religiosity is associated with affective and immune status in symptomatic HIV-infected gay men. *Journal of Psychosomatic Research, 46*(2), 165–176.

Chapter 9
Ayahuasca as an Addiction Treatment in Catalonia: Cognitive and Cultural Perspectives

Ismael Apud

Introduction

As in other countries, the cultural background that supported the arrival of aya-huasca to Spain is related to different traditions, usually connected to the so-called New Age networks (Hanegraaff 1996) or psychospiritual networks (Champion 1995). In the case of Catalonia, these networks are composed of a number of traditions and movements, which arrived in different periods of time: the psychedelic movement in the 1960s (Usó 2001), the alternative medicines of the 1970s (Perdiguero 2004), and new schools of psychology, such as bioenergetic therapy in the 1970s, gestalt associations in the 1980s, and transpersonal psychology in the 1990s (Apud 2017b). Nowadays, these networks include a variety of "cultural imageries," such as oriental spiritualities, natural and holistic therapies, esoteric tra-ditions, and neo-shamanic practices (Prat 2017; Prat et al. 2012). Furthermore, the popularization of ayahuasca, both in Spain and internationally, must be considered in the context of the "renaissance of psychedelic studies," occurring since the 1990s (Labate and Cavnar 2011; Sessa 2012). In this revival, Spanish therapists and researchers from several disciplines are playing an important role (Apud and Romaní 2017).

The first time ayahuasca arrived in Spain was at the end of 1980s, through a joint meeting organized by Claudio Naranjo (a follower of Fritz Perls' Gestalt psychol-ogy) and a Santo Daime Church (López-Pavillard 2008). Corbera (2012) identifies 17 groups in Catalonia, including shamans (e.g., Latin American healers), "neosha-mans" (e.g., Western "spiritual seekers," "wounded healers"), Brazilian churches (Santo Daime, União do Vegetal), schools of psychology (transpersonal psychol-ogy, Gestalt school), and different centers of alternative and holistic therapies.

I. Apud (✉)
Facultad de Psicología, Universidad de la Republica, Montevideo, Uruguay
e-mail: ismaelapud@psico.edu.uy

© Springer Nature Switzerland AG 2021
B. C. Labate, C. Cavnar (eds.), *Ayahuasca Healing and Science*,
https://doi.org/10.1007/978-3-030-55688-4_9

People get in touch with these groups for various reasons related to spiritual, religious, and existential quests and also for therapeutic needs. One of the medical conditions often presented in these centers is addiction. In the particular case of Catalonia, initiatives such as the Institute of Applied Amazonian Ethnopsychology (henceforth IDEAA; founded by the psychiatrist Josep Maria Fábregas, in the year 2000, and now closed) were important hallmarks for the boost of a new agenda interested in the therapeutic applications of psychedelics for these conditions.

The current chapter is a summary of the research, "Science, Medicine, Spirituality and Ayahuasca in Catalonia: Understanding Ritual Healing in the Treatment of Addictions from an Interdisciplinary Perspective." The qualitative study of the biographical narratives of former addicts who recovered using ayahuasca ritual as the main treatment was conducted in Spain in the period from December 2015 to July 2017. The entire research can be found in my doctoral dissertation (Apud 2017a) and in the book *Ayahuasca: Between Cognition and Culture*, published in the medical anthropology collection of the Universitat Rovira i Virgili (Apud 2020).

In the current chapter, I will focus on one topic of the research: the relation between experiences with ayahuasca and ritual healing. I will propose a novel approach to understanding how ayahuasca ritual healing works, using an interdisciplinary perspective that combines cultural and cognitive approaches. Ritual will be considered not only as a therapeutic device but also as a practice aimed at assorted purposes, such as cultural transmission of beliefs and enhancement of social commitment. The model stresses the capacity of ritual to produce "memories of the experience" that trigger new meanings in the biographical narratives of the participants, resulting in different psychological effects.

I will analyze how cultural context, ritual setting, and therapeutic demand are related to the medical outcome of the cases presented. Key concepts, especially, will be taken into consideration: the dynamic between "memory of the experience" and "narratives of the self" to understand what is usually called "integration" in these centers; the importance of social cognition in the cure, since narratives of the self are always a social drama; the belief in ayahuasca as a "superhuman agent" related to the effectiveness of the treatment; and the understanding of these new biographical narratives as "spiritual conversions" with a potential therapeutic effect.

The style of the directors of the center (from a psychotherapist's style to that of a charismatic authority) and the ideas of spirituality (from "free floating" beliefs to structured doctrines) play their role in the therapeutic process, and this will be analyzed. Finally, in the "Conclusions," I will stress the importance of connecting cognitive and cultural levels of explanation to understand not only the therapeutic ritual effects but also a variety of other effects. Ayahuasca ritual will be considered as a useful alternative strategy to cope with addiction disorders because it works not only with subjectivity but also with spiritual experiences and beliefs that are usually forgotten by orthodox Western therapies.

Antecedents

Treatment of addiction includes a variety of pharmacological, psychological, and communitarian approaches. The consensus is that treatment should consider different levels of the problem, including neurobiological, psychological, social, economic, institutional, and cultural factors (National Institute on Drug Addiction [NIDA] 2012).

In the particular case of ayahuasca, the compound is used in a variety of traditional, religious, and clinical settings that implicitly or explicitly combine pharmacological, psychological, communal, and spiritual or religious elements. Some novel approaches are related to the renaissance of psychedelic studies, through the emergence of several centers for the treatment of addiction, for example, Takiwasi (Peru), Runawasi (Argentina), and IDEAA (between Spain and Brazil). As Bouso points out (2012), clinical studies of ayahuasca are still few, but the evidence suggests it is safe in the appropriate ritual setting, with no deleterious effects on health, or even on potential therapeutic applications. In the academic literature, various clinical studies, testimonies, and ethnographic accounts suggest not only a low risk of developing dependence—a common feature with others psychedelics—but also its potential application for the treatment of addictions (Chiappe 1977; Da Silveira et al. 2005; Fábregas et al. 2010; Fernández et al. 2014; Grob et al. 1996; Halpern et al. 2008; Labate et al. 2008; Loizaga-Velder and Verres 2014; Mabit Bonicard and González Mariscal 2013; Mercante 2013; Santos Ricciardi 2008; Scuro 2016; Thomas et al. 2013).

Prickett and Liester (2014) describe various models used to understand how ayahuasca could work in the treatment of addiction: (a) the "biochemical hypothesis" (a normalization of the dopaminergic reward pathway through its connection with the serotonergic system), (b) the "physiological hypothesis" (the rewiring of neurons hijacked by addictive behavior), (c) the "psychological hypothesis" (the access to unconscious memories, repressed emotions, and unsolved traumas and the enhancing of psychological insights), and (d) the "transcendental hypothesis" (the production of transcendental experiences that cause radical changes in values, beliefs, and worldviews). Despite their differences, the four hypotheses could also be considered as overlapping each other. Maybe the resulting effect is a consequence of more than one of the processes described by the authors, as multilevel models of ayahuasca's therapeutic action propose (see, e.g., in this volume, Domínguez-Clavé et al., Chap. 1).

Furthermore, the phenomenological and subjective side of ayahuasca cannot be separated from a neurological level of analysis because biochemical and physiological effects are a necessary flipside of psychological and transcendental experiences. For example, Riba et al. (2006) show how the activation of brain regions are related to emotional and introspective processing: right anterior insula and its relation with the representations of bodily states and subjective feelings; medial prefrontal and anterior cingulate gyrus, related to motivational aspects of emotions and their processing; and left amygdala and parahippocampal gyrus related to negative emotional valence and the processing of memories. In a later article that specifically address the potential benefits of ayahuasca in the treatment of addictions, Bouso and Riba conclude that "It might be speculated that ayahuasca helps to bring to consciousness memories from the past, to re-experience associated emotions, and to reprocess them in order to make plans for the future" (Bouso and Riba 2014, p. 101).

Considering the qualitative research studies on ayahuasca, the testimonies of the subjective experiences of the participants seem to concur with these findings. For example, Western testimonies include cognitive alterations, such as change of thoughts, loss of volition, emotional changes, body image distortions, changes in meanings, increased capacity for insight, hypersuggestability, biographical revisions, and increased capacity for empathy (Apud 2013; Fericgla 2013; Fernández and Fábregas 2013, 2014; Grob 1999; Lewis 2008; Loizaga-Velder and Loizaga Pazzi 2014; Loizaga-Velder and Verres 2014; Shanon 2014). In the particular case of drug dependence, the experiences are the same but influenced by the specific therapeutic demand. Loizaga-Velder and Verres (2014) interviewed 14 patients treated with ayahuasca for an addiction disorder. As important subjective factors of their recovery, the patients mentioned: better understanding of the personal causes related to their addictive behavior, mobilization of positive resources, reduction of craving and withdrawal symptoms, and spiritual and transcendental experiences reinforcing their purpose in life. Loizaga-Velder and Loizaga Pazzi (2014) describe four main types of subjective experiences: body-oriented (e.g., body awareness, anti-craving), emotional/social (e.g., release of psychological burdens, forgiveness of self and others), insight-oriented/cognitive (e.g., insights into maladaptive psychological patterns and increased self-awareness), and transpersonal (e.g., spiritual healing, sense of meaning and purpose in life). Fernández and Fábregas (2014) analyzed the testimonies of 20 patients treated at IDEAA, describing six fundamental types of experiences: biographical revisions (e.g., memories of negative events related to the drug abuse, traumatic episodes), psychological insights (e.g., personal conflicts, patterns of psychological functioning), emotional experiences (e.g., shame, forgiveness), death-and-rebirth experiences, experiences with nature (e.g., reconciliation with nature, awareness of one's animal nature), and transcendental experiences (e.g., feelings of mystical union, spiritual experiences).

There are several similarities described in other chapters of the present volume: In Chap. 4, Diament and collaborators associate the therapeutic effect of ayahuasca with bodily, emotional, and other subjective experiences; in Chap. 8, Cavnar mentions the importance of visionary experiences, psychological insights, and transpersonal experiences to understand the psychotherapeutic effect of the brew; in Chap. 9, Mason and Kuypers describe several cognitive and experiential changes produced by ayahuasca—some of them are related to creativity, empathy, emotion regulation, and feelings of interconnectedness; lastly, in Chap. 12, Argento and collaborators describe their research using ayahuasca with indigenous Canadian participants who have addiction problems, noting how experiences of connectedness with oneself, others, nature, and spirits are usually considered as key elements in recovery.

Ritual Healing Using a Multilevel Model

To understand how ritual healing using ayahuasca works, we must consider not only the substance and its pharmacological effect on the brain and cognition but also the final psychological outcome. We must also consider the contextual elements mediating the effect. In the biomedical literature, they are usually labeled as "nonspecific factors" related to the placebo response, and in the psychedelic studies, they are grouped using the classic notion of "set and setting." Decades ago, Grob (1999) described how these extrapharmacological factors shape the visionary ayahuasca experience, comparing different cultural contexts from psychonautic Western practices to Amazonian shamanic ceremonies. More recently, Talin and Sanabria (2017) (and also Chap. 10 of this volume) analyzed the trajectories of seven Italian addicts who recovered in the Santo Daime church, stressing the therapeutic relevance of the semiotic and social world into which the participants are introduced. As I will show next, these extrapharmacological factors are essential to understanding the effects of ayahuasca.

The interdisciplinary approach proposed in the current chapter is part of the research I conducted in Catalonia, where I analyzed the therapeutic use of ayahuasca under the general framework of theory of ritual, using a perspective that integrates cultural and cognitive levels. Before my research in Catalonia, and during my fieldwork in Latin America, I proposed a theoretical ritual model, inspired by the distributed cognition model of Cole and Engeström (1993). The model (Fig. 9.1) disaggregates various contextual factors—the "setting"—that modulate the ayahuasca experience: the ritual design (rules, spatial order, and technologies used), community (relationships between the members of the group), subject (personal history, spiritual trajectory, cultural background), roles (assigned in the ritual to each of the participants), and artifacts (the set of instruments and techniques used by the healer to stimulate and manipulate the states of consciousness of the participants).

The model proposed in my research in Catalonia adds one further step (Fig. 9.2). It not only disaggregates the ritual and its variables but also the cognitive functions usually manipulated in it. For this, I used insights from interdisciplinary areas of research interested in ritual and symbolic healing: medical anthropology and

Fig. 9.1 Ritual design triangle (Apud 2013, 2015a, b)

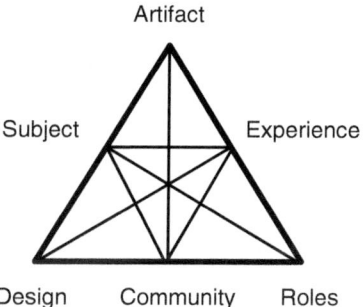

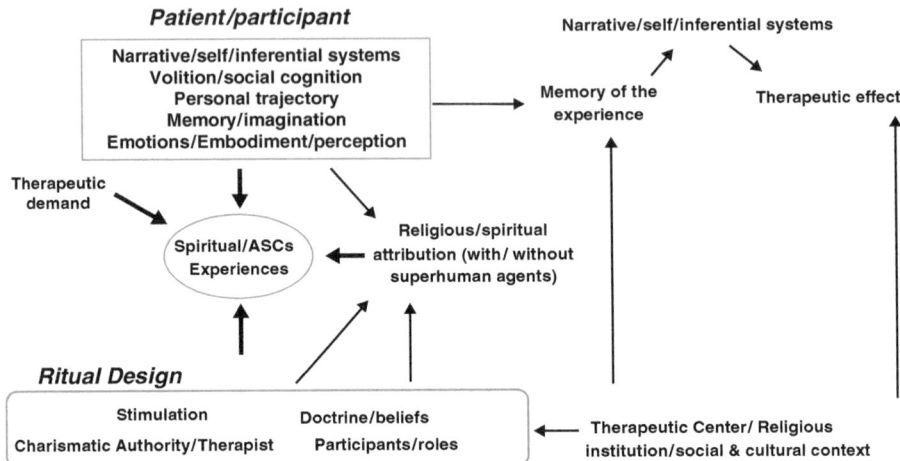

Fig. 9.2 Ritual healing model that integrates cultural and cognitive elements (Apud 2017a)

placebo agenda, ethnographic studies on trance and possession, cognitive science of religion, studies on religious experience, studies on religious conversion, psychedelic studies, and religious and health studies. The model was constructed between these theories and empirical data from my fieldwork in Catalonia and Latin America. The final categories presented are not exhaustive (other categories can be added) or necessarily relevant in different sets and settings (some categories could be excluded for certain cases). It is an open and emergent model, constructed in a "grounded theory" style.

In the model, ritual healing is a dramatic performance (Turner 1977) or a placebo drama (Kaptchuk 2002) where patient and practitioner interact in a therapeutic setting with the common goal of solving a therapeutic demand. The practitioner can be a therapist or a charismatic authority who uses different techniques of stimulation and who can manipulate mythological healing symbols and beliefs (Dow 1986). The ritual effectiveness is usually attributed to a "superhuman agent" who has "full access to strategic knowledge" (Boyer 2001) and is considered to produce a particular "super-permanent effect" (Lawson and McCauley 1990). The attribution of effectiveness to the ritual produces a "meaning response" (Moerman 2002) that could trigger a cascade of psychoneuroimmunological processes that give relief or healing (Geertz 2010; McClenon 1997; McNamara 2009; Winkelman 2010). The symbolic effectiveness does not require a common shared mythical structure, only certain conditions that induce the patient to search for new personal meanings. So, the patients are not mere passive agents; their personal predispositions and trajectories play an important role in the process.

We must also consider that the therapeutic outcome is not related to the experience itself but how it is remembered and re-signified later, in what Czachesz (2017) called "memory of the experience." It involves different psychological (e.g., agent-action representations) and cultural biases (e.g., narrative styles). The memory of the experience acts as potential "turning point" in the biography of the participants (Denzin 2014), producing various changes in the biographical narratives of the self (McNamara 2009).

From a cognitive point of view, biographical mental projection involves higher functions, such as self-relevant knowledge stored in the memory (episodic memories, semantic representations, metacognition, social identity), executive functions (valuation, learning, and cognitive homeostasis), reflexivity (interaction between executive functions and representational knowledge), and mental time travel (the capacity to reconstruct specific events of the past and engage in alternative mental simulations of future events), among others (Skoweonski and Sedikides 2007). From a cultural perspective, it involves a variety of narrative styles interpreting subjective experiences in a socially interactive process that, within these groups, is usually called "integration." The new narratives are not only a result of individual experiences but also of a social interpretive process in which patients evoke, assess, negotiate, legitimate, and give factuality to their experiences. The narrative mode of thought situates individuals as social agents within a cultural context, allowing subjects to dive through the cultural past and to project themselves from the present to the future (Bruner 1986).

In sum, biographical narratives have an important homeostatic function, both at psychological and social levels. They influence health, emotions, lifestyles, and social behavior and can also have psychoneuroimmunological effects. But, as was previously mentioned, both rituals and the construction of new narratives are related not necessarily to a therapeutic aim but to other goals, for example, religious conversion, which sometimes is overlapped with healing.

Method

This was a qualitative, case-focused study using a biographical approach. It is important to consider that, unlike quantitative research, qualitative sampling does not have the aim of inferring results to a whole population through statistical procedures but only to describe and analyze the cases in terms of processes and meanings. So, this study is not an assessment of the efficacy of ayahuasca in the treatment of addictions; it is only an analysis of certain cases where the treatment worked.

The units of analysis are former patients who recovered from an addiction disorder using ayahuasca ritual as the main treatment in the period of time between 2000 and 2016 in the geographical region of Catalonia and the surrounding autonomous communities. ICD-10 and DSM-V were used for the construction of the inclusion criteria for the definition of addiction, including features such as the following: at least 12 months of problematic use of the substance; significant social and psychological impairment caused by drug use; the inability to control the consumption and dosage administered; and at least 12 months without compulsively taking drugs. The final sample consisted of 12 subjects (10 males, 2 females) with a mean age of 41.8 (minimum 22, maximum 57) residing in Catalonia, Valencia, and the Balearic

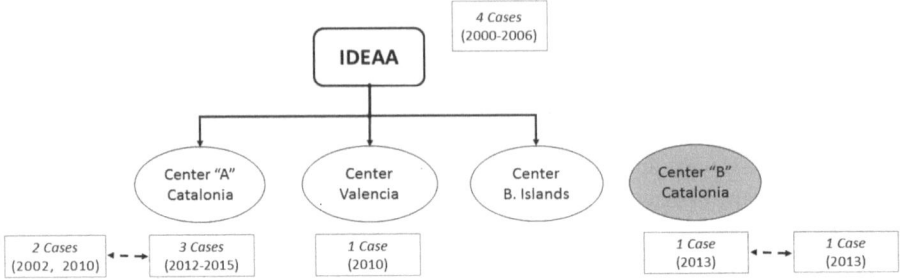

Fig. 9.3 Network of the groups and cases

Islands. The geographical zone was not a preestablished criterion but a consequence of the snowball sampling procedure.

The selection of the cases started with contacting therapists and patients who were part of IDEAA when it was still operating. Those informants gave access to current centers in the surrounding areas, most of them related to former therapists or patients of IDEAA who later became therapists. As shown in Fig. 9.3, four centers were finally contacted: two in Catalonia (one of them not related to IDEAA), one in Valencia, and one in the Balearic Islands (where I did not work with patients from the center but with two members who were previously treated in IDEAA). The final sample includes four former patients of IDEAA: three from Center A in Catalonia, one from Center B in Catalonia, one from a center in Valencia, and three cases recovered on their own, in ceremonies in Spain or Peru (two contacted during fieldwork in Center A and one in Center B).

All the centers are connected to the psychospiritual networks mentioned in the introduction of this chapter. Notably, the rituals at the centers are similar in their main features: they are usually performed at night; participants sit or lie in a circle on the floor; music is played live or on a computer; the rituals encourage self-reflection and insightful psychological or spiritual experiences. The sociocultural background is also similar in each of the centers. The 12 cases were from a middle-class context; all of them had completed secondary education; half of them had finished university studies. Most of them were polydrug users, with cocaine as the most common one, followed by heroin and alcohol (Fig. 9.4).

Results

The experiences of the participants are similar to those found in the previous qualitative studies mentioned, with biographical reviews and psychological insights usually considered to be of the most therapeutic value by the patients. These experiences include childhood remembrances, memories of negative events related to the addiction, traumatic episodes, insights into psychological patterns, and recognition of positive personal resources:

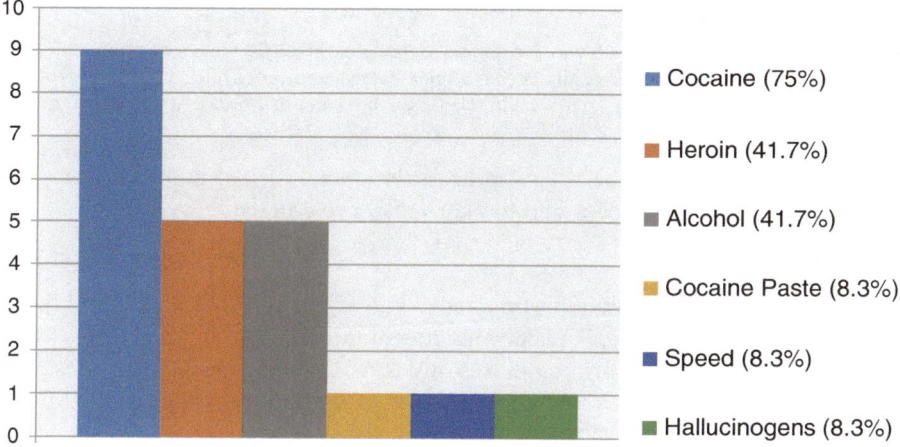

Fig. 9.4 Drugs used by the subjects

Case 1, from IDEAA

If it was for the enjoyment, I would have quit immediately because it was… It was about what you have been going through, the things you have done, the terrible condition you were in that moment. Flashbacks of bad episodes, decisions in which you could have taken a certain path and you finished in the wrong way … Lots of remembrances, of the problem that I had… [Memories] from my childhood also, but mostly of the problem that led me to the center…

Case 2, from IDEAA

In my case, and also in the case of other drug addicts, your life became a continuous lie. Everything was a lie. And the plant showed this to me. All the world of lies that was my life. And when I started to pull these things out, I saw that everything was a lie! And this was a critical moment when I went through a process with ayahuasca. In the moment when you contact with all of these, you aren't happy at all, the most common thing is to feel really afraid… because of the void that you suddenly see in your life, because everything is a lie, because I am a lie. It is a very delicate moment. And the person needs a lot of care, support, to go through this moment…

Another kind of experience is a variety of emotional states, such as happiness, relief, sadness, grief, rage, and love:

Case 3, from IDEAA

There was a huge fire, and I sat there, dizzy, but fine. And when I looked at the stars, some tears started to fall… but such tears! A feeling that is… very hard to describe. Because it was like a torrent… but not of sadness. The first time I drank it, ayahuasca did something to me. She taught me, showed me, made me see that she was capable of helping me…

Emotional experiences can also be accompanied by embodied experiences, such as crying, vomiting, and the sense of being cleansed or released from a psychological burden:

Case 1, from IDEAA

> The remembrance of my path was accompanied with continuous crying... of crying three hours in a row for example... and that was a release... The power of releasing is activated, and you are like a fountain... and I could release all the pain I had been carrying inside me ...

In certain parts of the therapeutic process, emotions such as guilt, empathy, forgiveness, and self-forgiveness start to play a major role in the recovery:

Case 4, from IDEAA

I closed my eyes and I saw monstrosities... knives, all very macabre. But I had a strong comprehension ... I understood lots of things ... What I had done, why I had behaved that way, my complexes, my fears ... What I focused on most was the topic of violence, and why I was engaged in drug trafficking and all those things ...

Case 1, from Center A

I put myself in the place of my father, in feeling as my father feels ... and I could reassess him, ayahuasca made me value him positively... seeing a little bit from his perspective, and consider him not only with a sense of resentment. To see his positive side ...

Finally, a variety of spiritual experiences are also present, including the following: traveling to other places, watching heavenly landscapes, a mystical feeling of union, out-of-body experiences, death-and-rebirth experiences, communication with spirits, and being possessed by entities:

Case 1, related to Center A

... visions... which later, searching in books, different cultures have this vision... of the *axis mundi*... like a column of energy connecting heaven and earth. And there I saw... snakes, crocodiles, low density spirits. And a world below was opened. I saw an amazing column of energy that was the whole universe.

Case 2, related to Center A

I had the sensation that ayahuasca... when I was at the peak of the experience... I opened my eyes and it was like the consciousness of the plant had possessed me ... and it was like she was studying the world from my perspective. And after that ... the plant travelled inside my body ... trying to clean it.

Spiritual experiences do not necessarily involve a superhuman agent but interacting with or seeing them is not an unusual thing. The contact with spirits may include dead or living relatives and friends, animals, and indigenous people. But the most important is the contact with ayahuasca and the belief in it as a teacher and healing plant. Ayahuasca is usually considered as the main agent behind the therapeutic process.

Dark experiences, such as visions of hell, sinister places, and evil entities, can occur. These experiences are sometimes counterproductive, causing strong fear in the participants and making them avoid ayahuasca for a certain period of time. But, in other cases, these negative experiences are considered as having a message with an important therapeutic value when decoded:

Case 2, related to Center A

I started seeing a monster, a huge, huge demon ... At the beginning, my thoughts were, "this is something wicked, from outside, that wants to enter me" ... But in the middle of the session, I saw an umbilical cord that came out of the monster and that was connected to me. So, I said, "this is not from the outside" ... I fell onto my knees and said to myself, "I have to recognize my dark side" ... I came out of the ceremony totally unstructured ... and it took me a lot of work to come to terms with what happened in that session...

In our 12 cases, the moderate, biographical, insightful experiences are the most common and also the ones most frequently considered important for recovery. But the transpersonal spiritual experiences are also considered a main catalyst for healing. Besides that, it is important to note that spiritual experiences cannot be divorced from psychological ones. There is usually an overlap or connection between both. As a paradigmatic example, the following case describes an experience where an emotional state, a repressed trauma, a death-and-rebirth experience, and a final spiritual awareness of life are chained together:

Case 2, from Center A

There was a turning point when I was 16... the death of my best friend. I was there with him, it was an accident, and after that, I rejected life. I said, "Life is shit and now I will burn it." And I started to take drugs to burn out my life ... I was wondering why, when I was at home and not taking drugs, I started to feel such a huge hole in my life, inside me, and I started to cry, and I did not know why, ok? ... During an ayahuasca ceremony, I asked about that hole. I saw a hole... like a giant black cloud, so, so big that I could not close my eyes. And then it came out, the death of my best friend, and I... cried for three hours, like I'd never cried before. ... I stopped crying and started laughing. ... I started to feel as if I was dying, that I was disintegrating, ok? ... I lay down, and I gave birth to myself, and felt as if I could not breathe because I was coming out of the uterus ... And I finally understood, and all the connections came into my mind. I understood that it was an unfinished mourning process. That life is wonderful, and death is part of it, and that there is no rebirth without death. I had to die and to have this experience of being reborn to understand death in another way. ... I was born again and I was cured of that emptiness. And since then, the void that I used to feel when I came home and made me start crying without knowing why, that void disappeared forever ... Pulling out the trauma of the death of my best friend was a turning point for me to start moving forwards...

An important therapeutic element mentioned by the patients is what they call "integration," that is, how the experiences are interpreted and psychologically assimilated after the ritual. In the cases studied, spiritual conversions could be

Table 9.1 Religious affiliation

Religious affiliation	Before ayahuasca	After ayahuasca
Agnostic/atheist	6 (50%)	1 (8.3 %)
Catholic	3 (25%)	1 (8.3 %)
Spiritual/no affiliation	3 (25%)	6 (50%)
Spiritual/Santo Daime	–	3 (25%)
Spiritual/Reiki	–	1 (8.3%)
Total	12 (100%)	12 (100%)

considered a recurrent final outcome of the process of integration of these experiences into their biographical narratives (Table 9.1).

The reflections on spirituality can be more or less generic, and the interpretations are diverse, particularly in those centers where there is no canonical view or doctrine. But, despite the differences, there are certain common beliefs "free floating" in these centers, related to different traditions within the psychospiritual networks: New Age ideas, transpersonal psychology, perennial philosophy, and Eastern ideas and practices, among others. The integration process will depend on what each participant found useful and how the patient interacted with the style of the center, the group, and the therapist. For example, Center A allowed participants to explore their psychological problems through different strategies. Some of them were learned in the center, but others came from other spiritual or therapeutic practices:

Case 2, from Center A

I had the session of ayahuasca, and later a session of reiki. And the woman with whom I was doing reiki integrated lot of things I had experienced in the sessions of ayahuasca, she helped me to order them…

In the case of Center B, a structured doctrine of a millennial spirituality could be identified, spinning around its controversial leader, who considers himself as a spiritual "guru" who had achieved wisdom:

Case 1, from Center B

When I went to retreats with him, he confronted me during the integrations. He really had a deep insight into what is happening to a person because he had worked on inner conquest, so what he says is pure intuition, not a projection. If he says something, it is because he is seeing it, because he has the ability to see into the depths of the person…

In this particular case, the therapist acted as a "charismatic authority." He considers himself as having access to spiritual relevant knowledge and teaches the patient how to "heal" and "evolve." In that way, the patient is introduced to a new worldview in an intellectual kind of conversion.

Discussion

As was already mentioned, the testimonies of the subjects are similar to the ones described in previous qualitative studies. This should not be a surprise, since context (a Western "psi" subculture focused on introspection), setting (the ritual as a psychotherapeutic device for self-knowledge), and therapeutic demand (the healing of an addiction disorder) are mostly the same. Biographical experiences were the most common ones. In addition, and as was previously mentioned, they are connected to other kinds of experiences, usually in a therapeutic plot.

Most of the biographical reviews are associated with emotions, such as guilt, shame, and, later, forgiveness and self-forgiveness. From a cognitive perspective, these emotions are related to social cognition, that is, our natural capacities to interact with others, to see from their perspectives, and to empathize. On one hand, ayahuasca, as well as other psychedelics, may trigger certain brain regions related to social cognition. On the other hand, the psychotherapeutic work of "integration" in these centers involves redefining a narrative of the self. Biographical narratives, too, are always a social drama. We are social beings and we construct our identity in a social scenario, with other agents, with social rules, moral values, and a sense of what is right and wrong (Bruner 1986).

When social dispositions such as empathy, altruism, and love are obstructed, different psychological disorders are more likely to appear (Cloninger and Kedia 2011). In the case of drug dependence, the individuals presented in this chapter connected both intellectually and emotionally with the social consequences of their disorder. Most of the patients described this as an essential part of the therapeutic process: to put themselves into the shoes of relatives and friends, to understand the damage they had done to them, and to find forgiveness in order to construct a fresh starting point for their new identity. Spiritual experiences are usually connected with other experiences in a psychotherapeutic plot; they are especially relevant, since they introduce the participant to a new worldview, producing diverse varieties of spiritual conversions, and the construction of a new narrative of the self. One relevant spiritual category is the contact with or the belief in ayahuasca as a teacher and healer. As was mentioned in the theoretical model, the connection with a "superhuman agent" is an important element in the effectiveness of the ritual.

Considering the empirical findings in disciplines such as neuroscience, psychology, and anthropology, it could be concluded that experiences with ayahuasca seem to have a profound psychological impact, affecting cognitive processes, such as perception, emotions, social cognition, corporal perception, and consciousness. It seems to act on mechanisms that are usually addressed by rituals, but the desired effects are easier to reach and also more intense. Despite this, to have a durable effect, the ritual must have an impact on the patient's self-identity. This effect involves a dialectic between the memory of the experience (how experience is stored, assessed, evoked, and re-signified) and the narrative of the self (identity expressed as a social story).

The experience should act as a significant episode in the biography of the subject, producing positive changes. I would like to stress, however, that these effects are not necessarily psychotherapeutic, and all these cognitive effects can be used in different ways, depending on the ritual design, the mindset of the participant, the cultural context, and the institutional and social background where the ritual is displayed. For example, in the traditional Amazonian shamanic context, the effects are directed to other purposes, such as fighting witchcraft, learning new spiritual knowledge, or healing different culture-bound syndromes (Apud 2013; Beyer 2009; Dobkin de Rios 1973; Luna 1986; MacRae 1992).

Conclusions

Clinical research on ayahuasca is usually interested in the pharmacological properties of the brew, that is, how it interacts at physiological and neurobiological levels. This an essential level of inquiry if we want to assess both positive and negative outcomes of its use. However, it is important not only to stay solely at one level but to also study the variations and influence of different factors from psychological, social, and cultural levels of analysis, and also try to connect those levels. The interdisciplinary perspective presented in this chapter is concerned with connecting cognitive and cultural levels of explanation. Ritual was considered as a cultural practice that produces certain psychological effects in the participants, including religious conversion, coping with mental disorders and culture-bound syndromes, enhancing social commitment, and as a rite of passage during different life's stages, among others. Religious rituals—and ayahuasca rituals are not an exception—are commonly used as cultural devices that help in the task of putting the new and old pieces of the biographical self together. This is a delicate moment, when the patient is vulnerable and suggestible, so negative effects are also possible. In this process, religious and spiritual conversion, adherence to certain community, and idealization of a charismatic leader can happen in both positive and negative ways.

In the case of addictions, and with the appropriate work of "integration," ayahuasca ritual seems useful, since it addresses different aspects related to the disorder at the neurological, psychological, social, and cultural levels involved in this complex problem. It can alter a participant's perspective on life, self-identity, lifestyles, moods, motivation, social relationships, and ways of understanding the past and projecting to the future. If the effect is to be long-lasting—in our case, the effect being the avoidance of a dependence behavior pattern—the ritual experience must leave a mark on the narrative of the self, a turning point in the biography of the individual that is usually expressed as a kind of spiritual conversion.

It is important to mention that the effectiveness of ayahuasca ritual varies from case to case, and some participants might not find this kind of approach useful at all. But, even if the treatment with ayahuasca works for only a few cases, we should consider the importance of it anyway, since sometimes these few cases might only find a solution in these kinds of heterodox strategies. For example, in this particular research, more than half of the cases had a long career of resistance to conventional treatments, and they finally overcame their problem through ayahuasca-related strategies; so, these alternative strategies should not be excluded from the range of potential therapies for addictions. This problem is addressed in the principles of effective treatment of NIDA (2012), where it is stated that no single treatment is appropriate for everyone and that different approaches and strategies should be considered. Some of these strategies should include spirituality as part of the treatment. This is important because most Western therapies do not specifically address spirituality, and the majority of the population has religious or spiritual backgrounds. As scientific literature suggests, spiritual or religious beliefs seem to have a positive effect on the health of their adherents, not only because of the sense of meaningfulness and hope that they usually produce but also because of the networks of support and the healthy lifestyles they promote (Koenig et al. 2001). In the cases studied, the recovery implied staying away from certain social contexts and social habits and starting a new social life, with other rewards, persons of reference, and cultural and spiritual motivations.

References

Apud, I. (2013). *Ceremonias de Ayahuasca: entre un centro holístico uruguayo y el curanderismo amazónico peruano* [Ayahuasca ceremonies: Between a Uruguayan holistic center and the Peruvian Amazon healer] [Unpublished master's thesis]. Universidad Nacional de Lanús.

Apud, I. (2015a). Ayahuasca, Contexto 1 a Partir de un Grupo Uruguayo de Terapias Alternativas [Ayahuasca, ceremonial context and subject: Set and setting analysis from a Uruguayan group of alternative therapies]. *Journal of Transpersonal Research, 7*(1), 7–18.

Apud, I. (2015b). Ayahuasca from Peru to Uruguay: Ritual design and redesign through a distributed cognition approach. *Anthropology of Consciousness, 26*(1), 1–27. https://doi.org/10.1111/anoc.12023.

Apud, I. (2017a). *Science, medicine, spirituality and ayahuasca in Catalonia. Understanding ritual healing in the treatment of addictions from an interdisciplinary perspective* [Unpublished doctoral dissertation]. Universitat Rovira i Virgili.

Apud, I. (2017b). Science, spirituality, and ayahuasca: The problem of consciousness and spiritual ontologies in the academy. *Zygon. Journal of Religion and Science, 52*(1), 100–123.

Apud, I. (2020). *Ayahuasca: Between cognition and culture. Perspectives from an interdisciplinary and reflexive ethnography.* Tarragona: Publicacions URV. http://llibres.urv.cat/index.php/purv/catalog/book/428.

Apud, I., & Romaní, O. (2017). Medicine, religion and ayahuasca in Catalonia: Considering ayahuasca networks from a medical anthropology perspective. *International Journal of Drug Policy, 39*, 28–36. https://doi.org/10.1016/j.drugpo.2016.07.011.

Beyer, S. V. (2009). *Singing to the plants: A guide to mestizo shamanism in the Upper Amazon.* Albuquerque: University of New Mexico Press.

Bouso, J. C. (2012). *Personalidad, psicopatología y rendimiento neuropsicológico de los consumidores rituales de ayahuasca* [Personality, psychopathology and neuropsychological performance of ritual ayahuasca consumers] [Unpublished doctoral dissertation]. Universitat Autonoma de Barcelona.

Bouso, J. C., & Riba, J. (2014). Ayahuasca and the treatment of drug addiction. In B. C. Labate & C. Cavnar (Eds.), *The therapeutic use of ayahuasca* (pp. 95–109). Berlin: Springer.

Boyer, P. (2001). *Religion explained: The evolutionary origins of religious thought*. New York City: Basic Books.

Bruner, J. (1986). *Actual minds, possible worlds*. Cambridge, MA: Harvard University Press.

Champion, F. (1995). Persona religiosa fluctuante, eclecticismo y sincretismos [Fluctuating religious person, eclecticism and syncretism]. In J. Delumenau (Ed.), *El Hecho Religioso. Enciclopedia de las grandes religiones* [The religious fact: Encyclopedia of the great religions] (pp. 235–243). Madrid: Alianza Editorial.

Chiappe, M. (1977). El empleo de alucinógenos en la psiquiatría folklórica [The use of hallucinogens in folk psychiatry]. *Boletín de La OPS, 81*(2), 176–186.

Cloninger, R. C., & Kedia, S. (2011). The phylogenesis of human personality: Identifying the precursors of cooperation, altruism, and well-being. In R. W. Sussman & R. C. Cloninger (Eds.), *Origins of altruism and cooperation* (pp. 63–107). New York City: Springer.

Cole, M., & Engeström, Y. (1993). Enfoque histórico-cultural de la cognición distribuída [Historical-cultural approach to distributed cognition]. In G. Salomon (Ed.), *Cogniciones distribuidas* [Distributed cognitions] (pp. 23–74). Buenos Aires: Amorrortu Editores.

Corbera, J. (2012). *L'Ayahuasca ara i Aquí: usos rituals de l'ayahuasca a Catalunya* [Ayahuasca now and here: Ritual uses of ayahuasca in Catalonia]. Barcelona: Universitat Oberta de Catalunya.

Czachesz, I. (2017). *Cognitive science and the new testament: A new approach to early Christian research*. Oxford: Oxford University Press.

Da Silveira, D. X., Grob, C. S., de Rios, M. D., Lopez, E., Alonso, L. K., Tacla, C., & Doering-Silveira, E. (2005). Ayahuasca in adolescence: A preliminary psychiatric assessment. *Journal of Psychoactive Drugs, 37*(2), 129–133. https://doi.org/10.1080/02791072.2005.10399792.

Denzin, N. (2014). *Interpretive autoethnography*. London: Sage Publications.

Dobkin de Rios, M. (1973). Curing with ayahuasca in an urban slum. In M. Harner (Ed.), *Hallucinogens and shamanism* (pp. 67–85). New York City: Oxford University Press.

Dow, J. (1986). Universal aspects of symbolic healing: A theoretical synthesis. *American Anthropologist, 88*(1), 56–69.

Fábregas, J. M., González, D., Fondevila, S., Cutchet, M., Fernández, X., Barbosa, P. C. R., et al. (2010). Assessment of addiction severity among ritual users of ayahuasca. *Drug and Alcohol Dependence, 111*(3), 257–261. https://doi.org/10.1016/j.drugalcdep.2010.03.024.

Fericgla, J. M. (2013). Cambios en el perfil de valores tras una experiencia con ayahuasca. Comparación de resultados del Test de Hartman administrado antes y después de una sesión de ayahuasca en un grupo de voluntarios [Changes in the profile of values after an experience with ayahuasca: Comparison of Hartman Test results administered before and after an ayahuasca session in a group of volunteers]. In B. C. Labate & J. C. Bouso Saiz (Eds.), *Ayahuasca y Salud* [Ayahuasca and health] (pp. 424–432). Barcelona: La Liebre de Marzo.

Fernández, X., & Fábregas, J. M. (2013). Experiencia de un tratamiento de ayahuasca para las drogodependencias en la amazonia brasileña [Experience of an ayahuasca treatment for drug addictions in the Brazilian Amazon]. In B. C. Labate & J. C. Bouso (Eds.), *Ayahuasca y Salud* [Ayahausca and health] (pp. 392–423). Barcelona: La Liebre de Marzo.

Fernández, X., & Fábregas, J. M. (2014). Experience of treatment with ayahuasca for drug addiction in the Brazilian Amazon. In B. C. Labate & C. Cavnar (Eds.), *The therapeutic use of ayahuasca* (pp. 161–182). Berlin: Springer.

Fernández, X., dos Santos, R. G., Cutchet, M., Fondevila, S., González, D., Alcázar, M. A., et al. (2014). Assessment of the psychotherapeutic effects of ritual ayahuasca use on drug dependency: A pilot study. In B. C. Labate & C. Cavnar (Eds.), *The therapeutic use of ayahuasca* (pp. 183–196). Berlin: Springer.

Geertz, A. (2010). Brain, body and culture: A biocultural theory of religion. *Method & Theory in the Study of Religion, 22*(4), 304–321.

Grob, C. S. (1999). The psychology of ayahuasca. In R. Metzner (Ed.), *Sacred vine of spirits: Ayahuasca*. Rochester: Park Street Press.

Grob, C. S., Mckenna, D. J., Callaway, J. C., Brito, G. S., Neves, E. S., Oberlaender, G., Saide, O. L., Labigalini, E., et al. (1996). Human psychopharmacology of hoasca, a plant hallucinogen used in ritual context in Brazil. *The Journal of Nervous & Mental Disease, 184*(2), 86–94.

Halpern, J. H., Sherwood, A. R., Passie, T., Blackwell, K. C., & Ruttenber, A. J. (2008). Evidence of health and safety in American members of a religion who use a hallucinogenic sacrament. *Medical Science Monitor, 14*(8), 15–22.

Hanegraaff, W. J. (1996). *New age religion and Western culture. Esotericism in the mirror of secular thought.* Leiden: Brill.

Kaptchuk, T. J. (2002). The placebo effect in alternative medicine: Can the performance of a healing ritual have clinical significance? *Annals of Internal Medicine, 136*(11), 817–825.

Koenig, H. G., McCullough, M. E., & Larson, D. B. (2001). *Handbook of religion and health.* New York City: Oxford University Press.

Labate, B. C., & Cavnar, C. (2011). The expansion of the field of research on ayahuasca: Some reflections about the ayahuasca track at the 2010 MAPS "Psychedelic Science in the 21st Century" conference. *The International Journal of Drug Policy, 22*(2), 174–178.

Labate, B. C., dos Santos, R. G., Mercante, M. S., & Ribeiro Barbosa, P. C. (2008, July). *Considerações sobre o tratamento da dependência por meio da ayahuasca* [Considerations on the treatment of addiction through ayahuasca]. Retrieved from http://neip.info/novo/wp-content/uploads/2015/04/labate_et_all_tratamento_dependencia_ayahuasca_final.pdf

Lawson, E. T., & McCauley, R. N. (1990). *Rethinking religion: Connecting cognition and culture.* Cambridge: Cambridge University Press.

Lewis, S. E. (2008). Ayahuasca and spiritual crisis: Liminality as space for personal growth. *Anthropology of Consciousness, 19*(2), 109–133. https://doi.org/10.1111/j.1556-3537.2008.00006.x.

Loizaga-Velder, A., & Loizaga Pazzi, A. (2014). Therapist and patient perspectives on ayahuasca-assisted treatment for substance dependence. In B. C. Labate & C. Cavnar (Eds.), *The therapeutic use of ayahuasca* (pp. 133–152). Berlin: Springer.

Loizaga-Velder, A., & Verres, R. (2014). Therapeutic effects of ritual ayahuasca use in the treatment of substance dependence: Qualitative results. *Journal of Psychoactive Drugs, 46*(1), 63–72. https://doi.org/10.1080/02791072.2013.873157.

López-Pavillard, S. (2008). *Recepción de la ayahuasca en España* [Reception of ayahuasca in Spain]. Madrid: Universidad Complutense de Madrid.

Luna, L. E. (1986). *Vegetalismo: Shamanism among the mestizo population of the Peruvian Amazon.* Stockholm: Almqvist & Wiksell.

Mabit Bonicard, J., & González Mariscal, J. M. (2013). Hacia una medicina transcultural: Reflexiones y propuestas a partir de la experiencia en Takiwasi [Towards a transcultural medicine: Reflections and proposals from the experience in Takiwasi]. *Journal of Transpersonal Research, 5*(2), 49–76.

MacRae, E. (1992). *Guided by the moon. Shamanism and the ritual use of ayahuasca in the Santo Daime Religion in Brazil.* São Paulo: Brasiliense.

McClenon, J. (1997). Shamanic healing, human evolution, and the origin of religion. *Journal for the Scientific Study of Religion, 36*(3), 345–354. Retrieved from http://www.jstor.org/stable/1387852?origin=JSTOR-pdf.

McNamara, P. (2009). *The neuroscience of religious experience.* New York City: Cambridge University Press.

Mercante, M. S. (2013). A ayahuasca e o tratamiento da dependencia [Ayahuasca and treatment of dependency]. *Mana, 19*(3), 529–558.

Moerman, D. E. (2002). Explanatory mechanisms for placebo effects: Cultural influences and the meaning response. In H. A. Guess, A. Kleinman, J. W. Kusek, & L. W. Engel (Eds.), *The science of the placebo: Toward an interdisciplinary research agenda* (pp. 77–107). London: BMJ Books.

National Institute of Drug Addiction (NIDA). (2012). *Principles of drug addiction treatment: A research-based guide* (3rd ed.). Bethesda: National Institute of Health Publication.

Perdiguero, E. (2004). El fenómeno del pluralismo asistencial: una realidad por investigar [The phenomenon of healthcare pluralism: A reality to investigate]. *Gaceta Sanitaria, 18*(Supl.1), 140–145. https://doi.org/10.1157/13062263.

Prat, J. (2017). *La Nostalgia de los Orígenes. Chamanes, gnósticos, monjes y místicos* [The nostalgia of the origins. shamans, gnostics, monks and mystics]. Barcelona: Kairós.

Prat, J., Anguera, M., Cuadet, F., Dittwald, D., Reche, J., Tomas, I., & Vivancos, I. (2012). *Els Nous Imaginaris Culturals. Espiritualitats orientals, therapies naturals I sabers esoterics* [The new cultural imaginaries. Oriental spiritualities, natural therapies and esoteric knowledge]. Tarragona: Publicaciones URV.

Prickett, J. I., & Liester, M. B. (2014). Hypotheses regarding ayahuasca's potential mechanisms of action in the treatment of addiction. In B. C. Labate & C. Cavnar (Eds.), *The therapeutic use of ayahuasca* (pp. 111–132). Berlin: Springer.

Riba, J., Romero, S., Grasa, E., Mena, E., Carrió, I., & Barbanoj, M. J. (2006). Increased frontal and paralimbic activation following ayahuasca, the pan-Amazonian inebriant. *Psychopharmacology, 186*(1), 93–98. https://doi.org/10.1007/s00213-006-0358-7.

Santos Ricciardi, G. (2008). *O uso da Ayahuasca e a experiencia de transformacao, alivio e cura, na Uniao do Vegetal* [The use of ayahuasca and the experience of transformation, relief and healing in the Uniao do Vegetal] [Unpublished master's thesis]. Universidade Federal da Bahia.

Scuro, J. (2016). *Neochamanismo en América Latina. Una cartografía desde el Uruguay* [Neo-shamanism in Latin America. A cartography from Uruguay] [Unpublished doctoral dissertation]. Universidade Federal do Rio Grande do Sul.

Sessa, B. (2012). Shaping the renaissance of psychedelic research. *The Lancet, 380*(9839), 200–201.

Shanon, B. (2014). Moments of insight, healing, and transformation: A cognitive phenomenological analysis. In B. C. Labate & C. Cavnar (Eds.), *The therapeutic use of ayahuasca* (pp. 59–76). Berlin: Springer.

Skoweonski, J., & Sedikides, C. (2007). Temporal knowledge and autobiographical memory: An evolutionary perspective. In R. Dunbar & L. Barrett (Eds.), *Oxford handbook of evolutionary psychology* (pp. 505–517). New York City: Oxford University Press.

Talin, P., & Sanabria, E. (2017). Ayahuasca's entwined efficacy: An ethnographic study of ritual healing from "addiction.". *International Journal of Drug Policy, 44*, 23–30.

Thomas, G., Lucas, P., Capler, N. R., Tupper, K. W., & Martin, G. (2013). Ayahuasca-assisted therapy for addiction: Results from a preliminary observational study in Canada. *Current Drug Abuse Reviews, 6*(1), 30–42. https://doi.org/10.2174/15733998113099990003.

Turner, V. (1977). *The ritual process: Structure and anti-structure*. Ithaca: Cornell University Press.

Usó, J. C. (2001). *Spanish trip. La aventura psiquedélica en España* [Spanish trip: The psychedelic adventure in Spain]. Barcelona: La Liebre de Marzo.

Winkelman, M. (2010). *Shamanism. A biopsychosocial paradigm of consciousness and healing* (2nd ed.). Santa Barbara: Praeger.

Chapter 10
Perspectives on Healing and Recovery from Addiction with Ayahuasca-Based Therapy Among Members of an Indigenous Community in Canada

Elena Argento, Rielle Capler, Gerald Thomas, Philippe Lucas, and Kenneth W. Tupper

Introduction

This chapter is based on an observational research project that assessed the impact of ayahuasca-based therapy on addiction and other substance use-related outcomes among members of a rural Indigenous community in Western Canada. These quantitative and qualitative findings summarize the lived experiences of participants and shed light on how and why ayahuasca delivered in a ritualized retreat setting influenced healing and recovery from addiction and overall psychosocial well-being.

Indigenous peoples in Canada and globally experience a disproportionate burden of social and health-related problems that are the legacies of Euro-American colonialism, racialized policies, and consequent territorial and cultural dislocation. Experiences of multigenerational trauma, including concurrent mental health and substance use disorders—with significant heterogeneity across settings and sub-populations—remain of critical concern, despite decades of ongoing intervention efforts (Chansonneuve 2007; Gracey and King 2009; Health Canada and Assembly of First Nations 2015). Conventional approaches to treating addictions have had limited success in breaking the formidable barriers to health and well-being faced by many Indigenous people who continue to experience the consequences of devastating disconnection from traditions, culture, language, and spirituality (Chansonneuve 2007; Gracey and King 2009).

E. Argento (✉) · R. Capler · K. W. Tupper
University of British Columbia, Vancouver, BC, Canada
e-mail: elena.argento@bccsu.ubc.ca; rielle.capler@bccsu.ubc.ca

G. Thomas · P. Lucas
Centre for Addictions Research of British Columbia, University of Victoria, Victoria, BC, Canada
e-mail: gthomas@okanaganresearch.com; philippe@tilray.ca

© Springer Nature Switzerland AG 2021
B. C. Labate, C. Cavnar (eds.), *Ayahuasca Healing and Science*,
https://doi.org/10.1007/978-3-030-55688-4_10

Ayahuasca is among various psychedelic substances demonstrated to have potential therapeutic benefits in both nonclinical and clinical settings (dos Santos et al. 2016; Johnson et al. 2017; Sessa 2012; Tupper et al. 2015). The Amazonian plant brew, typically made from plants containing harmala alkaloids (*Banisteriopsis caapi*) and dimethyltryptamine (*Psychotria viridis*), has a vibrant history of social, cultural, and medical uses.

Ayahuasca and other psychedelic plants have been used for centuries across diverse settings for a wide range of spiritual and therapeutic purposes, and in recent years, the traditional uses of ayahuasca among Indigenous peoples in South America and Brazilian folk religions have expanded rapidly into mainstream global culture (Labate and Cavnar 2018; McKenna and Riba 2015; Sánchez and Bouso 2015; Tupper 2008). A renewed interest in ayahuasca and other psychedelic-assisted therapies has reemerged among health researchers, receiving mounting and far-reaching attention for potentially powerful therapeutic benefits through the induction of altered states of consciousness (Mithoefer et al. 2016; Pollan 2015; Sessa 2012), including improving problematic substance use (Bogenschutz et al. 2015; Johnson et al. 2014; Krebs and Johansen 2012; Thomas et al. 2013), addressing trauma (Mithoefer et al. 2016), recovering from eating disorders (Lafrance et al. 2017), and promoting psychological well-being (Gasser et al. 2015; Griffiths et al. 2016; Hendricks et al. 2015; Osório et al. 2015; Ross et al. 2016).

Specifically, ceremonial and ritualistic use of ayahuasca has been associated with reductions in various substance use problems (Fábregas et al. 2010; Grob et al. 1996), and a recent open-label trial in Brazil demonstrated rapid and sustained antidepressant effects following a single dose of ayahuasca administered to individuals suffering from treatment-resistant depression (Sanches et al. 2016).

The primary psychoactive constituent in ayahuasca is dimethyltryptamine (DMT), which is an agonist for sigma-1 receptors and 5-HT$_{2A}$ serotonin receptors in the human brain (Fontanilla et al. 2009; McKenna and Riba 2015). Similar to other classical psychedelic substances (e.g., lysergic acid diethylamide (LSD), psilocybin, and mescaline), ayahuasca is believed to influence a group of interconnected brain regions known as the default mode network (DMN), which plays a key role in mediating cognitive states, such as daydreaming and mind-wandering (Carhart-Harris et al. 2014; Palhano-Fontes et al. 2015). The DMN has attracted considerable attention among cognitive neuroscientists, whereby neuroimaging research postulates that the DMN may be the physical counterpart to the psychological construct of the ego (Carhart-Harris et al. 2012; Carhart-Harris et al. 2014). The potential and tendency of ayahuasca to generate profound spiritual or "mystical-type" experiences, in which meaning, spirituality, connectedness, and recognition of the sacredness of life are enhanced and validated (Griffiths et al. 2006; Nour et al. 2017; Watts et al. 2017), may have utility in interventions to reduce problematic substance use (Garcia-Romeu et al. 2014).

Generating the Research

This research project developed after Gabor Maté, a Canadian physician specialized in addictions medicine with extensive experience working with Indigenous people, began using ayahuasca as an adjunct to group therapy in 2009. In partnership with a Peruvian Shipibo ayahuasquero and three of his Canadian apprentices, Maté's "Working with Addiction and Stress" retreats were being conducted to address various psychological health conditions, when a rural Coast Salish First Nations band learned of and became interested in the work being done with ayahuasca. After approval from the Band Council and Elders, the band invited the retreat team to hold the retreats in the community and offered our research group the opportunity to conduct an observational study to document and assess the impacts and outcomes of the retreats. Subsequently, two retreats were conducted in 2011, one in June and one in September.

The retreats employed multiple healing modalities, including group and individual counseling and two ayahuasca ceremonies. The retreat team facilitated group talk therapy sessions over a period of 4 days to elicit personal insights about historical experiences of trauma and related emotional or psychological issues, including problematic substance use and addictions. The ayahuasca ceremonies were presided over by a Shipibo master ayahuasquero alongside three (non-Indigenous) Canadian apprentice ayahuasqueros.

The research team obtained funding for the observational study through philanthropic donations to the Multidisciplinary Association of Psychedelic Studies (MAPS), the River Styx Foundation, and through an anonymous donor. Ethics approval was provided by the Institutional Review Board Services (IRBS), an independent research ethics review board in Ontario, Canada.

Description of the Study

The band's health office notified the community of the ayahuasca retreats and recruited potentially eligible participants. A total of 18 participants from the Coast Salish band were recruited for the two retreats. Eligibility criteria included being age 18 or older and having the ability to communicate in English. Participants were excluded if they had drunk ayahuasca in the past, were taking selective serotonin reuptake inhibitors or monoamine oxidase inhibitor medications, or were currently experiencing psychosis or had experienced a psychotic break in the recent past.

Initial self-administered surveys were conducted by the research team to collect baseline data on a number of psychological and behavioral factors related to problematic substance use. The specific psychometric instruments used in this study included the following: the *Difficulty in Emotion Regulation Scale* (DERS), the

Philadelphia Mindfulness Scale (PHLMS), the *Empowerment Scale* (ES), the *Hope Scale* (HS), the *McGill Quality of Life* survey (MQL), and the *4 Week Substance Use Scale* (4WSUS). Scale scores were calculated for the following measures: *mindfulness, empowerment, emotional regulation, hopefulness,* and *quality of life* (which included five subscales: overall quality of life, physical symptoms, psychological symptoms, outlook, and meaning). The instruments were selected based on a holistic interpretation of the biopsychosocial-spiritual model of human behavior, trauma, and substance dependence. More detail on the instruments, explicit rationales for their inclusion, and statistical analyses can be found in a previously published paper (Thomas et al. 2013).

Follow-up data collection was conducted in a group setting at the band's health office at 2 weeks and 4 weeks after the retreats and monthly thereafter for 6 months. For those unable to attend in-person follow-ups, sessions were conducted by telephone, and occasionally the instruments were completed by the participant on their own time and returned to the research team. A $20 gift certificate to a grocery store was offered to participants at each follow-up session, and meals were provided at in-person sessions.

At the request of several study participants to share their experiences in their own words, the research team added a short semi-structured interview as part of the final (6-month) follow-up session of the study to collect qualitative data and provide context around participants' experiences and impressions during and after the retreats. The interviews were conducted either in person or over the phone and were audio recorded and transcribed. Thematic analysis was performed to identify key themes (Argento et al. 2019). The questions included in the semi-structured interviews were as follows:

1. Did the addiction and stress retreat (including the two ayahuasca ceremonies) have any impact on your life? If "yes" to question 1, please answer questions 2 and 3.
2. On a scale of 1–10, with 1 being extremely negative and 10 being extremely positive, how would you rank the impact of the addiction and stress retreat on your life?
3. Please describe the impact that the addiction and stress retreat (and ayahuasca ceremonies) had on your life, for example, your substance use, personal relationships, sense of self, and your connection with nature and/or spirit.

Quantitative Analysis of Therapeutic Effects

Of the total 18 retreat participants, quantitative analyses were conducted for 12 participants who attended at least one retreat and had no missing data. Results demonstrated statistically significant improvements ($p < 0.05$) over time for measures relating to *mindfulness, empowerment, hopefulness,* and two *quality of life* measures: meaning and outlook (Table 10.1). Among participants who completed the final 6-month follow-up (n = 11), data suggest that fewer participants used alcohol,

Table 10.1 Repeated measures ANOVA for psychometric scales

Scale	F score	P value	N	Cronbach's alpha
Hope	3.14	0.023	12	0.456
Empowerment*	5.07	0.002	10	0.702
Mindfulness*	2.76	0.041	11	0.408
Emotional regulation*a	1.96	0.124	9	0.929
Quality of life—overall*	2.05	0.105	11	n/a
Quality of life—psychological*	2.52	0.055	12	0.736
Quality of life—meaning*	4.36	0.005	12	0.879
Quality of life—outlook*	4.43	0.004	12	0.925

*Sphericity is assumed as Mauchly's W is >0.05 for all measures
aLower scores equal greater emotional regulation

Table 10.2 Proportion of participants who reported substance use at baseline compared with final follow-up ($n = 11$)

Substance	Proportion that used at baseline (%)	Proportion that used at final follow-up (%)
Tobacco	81.8	63.6
Alcohola	50	20
Cannabis	45.5	45.5
Cocainea	60	0
Amphetamines	0	0
Inhalants	0	0
Sedatives	18.2	18.2
Hallucinogens	0	9.1
Opioids	9.1	9.1

aN = 10; 1 participant did not answer the question at one phase

tobacco, and cocaine in the 4 weeks preceding the final follow-up than in the 4 weeks preceding the retreats at baseline (Table 10.2).

Figure 10.1 displays changes in the 4-Week Substance Use Scale (4WSUS) scores, measuring changes in participants' levels and patterns of use, cravings, and substance use-related harm for the substances used by at least 40% of the participants at baseline: tobacco, alcohol, cannabis, and cocaine. All 4WSUS scores decreased except for cannabis. Statistically significant reductions were demonstrated for problematic cocaine use over the study period (Thomas et al. 2013).

Perspectives on Healing and Recovery from Addiction

Qualitative interviews were conducted with 11 of the 12 participants who participated in the retreats and quantitative study. Participants included six men and five women with a mean age of 38 (ranging from 19 to 56 years of age). The primary objective of the semi-structured interviews was to elicit participants' perspectives

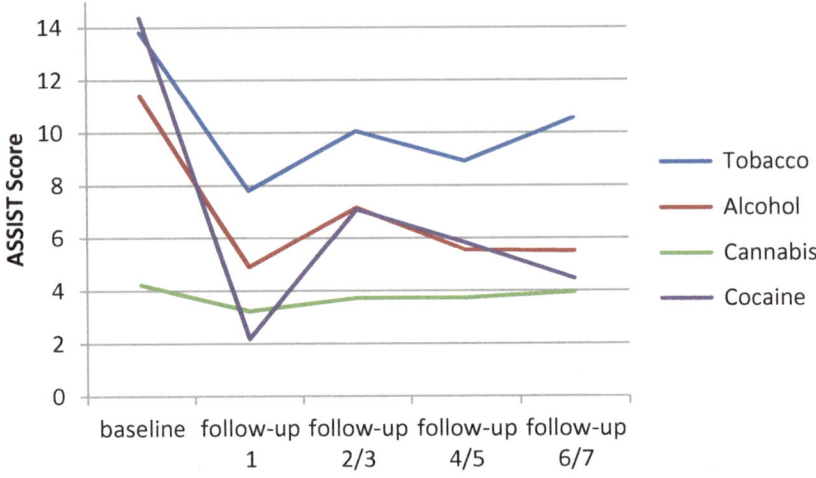

Fig. 10.1 Changes in 4WSUS scores for tobacco, alcohol, cannabis, and cocaine

and deeper understanding of their experiences with the ayahuasca retreats. In addition to eliciting responses around changes in substance use and addiction, the interviews explored how the retreats influenced participants' sense of connectedness with self, spirit and nature, and personal relationships with others (family, friends, and community). The qualitative findings expanded upon the prior quantitative analysis to provide a deeper understanding of the emotional, physical, psychological, spiritual, and interpersonal contexts around ayahuasca-based therapy for addiction (Argento et al. 2019).

Reductions in Substance Use and Addiction All participants described the retreats as having a positive impact on their substance use, coinciding with significant life improvements and healthier relationships with themselves and others. The retreats resulted in considerable reductions in substance use and the desire to use drugs and, for some, complete abstinence from alcohol and other drug use. Some participants explained that ayahuasca-based therapy helped them to better understand the underlying pain or trauma behind their substance use (e.g., losing loved ones), allowing them to address the causes of their addiction:

Before the ceremony I was struggling with my addiction, crack cocaine, for many years. When I went to this retreat, it more or less helped me release the hurt and pain that I was carrying around and trying to bury…with drugs and alcohol. Ever since this retreat I've been clean and sober. So, it had a major impact on my life in a positive way. It affected my life in giving me another chance at life. (P4, 41-year-old woman)

It [the retreats] put my substance [use] in my face and I just faced the problems that I was having that was keeping me in my abuse, and it… mostly had to do with my grief and loss of my twin brother and my… infant daughter that passed away. I just kept living it over and over again until I finally faced it and I just feel that there was closure. I don't use any more. I don't use anything. I don't smoke or anything. (P8, 34-year-old man)

The impact of the retreats on reducing substance use and cravings was a powerful theme to emerge from the narratives. Some participants reported a complete absence of cravings following the retreats, alongside a visceral, physical rejection by their bodies of any substance that was previously being used problematically, including tobacco, alcohol, pharmaceutical painkillers, and cocaine:

> When I left ayahuasca... I bought a pack of smokes, and I pretty much vomited... It was literally like my body would not absorb the nicotine... So that was really amazing... because I had this old habit of going to nicotine when overwhelmed with stress.... So, it helped me... be with all of the emotions and...not need a substance...helped me eliminate an old behavior... it's quite the journey. Whew [laughter]. (P3, 22-year-old woman)

> I'm off all my painkillers. I was on 50 mg of Fentanyl patches, I was on...oxycodone, you name it, I was on it. But I'm off everything...I believe that...my ayahuasca experience, won't let me put any more drugs in my body... It's rejecting it. Because it's making me physically ill... I was drinking, I was drugging, I was cracking out, I was IV drug using. I was a hard-core drug addict. And now I'm just down to maybe one or two medicinal marijuana joints a day. I'm off everything. (P1, 49-year-old woman)

> No cravings whatsoever for the crack cocaine or drinking. It's pretty strong, that ayahuasca, as far as removing that craving, that desire, that habit, or however you want to describe it. For me, it's not even there. (P7, 56-year-old man)

The narratives indicated that the retreats allowed some participants to reflect upon their addiction and elicited conscious contemplation of their substance use, leading to more acceptance of the nature of their addiction and a better ability to cope:

> It [the retreat experience] definitely makes me think more. Like, I am, like, I have cravings... I think I have more of a logical approach to it instead of just getting frustrated with myself... You know, I can say yeah, okay, I'm having a craving... I'm able to deal with it, cope with it easier... I definitely feel stronger. (P11, 30-year-old man)

> Overall, participants reported that the retreats had an overwhelmingly positive impact on reducing problematic substance use and addiction. The narratives portrayed considerable shifts in the ways participants perceived and experienced their substance use, and participants attributed these changes to the insights they gained into the reasons behind their addiction following the retreats. All participants noted that the retreats led to a reduction or, in some instances, a complete elimination of substance use. One participant explained: "I felt free of my addiction, that's for sure... I stopped completely. I had no desire to use" (P5, 19-year-old woman).

Comparison of Ayahuasca-Based Therapy with Other Treatment Approaches Participants reported that the retreats were more effective in treating addiction than their other treatment experiences. Some explained that other forms of therapy were often unsuccessful in addressing the underlying emotional issues related to their addiction. The narratives indicated that the success of the retreats was largely attributed to the ways in which ayahuasca-based therapy addressed unresolved trauma and emotional issues. Many of the participants' accounts provided greater insights into how and why the retreats and ayahuasca ceremonies were so successful at diminishing addiction and cravings:

I'm on methadone and that did not work… and after that [retreat] I had no desire… I don't know what it is about that, but it really, it is very life changing. (P5, 19-year-old woman)

This treatment is better, man. I know it is. Maybe half a dozen people that have gone to treatment over and over and over again… are now [following the retreats] …finally clean. (P6, 32-year-old man)

I went to other treatments and I thought I dealt with it and it kept coming back to me. This [retreat] got into my mind and into me better. I felt I got more out of it and it was… a short period of time. It's exactly what I needed… things have changed now. I feel a lot better… healthier. My mind's more clear. (P8, 34-year-old man)

Other treatments sort of like scraped the surface, as they say. This one got me… deep, deep into myself which I've never… confronted, I guess you could say, in the other treatments. This was just a mind-bending experience, boy (laughter). I can't believe what I seen and who I talked to, like, my mom and my dad and my granddaughter who are in the next world there [i.e., deceased]. And it was really, really touched me deeply and I think about that every day. So, there you go. (P9, 55-year-old man)

The narratives elucidated that the retreats cultivated an environment, whereby participants could access parts of the psyche that had been inaccessible to them with other treatment approaches, facilitating a healing of the foundational emotional issues underlying addiction:

It was a whole new experience for me. I don't think I could compare it to anything else… this one was like… just going inside yourself and, like, in a different way… it was way more spiritual than any other treatment because the plant [ayahuasca] intensified it. (P2, 28-year-old woman)

Enhanced Connectedness with Self, Spirit, and Nature   One of the most profound and consistent themes to emerge from the narratives was an enhanced sense of connectedness with self, spirit, and nature following the retreats, which was described as a key element associated with reduced substance use and cravings. The narratives suggested that participating in the ayahuasca retreats facilitated a connection to a "higher self" and provided meaningful insights into the psyche or emotional and psychological states related to addiction:

Ayahuasca connected me to my higher self in a way… phenomenal changes, it's indescribable. It's just profound, big changes and big strides in areas of my life. Being free from myself… holding myself hostage through my addiction. (P7, 56-year-old man)

The impact was huge on my spiritual and my emotional side. I opened up hugely in the emotional department… I feel stronger… my last experience with ayahuasca, I really faced myself. Like, my fear, my anger. Which, really, I think is a big part of my addictions. (P11, 30-year-old man)

Some participants indicated that the retreat experiences altered negative thought patterns to help gain important insights into psychological barriers to healing, such as relieving feelings of grief and facilitating self-acceptance and self-love. Others described feeling an emotional opening or release of repressed emotions after drinking ayahuasca:

It relieved... a lot of negative thoughts within my body. Thoughts of what my life was like. It... opened my eyes to see where my stress and conflict is coming from... It's hard to explain but... it just brought a lot of grief up that I had inside me; it brought it out and I got rid of a lot of grief. (P9, 55-year-old man)

It really opened my eyes...taking ayahuasca... It brought me back to my childhood and what I didn't like growing up, alcohol at home and abuse of home... and it finally opened my eyes to a realization of what I was doing to myself and my daughter. (P4, 41-year-old woman)

It's actually helping me to do my psychological work... it made me look at myself and see how I react to the world, and how I can shift in the way that I actually, truly desire which is, you know, unconditional love. That's where I want to go. And so, ayahuasca has helped me, the whole retreat, to eliminate those barriers so I can... keep on healing and go forward and accomplish what I want to. (P3, 22-year-old woman)

I felt like, just like a whole new reborn person. And yeah, I was really happy. I hadn't felt that happy in a long, long time. I felt way better about myself (P5, 19-year-old woman)

Ayahuasca-based therapy was also described as helping some participants to achieve more balance in life, gain a deeper understanding of the importance of stillness and mindfulness, and harness courage to confront emotions with acceptance and patience in the process of healing:

It gave me the courage to go deeper... I found out why I didn't feel worthy, or didn't feel good enough to do anything... before I ever did ayahuasca, I had absolutely no balance in my life... I'm not on stimulants anymore... I always get up at five in the morning... because it's the only time that it's quiet in my house. I go in my lazy boy and just stare at my wolves... or I just sit there and meditate. (P1, 49-year-old woman)

I realize that I deserve a better life and I love myself... After the retreat I felt like a brick that was lifted off of my shoulders and just feeling free. (P4, 41-year-old woman)

Participants further reported that the retreat experiences elicited a profound sense of spirituality and appreciation for the natural world. One participant reported that the impact of the retreat was "spiritually huge" (P11, 30-year-old man). Most participants described a heightened awareness of beauty in nature and a deeper sense of connection and gratitude:

About a week or two after [the retreat] I was just waking up every morning at like five, six and going outside... to have a big connection with nature and just the trees moving, I just sat and stared at the trees and the wind for, like, 2 h. I would sit outside and it was just beautiful. I've never noticed it that much, ever, in my life. And, after I had the ayahuasca, it was just amazing, the connection with nature. (P5, 19-year-old woman)

I had no sense of spirituality before, really, coming clean and sober even while I was going through, like, AA and NA. They tell you to reach your higher power or whatever. I thought that was a bunch of bull. But, after the retreats, I've really opened up to spirituality big time. I smudge every night before bed. I pray... I say thanks to whatever is out there, you know? (P11, 30-year-old man)

While some participants expressed having a well-established sense of spirituality prior to the retreats, narratives suggest that the connectedness to spirit and nature was intensified or renewed after participating in ayahuasca-based therapy:

I just feel that definitely it [ayahuasca] got me more in touch with my spirit... I already was in touch with it. It just seems like it's more intensified, where everything is, like, has meaning and there is a way of life...Well, it seemed like it brought it out more 'cause it was there. But now it's there even more. (P2, 28-year-old woman)

I've always been connected to nature and spirit bathing [a traditional Coast Salish spiritual ritual], but, upon ayahuasca, it's a far greater connection... to all living things. It's grounding, like feeling cleansed and purified. Grounded to Mother Earth. (P7, 56-year-old man)

Transformations in Personal Relationships Many participants reported that the retreats had a positive impact upon their personal relationships and that subsequent communication with close friends and family became easier and more constructive. For one participant, the retreats facilitated feelings of joy and happiness in reconnecting with his family, allowing him to spend time with his grandchildren again. He also reported that his friends could see a change in him and noted that he was "glowing" (P10, 51-year-old man). A common theme to emerge from the interviews was that the retreats facilitated more open and less judgmental interactions between participants and their friends and family, fostering more feelings of love for others:

Since doing the ceremonies... I've grown hugely in that I can accept more. I'm more open to caring for other people... I think I have more love and respect for the people in my life. You know, more gratitude. (P11, 30-year-old man)

It helped me see that they [friends] are actually trying and to eliminate judgment and have only love for them. So, it's helped me in a lot of ways to love, to love people on so many levels. (P3, 22-year-old woman)

There have been a lot of changes, but all for the positive. There was a little rough spell there for a while, but, after the ayahuasca ceremonies, it allowed me to address things that were holding me back and upon my experiences that I had, showed me a way, a healthier way of [being] in my relationships... opened up where I felt I had a door closed to allow... my close family members inside. (P7, 56-year-old man)

One participant described how the retreats cultivated new friendships with other study participants and helped to build mutual trust and understanding for the challenges faced by people in the community. It was reported that the opportunity to share stories with other study participants made it easier to be nonjudgmental and enhanced a sense of community:

As for relationships with the participants... we've become friends and it's really nice to actually have friends in the community. In First Nations communities, the norm is, usually, we are all separated. The retreat brought in different people and I've been able to connect with them and it's really nice... I got a sense of where the community is at and... I got to hear their stories. They shared their past and I was like, okay, they are the way they are because of multiple reasons. (P3, 22-year-old woman)

Overall Reflections on the Retreat Experiences The narratives demonstrated that participants' overall experiences with the retreats were predominantly positive and transformative. Participants were asked if the retreats had any impact on their lives and to rank the experience on a scale of 1 (extremely negative) to 10 (extremely positive). The mean ranking was 7.95, and most participants (n = 8) gave their experience a ranking of 8 or higher (range = 5–10).

Overall, the narratives demonstrated a great appreciation for the entire retreat experience and admiration for all who guided the participants through the healing and recovery process, including the spirit of ayahuasca itself, the shamans, and the physician who facilitated the retreats:

> I admire the shamans… they were given the gift by the Creator, and to see people wanting to help First Nations individuals who live in addiction… to have this group really open-minded and willing to come to the community and help, that was an honor. As for the doctor… I admire him… it was just amazing to see the patience… and determination… his being… so intelligent and understanding human behavior, past history, present, future… oh man… I'm floored… I'm really shy… so to be able to interact with people and feel comfortable around them… was really nice… It's just real. (P3, 22-year-old woman)

> I felt honored and proud… I feel like I'm a success through the program, right? (P1, 49-year-old woman)

> I wish I was introduced to it [ayahuasca], like, 20 years ago. It could have saved me a lot of time and trouble. (P11, 30-year-old man)

Participants conveyed deep gratitude for the retreat experiences and an enhanced appreciation for life, highlighting how ayahuasca-based therapy led to reduced substance use issues and benefited overall health and well-being. Participants reported that valuable tools and knowledge were gained from the retreats that could be applied to aspects of daily life:

> The ayahuasca retreat was overwhelmingly educational. Like, from the research, to the research committee, to the ayahuasca group, to the doctor, even to the participants. Like, everybody educated me in one way or another and I left with an immense amount of knowledge that I was able to attempt to apply to my daily life. (P3, 22-year-old woman)

> I'm just grateful that this has happened because it made me realize how much my life is to me now and how much my daughter's life is to me, my husband, so much better without the drugs and alcohol. (P4, 41-year-old woman)

> I just enjoyed it and I'm glad I went to it. I feel a lot better and grateful for what I have today. (P8, 34-year-old man)

One participant described the retreat as the "best experience I've had in my life" (P5, 19-year-old woman). Another participant described it as "the most awesome experience I've ever experienced in my life" (P1, 49-year-old woman). Some participants also expressed that they would like to see ayahuasca-based therapy become available to others who suffer from mental and substance use disorders and that the overall experience is one that they would highly recommend to others:

> I'm looking forward to having that experience again. It was probably the best experience I've had in my life. And I haven't had a very good life. I've lived a very shitty life around people in addiction in my family and everything… doing that, being around all the people and after the feeling, it was just amazing. I don't know what else to really say… that lifestyle is very hard to get out of. And, I didn't think it was possible. I've tried to do it on my own and I don't know what it was but it actually it worked. (P5, 19-year-old woman)

I think it [ayahuasca] is a good medicine. Why? Because it can shift energy and help people clear up some things so they can continue on their journey... Because we are so unaware about if we are holding things in or not... We are almost naïve to the expansiveness of the Creator and all living creatures... So, I'm really happy that... they can shift energies of people and help them on their journey. I highly advise it to be used. It [ayahuasca]'s a nice steppingstone to another level. (P3, 22-year-old woman)

The narratives unveiled another emergent theme around intercultural aspects of ayahuasca-based therapy. Ayahuasca was described as an important plant medicine that has therapeutic utility across Indigenous communities. One participant explained:

It's an Aboriginal plant and we are Aboriginal people. I pray every morning that they let it be practiced within our communities... there is nothing wrong with Aboriginal medicine bringing in to another Aboriginal community. (P1, 49-year-old woman)

Limitations

The study design was limited to a small convenience sample of those who self-identified as having an interest in participating in the ayahuasca retreats, introducing the potential for self-selection bias. It is not possible to make any claims about whether the observed positive effects of ayahuasca-based therapy may be generalizable to other Indigenous peoples in Canada or elsewhere in the world. The study was not designed to assess the relative effects of group therapy work, other ancillary potentially therapeutic elements (e.g., spirit baths, sweat lodges), the pharmacological action of the ayahuasca, or the psychodynamic context of the retreats combining these various elements and did not track whether participants were involved in other forms of treatment during the follow-up period. Given the complex pharmacology of ayahuasca and unknown chemical composition of the brew drunk during the ceremonies, it is not possible to know whether the outcomes were dependent upon varying amounts or relative concentrations of the psychoactive components ingested. Nevertheless, as the narratives suggest, the effects of the brew were characteristic of ayahuasca phenomenology and therefore do not suggest any concerns about its composition or potency. Given the small number of participants and that some participated in two retreats while others participated in only one retreat, no conclusion can be drawn from this study on the potential harms/benefits of additional treatments.

There are notable challenges to conducting research on the effects of ayahuasca, some of which are ubiquitous to the broader field of research on illicit substance use and some that are unique to the study of plant medicines and psychedelics more generally. Firstly, the stigmatization and legal restrictions around ayahuasca and other psychedelics have made it difficult to obtain funding and government approval for research. Substantial barriers to obtaining financial and sociopolitical support for this work have considerably hampered the progress of academic and scientific efforts. Randomized clinical trials, which have become the gold standard, are somewhat at odds with investigating the therapeutic potential of ayahuasca and other psychedelic substances. With respect to the complex nature of ayahuasca's plant brews and mixtures, as well as the associated ceremony, the stringent controls on the set and setting inherent to clinical trials, which are designed to isolate the effects of the drug, result in poor generalizability to populations outside the study. Clinical trials are further challenged with respect to ensuring double blinding, although a recent trial in Brazil implemented a series of measures in a double-blind placebo-controlled study of ayahuasca (Palhano-Fontes et al. 2018). Further, given the high cost of running clinical trials, they are liable to falling short on therapeutic preparation and integration of the ayahuasca experience. As such, mixed methods and qualitative approaches are needed in the field of research on ayahuasca. The qualitative aspect of this research added near the end of the project at the urging of participants allowed for a deeper and more personal examination of the role of internal and external connectedness in the healing journey, a crucial factor that was missed in the initial conceptualizations of the project.

Specific to studying ancient plant medicines such as ayahuasca, researchers are faced with extra challenges posed by the traditional cultural and spiritual contexts from which these plant medicines have emerged (Fotiou 2016). Of note, the ayahuasca-based therapy retreat in the present study entailed introducing a geographically and culturally distinct plant spirit and ceremony into an Indigenous community. The potential impact of the intermingling of spirits and ceremonies had to be carefully considered by the Coast Salish First Nation's Band Council and Elders before they extended their approval for the retreats to take place in their community.

Another important issue inherent to research on traditional plant medicines involves the inclusion of Indigenous knowledge. While a more in-depth discussion is beyond the scope of this chapter, Indigenous cultures have been working with ayahuasca for thousands of years, and critical questions remain around cultural appropriation, ecological sustainability, and how to best integrate the deep well of Indigenous knowledge and experience within research and treatment approaches (Labate and Cavnar 2018).

As with other research on substance use among marginalized populations, community-based and trauma-informed approaches are critical to addressing the needs faced by these populations (Substance Abuse and Mental Health Services Administration [SAMHSA], 2014). In the case of working with Indigenous communities, an aspect of primary importance in conducting our research was that it was guided by the principles of ownership, control, access, and possession (OCAP®), developed by the National Steering Committee of the First Nations Regional Health Survey (Schnarch 2004). Research by Indigenous scholars has further highlighted culture-based approaches and decolonizing methodologies as necessary to understanding Indigenous peoples' research experiences and ensuring benefit to their communities (Goodman et al. 2017).

Final Thoughts

Findings from this study illuminate the emotional, physical, psychological, spiritual, and interpersonal contexts around ayahuasca-based therapy for treating addiction and addressing other substance use-related outcomes, with a central theme describing enhanced connectedness with sense of self, spirit and nature, and improved personal relationships. The results from this study corroborate and expand upon findings from previous studies and a growing body of research describing the therapeutic potential of ayahuasca and other psychedelics. As evidenced in this study, ayahuasca-based therapy has the potential to significantly alter one's relationship with self, spirit, and the natural world in ways that can facilitate healing and recovery from addiction. Experiencing enhanced connectedness was characterized in the narratives as a key element of the psychosocial processes associated with reducing substance use and addiction issues. A recent qualitative analysis further highlights the importance of transitioning from disconnection to enhanced connectedness in benefiting patients in the context of psilocybin-assisted therapy for treatment-resistant depression (Watts et al. 2017). Findings are also aligned with the well-developed literature on the protective effects of meaningful connection with others, as extensively documented in the youth prevention literature (Centres for Disease Control and Prevention 2009; Sieving et al. 2017).

The potential and tendency of psychedelics to generate "mystical-type" experiences, whereby personal meaning and the sacredness of life is enhanced and validated (Griffiths et al. 2006), may play a key role in reducing problematic substance use, by introducing much-desired meaning and purpose into the lives of those suffering from addiction (Garcia-Romeu et al. 2014). Furthermore, a core belief among Indigenous people in Canada, and elsewhere, is that there exists an interconnectedness of all things (e.g., people, animals, the earth, nature) and that a holistic approach to healing and recovery from addiction is one that encompasses a renewal of spirituality, a recovery of awareness, and the therapeutic power of the natural world (Chansonneuve 2007). The foundation of recovery from addiction is comprised of overlapping and interdependent dimensions of the psyche and the spirit (Mustain and Helminiak 2015), which is reiterated within Indigenous belief systems, whereby strengthening connections with self, spirit, and nature, alongside self-actualization, helps to achieve balance and attain healing (Chansonneuve 2007). Supportive findings from research using neuroimaging suggest psychedelics may allow for vivid recollection of autobiographical memories and may alter neural pathways to shift cognitive biases (i.e., habits of thought) and facilitate the positive reprocessing and reconciliation of traumatic memories (Bogenschutz and Pommy 2012; Carhart-Harris et al. 2012).

Notably, most participants in the present study ranked their ayahuasca-based therapy experience very highly in terms of the impact it had on their lives, and the majority emphasized the significance of the spiritual component of the experience. Narratives communicated deep gratitude for ayahuasca-based therapy and reiterated that the emotional clarity, self-acceptance, and positive transformations that occurred within participants, and in their relationships with others and the world around them, had not occurred with other forms of addictions treatment.

Overall, this study highlights the therapeutic potential of ayahuasca-based therapy to facilitate deep and sustained healing and recovery from addiction for members of an Indigenous community in Canada, including some who had experienced highly traumatic histories and several failed treatment attempts. Findings describe the ways in which ayahuasca delivered in a ritualized and therapeutic retreat setting led to significant reductions in problematic substance use and cravings, as well as important enhancements to how participants relate to themselves, nature and spirit, and key individuals in their social support networks. Given the heterogeneity of Indigenous populations and distinct vulnerabilities associated with living in rural and remote settings, this study encourages further mixed-methods research tailored to the needs of marginalized populations. Future research should explore how ayahuasca and other psychedelics affect connectedness and other factors that may improve well-being and facilitate recovery from addiction.

References

Argento, E., Capler, R., Thomas, G., Lucas, P., & Tupper, K. (2019). Exploring ayahuasca-assisted therapy for addiction: A qualitative analysis of preliminary findings among an Indigenous community in Canada. *Drug and Alcohol Review, 38*(7), 781–789. https://doi.org/10.1111/dar.12985.

Bogenschutz, M. P., & Pommy, J. M. (2012). Therapeutic mechanisms of classic hallucinogens in the treatment of addictions: From indirect evidence to testable hypotheses. *Drug Testing and Analysis, 4*(1), 543–555. https://doi.org/10.1002/dta.1376.

Bogenschutz, M. P., Forcehimes, A. A., Pommy, J. A., Wilcox, C. E., Barbosa, P. C. R., & Strassman, R. J. (2015). Psilocybin-assisted treatment for alcohol dependence: A proof-of-concept study. *Journal of Psychopharmacology, 29*(3), 289–299. https://doi.org/10.1177/0269881114565144.

Carhart-Harris, R. L., Leech, R., Williams, T. M., Erritzoe, D., Abbasi, N., Bargiotas, T., Hobden, P., Sharp, D. J., Evans, J., Feilding, A., Wise, R. G., & Nutt, D. J. (2012). Implications for psychedelic-assisted psychotherapy: Functional magnetic resonance imaging study with psilocybin. *British Journal of Psychiatry, 200*(3), 238–244. https://doi.org/10.1192/bjp.bp.111.103309.

Carhart-Harris, R. L., Leech, R., Hellyer, P. J., Shanahan, M., Feilding, A., Tagliazucchi, E., Chialvo, D. R., & Nutt, D. J. (2014). The entropic brain: A theory of conscious states informed by neuroimaging research with psychedelic drugs. *Frontiers in Human Neuroscience, 8*, 20. https://doi.org/10.3389/fnhum.2014.00020.

Centers for Disease Control and Prevention. (2009). *School connectedness: Strategies for increasing protective factors among youth*. Atlanta. https://www.cdc.gov/healthyyouth/protective/pdf/connectedness.pdf.

Chansonneuve, D. (2007). *Addictive behaviours among aboriginal people in Canada* (The aboriginal healing foundation research series). Ottawa. http://www.ahf.ca/downloads/addictive-behaviours.pdf.

dos Santos, R. G., Osorio, F. L., Crippa, J. A. S., Riba, J., Zuardi, A. W., & Hallak, J. E. C. (2016). Antidepressive, anxiolytic, and antiaddictive effects of ayahuasca, psilocybin and lysergic acid diethylamide (LSD): A systematic review of clinical trials published in the last 25 years. *Therapeutic Advances in Psychopharmacology, 6*(3), 193–213. https://doi.org/10.1177/2045125316638008.

Fábregas, J. M., González, D., Fondevila, S., Cutchet, M., Fernández, X., Barbosa, P. C. R., Alcázar-Córcoles, M. Á., Barbanoj, M. J., Riba, J., & Bouso, J. C. (2010). Assessment of addiction severity among ritual users of ayahuasca. *Drug and Alcohol Dependence, 111*(3), 257–261. https://doi.org/10.1016/j.drugalcdep.2010.03.024.

Fontanilla, D., Johannessen, M., Hajipour, A., Cozzi, N., Jackson, M., & Ruoho, A. (2009). The hallucinogen N,N-Dimethyltryptamine (DMT) is an endogenous sigma-1 receptor regulator. *Science, 323*(5916), 934–937. https://doi.org/10.1126/science.1166127.The.

Fotiou, E. (2016). The globalization of ayahuasca shamanism and the erasure of Indigenous shamanism. *Anthropology of Consciousness, 27*(2), 151–179. https://doi.org/10.1111/anoc.12056.

Garcia-Romeu, A., Griffiths, R. R., & Johnson, M. W. (2014). Psilocybin-occasioned mystical experiences in the treatment of tobacco addiction. *Current Drug Abuse Reviews, 7*(3), 157–164. http://www.drugandalcoholdependence.com/article/S0376-8716(14)00255-5/fulltext.

Gasser, P., Kirchner, K., & Passie, T. (2015). LSD-assisted psychotherapy for anxiety associated with a life-threatening disease: A qualitative study of acute and sustained subjective effects. *Journal of Psychopharmacology (Oxford, England), 29*, 1. https://doi.org/10.1177/0269881114555249.

Goodman, A., Fleming, K., Markwick, N., Morrison, T., Lagimodiere, L., & Kerr, T. (2017). "They treated me like crap and I know it was because I was Native": The healthcare experiences of Aboriginal peoples living in Vancouver's inner city. *Social Science and Medicine, 178*, 87–94. https://doi.org/10.1016/j.socscimed.2017.04.050.

Gracey, M., & King, M. (2009). Indigenous health part 1: Determinants and disease patterns. *The Lancet, 374*(9683), 65–75. https://doi.org/10.1016/S0140-6736(09)60914-4.

Griffiths, R. R., Richards, W. A., McCann, U., & Jesse, R. (2006). Psilocybin can occasion mystical-type experiences having substantial and sustained personal meaning and spiritual significance. *Psychopharmacology, 187*(3), 268–283. https://doi.org/10.1007/s00213-006-0457-5.

Griffiths, R. R., Johnson, M. W., Carducci, M. A., Umbricht, A., Richards, W. A., Richards, B. D., Cosimano, M. P., & Klinedinst, M. A. (2016). Psilocybin produces substantial and sustained decreases in depression and anxiety in patients with life-threatening cancer: A randomized double-blind trial. *Journal of Psychopharmacology, 30*(12), 1181–1197. https://doi.org/10.1177/0269881116675513.

Grob, C. S., McKenna, D. J., Callaway, J. C., Brito, G. S., Neves, E. S., Oberlaender, G., Saide, O. L., Labigalini, E., Tacla, C., Miranda, C. T., Strassman, R. J., & Boone, K. B. (1996). Human psychopharmacology of hoasca, a plant hallucinogen used in ritual context in Brazil. *The Journal of Nervous and Mental Disease, 184*(2), 86–94. https://doi.org/10.1097/00005053-199602000-00004.

Health Canada and Assembly of First Nations. (2015). *First nations mental wellness continuum framework – summary report*. Ottawa. https://www.canada.ca/en/health-canada/services/first-nations-inuit-health/reports-publications/health-promotion/first-nations-mental-wellness-continuum-framework-summary-report.html.

Hendricks, P., Thorne, C. B., Clark, C. B., Coombs, D. W., & Johnson, M. W. (2015). Classic psychedelic use is associated with reduced psychological distress and suicidality in the United States adult population. *Journal of Psychopharmacology, 29*(3), 280–288. https://doi.org/10.1177/0269881114565653.

Johnson, M. W., Garcia-Romeu, A., Cosimano, M. P., & Griffiths, R. R. (2014). Pilot study of the 5-HT2AR agonist psilocybin in the treatment of tobacco addiction. *Journal of Psychopharmacology, 28*(11), 983–992. https://doi.org/10.1177/0269881114548296.

Johnson, M. W., Garcia-Romeu, A., Johnson, P. S., & Griffiths, R. R. (2017). An online survey of tobacco smoking cessation associated with naturalistic psychedelic use. *Journal of Psychopharmacology, 31*(7), 841–850. https://doi.org/10.1177/0269881116684335.

Krebs, T. S., & Johansen, P.-Ø. (2012). Lysergic acid diethylamide (LSD) for alcoholism: Meta-analysis of randomized controlled trials. *Journal of Psychopharmacology, 26*(7), 994–1002. https://doi.org/10.1177/0269881112439253.

Labate, B. C., & Cavnar, C. (Eds.). (2018). *The expanding world ayahuasca diaspora: Appropriation, integration and legislations*. London: Routledge.

Lafrance, A., Loizaga-Velder, A., Fletcher, J., Renelli, M., Files, N., & Tupper, K. W. (2017). Nourishing the spirit: Exploratory research on ayahuasca experiences along the continuum of recovery from eating disorders. *Journal of Psychoactive Drugs, 45*(5), 427–435. https://doi.org/10.1080/02791072.2017.1361559.

McKenna, D., & Riba, J. (2015). New world tryptamine hallucinogens and the neuroscience of ayahuasca. *Current Topics in Behavioral Neurosciences, 36,* 283–311. https://doi.org/10.1007/7854_2015_368.

Mithoefer, M. C., Grob, C. S., & Brewerton, T. D. (2016). Novel psychopharmacological therapies for psychiatric disorders: Psilocybin and MDMA. *The Lancet Psychiatry, 3*(5), 481–488. https://doi.org/10.1016/S2215-0366(15)00576-3.

Mustain, J. R., & Helminiak, D. A. (2015). Understanding spirituality in recovery from addiction: Reintegrating the psyche to release the human spirit. *Addiction Research & Theory, 23*(5), 364–371. https://doi.org/10.3109/16066359.2015.1011623.

Nour, M. M., Evans, L., & Carhart-Harris, R. L. (2017). Psychedelics, personality, and political perspectives. *Journal of Psychoactive Drugs, 49*(3), 182–191. https://doi.org/10.1080/02791072.2017.1312643.

Osório, F. L., Sanches, R. F., Macedo, L. R., dos Santos, R. G., Maia-de-Oliveira, J. P., Wichert-Ana, L., Araujo, D. B., Riba, J., Crippa, J. A., & Hallak, J. E. (2015). Antidepressant effects of a single dose of ayahuasca in patients with recurrent depression: A preliminary report. *Revista Brasileira de Psiquiatria, 37*(1), 13–20. https://doi.org/10.1590/1516-4446-2014-1496.

Palhano-Fontes, F., Andrade, K. C., Tofoli, L. F., Jose, A. C. S., Crippa, A. S., Hallak, J. E. C., Ribeiro, S., & De Araujo, D. B. (2015). The psychedelic state induced by ayahuasca modulates the activity and connectivity of the default mode network. *PLoS One, 10*(2), 1–13. https://doi.org/10.1371/journal.pone.0118143.

Palhano-Fontes, F., Barreto, D., Onias, H., Andrade, K. C., Novaes, M., Pessoa, J., Osório, F. L., Sanches, R., Dos Santos, R. G., Tófoli, L. F., de Oliveira Silveira, G., Yonamine, M., Riba, J., Santos, F. R., Silva-Junior, A. A., Alchieri, J. C., Galvão-Coelho, N. L., Lobão-Soares, B., et al. (2018). Rapid antidepressant effects of the psychedelic ayahuasca in treatment-resistant depression: A randomized placebo-controlled trial. *Psychological Medicine, 49*(4), 655–663. https://doi.org/10.1101/103531.

Pollan, M. (2015, February 2). The trip treatment. *The New Yorker.* http://www.newyorker.com/magazine/2015/02/09/trip-treatment

Ross, S., Bossis, A., Guss, J., Agin-Liebes, G., Malone, T., Cohen, B., Mennenga, S. E., Belser, A., Kalliontzi, K., Babb, J., Su, Z., Corby, P., & Schmidt, B. L. (2016). Rapid and sustained symptom reduction following psilocybin treatment for anxiety and depression in patients with life-threatening cancer: A randomized controlled trial. *Journal of Psychopharmacology, 30*(12), 1165–1180. https://doi.org/10.1177/0269881116675512.

Sanches, R. F., de Lima Osorio, F., dos Santos, R. G., Macedo, L., Maia-de-Oliviera, J., Wichert-Ana, L., de Araujo, D. B., Riba, J., Crippa, J. A., & Hallak, J. (2016). Antidepressant effects of a single dose of ayahuasca in patients with recurrent depression: A SPECT study. *Journal of Clinical Psychopharmacology, 36*(1), 77–81.

Sánchez, C., & Bouso, C. (2015). *Ayahuasca: From the Amazon to the global village.* Amsterdam: ICEERS; Transnational Institute (TNI).

Schnarch, B. (2004). Ownership, control, access, and possession (OCAP) or self-determination applied to research. *Journal of Aboriginal Health, 1*(1), 80–95.

Sessa, B. (2012). Shaping the renaissance of psychedelic research. *The Lancet, 380*(9838), 200–201. https://doi.org/10.1016/S0140-6736(12)60600-X.

Sieving, R. E., McRee, A.-L., McMorris, B. J., Shlafer, R. J., Gower, A. L., Kapa, H. M., Beckman, K. J., Doty, J. L., Plowman, S. L., & Resnick, M. D. (2017). Youth–adult connectedness: A key protective factor for adolescent health. *American Journal of Preventive Medicine, 52*(3), S275–S278. https://doi.org/10.1016/j.amepre.2016.07.037.

Substance Abuse and Mental Health Services Administration (SAMHSA). (2014). *SAMHSA's concept of trauma and guidance for a trauma-informed approach.* Rockville. https://store.samhsa.gov/

product/SAMHSA-s-Concept-of-Trauma-and-Guidance-for-a-Trauma-Informed-Approach/
SMA14-4884.

Thomas, G., Lucas, P., Capler, N. R., Tupper, K. W., & Martin, G. (2013). Ayahuasca-assisted
therapy for addiction: Results from a preliminary observational study in Canada. *Current Drug
Abuse Reviews, 6*(1), 30–42. https://doi.org/10.2174/15733998113099990003.

Tupper, K. W. (2008). The globalization of ayahuasca: Harm reduction or benefit maximization? *International Journal of Drug Policy, 19*(4), 297–303. https://doi.org/10.1016/j.
drugpo.2006.11.001.

Tupper, K. W., Wood, E., Yensen, R., & Johnson, M. W. (2015). Psychedelic medicine: A re-
emerging therapeutic paradigm. *Canadian Medical Association Journal, 187*(14), 1054–1059.
https://doi.org/10.1503/cmaj.141124.

Watts, R., Day, C., Krzanowski, J., Nutt, D. J., & Carhart-Harris, R. L. (2017). Patients' accounts
of increased "connectedness" and "acceptance" after psilocybin for depression. *Journal of
Humanistic Psychology, 57*(5), 520–564. https://doi.org/10.1177/0022167817709585.

Chapter 11
Ayahuasca as a Healing Tool Along the Continuum of Recovery from Eating Disorders

Adele Lafrance, Marika Renelli, Jenna Fletcher, Natasha Files, Kenneth W. Tupper, and Anja Loizaga-Velder

Eating disorders (EDs), including anorexia nervosa and bulimia nervosa, are serious mental health disorders that involve disturbances in both eating behavior and perception of body image (American Psychiatric Association 2013). They are associated with a wide spectrum of symptoms and associated conditions that can have a devastating impact on physical, psychological, and social well-being (Löwe et al. 2001). For example, EDs are associated with a range of serious health problems including dermatological, gastrointestinal, cardiovascular, and endocrine issues (Mehler et al. 2010). They are notoriously comorbid with disorders of anxiety, depression, and substance use (Blinder et al. 2006) and can significantly and negatively impact meaningful interpersonal relationships as well (Hartmann et al. 2010). Moreover, EDs have high rates of morbidity and premature mortality (Arcelus et al. 2011) and are among the most challenging psychiatric illnesses to treat. Current treatment approaches for adults offer limited success with elevated dropout and relapse rates (Fassino et al. 2009; Waller 2016). For these reasons, new treatments must be explored.

A. Lafrance (✉) · M. Renelli
Laurentian University, Sudbury, ON, Canada
e-mail: adele@dradelelafrance.com

J. Fletcher
Private Practice, Ottawa, ON, Canada

N. Files
Mental Health Foundations, Vancouver, BC, Canada

K. W. Tupper
University of British Columbia, Vancouver, BC, Canada

A. Loizaga-Velder
National Autonomous University of Mexico, Mexico City, Mexico

Nierika Institute for Intercultural Medicine, Chalmita, Mexico

© Springer Nature Switzerland AG 2021
B. C. Labate, C. Cavnar (eds.), *Ayahuasca Healing and Science*,
https://doi.org/10.1007/978-3-030-55688-4_11

Ayahuasca has been used for centuries by indigenous Amazonian peoples in folk healing and spiritual contexts (Stephan 2010). Ayahuasca drinking has been gaining popularity worldwide over the past few decades for purposes including spiritual, therapeutic, and spiritual growth (Fotiou 2016; Labate et al. 2016; Tupper 2008). A growing body of research has pointed to the potential for ayahuasca to assist individuals in the improvement of mental health issues, such as problematic substance use and depression (dos Santos et al. 2016; Loizaga-Velder and Verres 2014; Osório et al. 2015; Thomas et al. 2013). More recently, the potential for ayahuasca to offer healing to individuals along the continuum of recovery from an ED has begun to be explored (Lafrance et al. 2017; Renelli et al. 2018, 2020). This chapter will review and deepen these investigations on the basis of retroactive qualitative studies exploring perceptions of the outcomes associated with ritual ayahuasca use among individuals with a history of a diagnosed ED. New themes and quotes will be presented for illustrative purposes. Data from a subset of participants who engaged in conventional ED treatment in North America and ceremonial ayahuasca use in both underground circles and ayahuasca retreats in South America will also be discussed. Finally, perceived and experienced risks and challenges associated with ceremonial ayahuasca use will be described.

Method

Participants The total sample consisted of 14 women and two men with a mean age of 34 years (range 21–55). In terms of diagnosis, ten of the participants reported suffering from anorexia nervosa and six from bulimia nervosa. Thirteen of the 16 participants had at some point engaged in conventional ED treatment in inpatient, day hospital, and outpatient settings with various psychotherapeutic approaches (Renelli et al. 2020). A history of trauma was shared by 14 participants, and these experiences were related to emotional, physical, or sexual abuse, or learning of the violent death of a loved one ($n = 4$). Frequency of participation in ceremonial ayahuasca use ranged from 1 to 30 times. Prior to their experiences with ayahuasca, six of the participants were psychedelic naïve, and ten participants reported previous experience with the following substances: psilocybin ($n = 9$), lysergic acid diethylamide (LSD: $n = 5$), methylenedioxymethamphetamine (MDMA: $n = 3$), salvia ($n = 2$), phencyclidine ($n = 1$), mescaline ($n = 1$), and N, N-dimethyltryptamine (DMT: $n = 1$).

Interview Procedure and Analysis Interviews were conducted via telephone, recorded and transcribed, and were 75–180 min in length. Questions covered a broad range of topics. Using the methodology of thematic analysis (Braun and Clarke 2006), the interviews were analyzed by at least two separate raters, and common themes were identified. A theme was identified as major when 50% or more of the sample endorsed the theme. Themes were rated as minor when they did not meet the 50% threshold but were considered of theoretical or clinical importance.

Qualitative descriptive analysis (Sandelowski 2000) was utilized for the analysis of perceived challenges and risks of ceremonial ayahuasca. This method was used to capture all perspectives relating to real and potential risks as well as any hypothetical concerns and actual challenges experienced firsthand by participants.

Results

In terms of outcomes, all but one participant reported that the ayahuasca-induced experiences contributed to a greater sense of well-being and were beneficial to them with respect to ED thoughts and symptoms. Most participants perceived benefits in an integral way, which we have conceptualized using a biopsychosocial-spiritual framework (Renelli et al. 2018). Specifically, the major themes reported related to positive shifts that were psychological, physical or body-oriented, relational, or spiritual in nature. Table 11.1 includes an overview of each of these themes in order of frequency in reporting along with new quotes for illustration purposes.

As Table 11.1 suggests, participants reported therapeutic outcomes that were related to several key themes, including psychological changes, body perception and physical sensations and effects, relational effects and experiences, and spiritual or transpersonal effects and experiences. The psychological changes ranged from emotion processing and increased self-love or self-acceptance to feeling that ayahuasca helped address the root cause of the ED or predisposing underlying trauma. A number of subjects also reported reduced symptoms of other co-occurring mental health issues, including anxiety, depression, suicidality, and problematic substance use. Bodily perceptions or physical sensations that were noted include insights into purging as a symptom of ED (vs. purging in ceremony as an acute effect of drinking ayahuasca), improved relationship with the body, with food and eating, and overall better physical health, including weight regulation. Relational effects included improved relationships with family or loved ones and intergenerational or relationship insights from the ayahuasca experience. Spiritual or transpersonal experiences included greater awareness of spiritual connections and, in some cases, a deepening of spiritual or religious practices.

One particularly interesting finding that perhaps makes sense of these positive outcomes was reflected in a theme relating to participants' newfound sense of self-efficacy with respect to their own healing. Several participants ($n = 7$) reported that their ayahuasca experiences facilitated the activation of inner psychological resources and instilled within them a belief in their innate capacity to heal. Many spoke of discovering and mobilizing these internal resources to heal psychologically, physically, relationally, and spiritually. They also spoke of an ability to maintain a greater sense of health and wellness in these domains post-ayahuasca. This recognition and belief in one's ability to heal is expressed by participant 2:

> Just using myself as a resource and to know that I really do possess everything I do need to treat and to manage whatever it is that I'm dealing with. More specifically, she [ayahuasca] gave me a reference point, an idea of where to go during my own meditation sessions to

Table 11.1 Overview of themes related to outcomes

Theme	Subtheme	Quote	Participant endorsement (%)
Psychological changes	Improved ability to respond to and process emotion	I choose [my emotions], they don't choose me. Sometimes I choose anger, sometimes I choose disappointment, and sometimes I even choose sadness.	87.5
		My emotional awareness is heightened and with that I feel that I can sit with my emotions longer without them taking over.	
	Validated or changed personal theory of the ED	The eating disorder was a form of self-care, although it was not an optimal form of self-care. It was a way of trying to manage the trauma.	81.3
		I felt a very deep sense of meaning, like [the ED] just really made sense to me, and that was something I had been searching for a really long time. I began to make meaning of it, and it was just a deep felt "I get it now." That in itself was very helpful for me.	
	Developed more self-love, self-acceptance, self-esteem, self-forgiveness, and self-compassion	I felt like I needed to have acceptance for myself and having love for myself... I learned that once I could learn to fully love myself unconditionally, I wouldn't need the eating disorder.	81.3
		I seem to think about myself and talk about myself a lot more kindly than I previously did. And I'm a lot gentler [to myself].	
	Decreased or stopped ED symptoms	It was a 100% reduction of any bulimic-related purging	68.8
		I also haven't done any over-exercising whatsoever... that used to be a daily ritual for me. It just doesn't exist.	
		I would say that I've been tempted. It's been a really, really tough year, it's been a really, really hard year... And somehow, I have managed not to engage in eating disorder behavior.	

(continued)

Table 11.1 (continued)

Theme	Subtheme	Quote	Participant endorsement (%)
	Healed root cause of the ED and/or trauma	You know how there's different tiers of vulnerability, like psychological vulnerabilities? And how some people tend towards anxiety, some people tend towards depression, depending on other personality features? But we all have a certain underlying psychological vulnerability. I see that ayahuasca gets at those underlying psychological vulnerabilities, for they could manifest themselves in all different kinds of ways, whether it's an eating disorder, whether it's anxiety, or other things.	62.5
		What I noted from the group therapy, the ayahuasca experiences were, they said, ayahuasca seems to tackle the trauma, it focuses in on trauma.	
	Reduced anxiety, depression/self-harm, suicidality	I still experience periods of feeling anxiety, but I feel like they don't last as long, whereas before I would spiral downward and get depressed and then start to restrict and start to purge and binge and all of that. I feel like I can notice when my energy is changing, and then I am more able to be with it and sort of resist it and then it moves after.	56.3
		Since ayahuasca, I don't even imagine being dead. I mean I know it's going to happen, but I don't want to die. I want to live.	

(continued)

Table 11.1 (continued)

Theme	Subtheme	Quote	Participant endorsement (%)
	Reduced cravings and/or use of psychoactive substances	I don't do any pharmaceuticals at all anymore, and I don't even do alcohol. There have been a few times when I'm like "hmm" and no, not really, don't even really want it.	50.0
		My experience with ayahuasca has very much taught me how to interact with these sacred medicines and how to hold them with such reverence and respect and intention and prayer, and certainly I am not wanting to use any psychedelic substances recreationally even... I definitely would say I don't have any drive to use substances recreationally, and that's been since my relationship with ayahuasca.	
	Increased self-efficacy	I think that's just the biggest thing for me, was that it showed me how I could heal myself in a deeper way and that I could continue doing that in every part, every aspect of my life.	43.8
		So, she made it really clear to me that... she isn't my own innate power to heal, she's just there to help reveal that truth in me. Yeah, so I see it as this potent way to help us become aware not only of our wounding and its origins but also of our inherent ability to heal our inherent wholeness.	
	Increased mindfulness	Just to be. The three key points that I took away from the experience were to stay humble, to stay grateful, and to stay mindful. And these were points that were very well made and are exceptionally good at serving to manage negative mind chatter, which is still present, though the voice has been identified.	37.5
		I think it goes hand in hand with the kind of skills that I've been learning [in the ED program]. It put a big emphasis on being mindful and being in the present and think that kind of is what I got from [an experience I had].	

(continued)

Table 11.1 (continued)

Theme	Subtheme	Quote	Participant endorsement (%)
Body perception and physical sensations and effects	Received insight into purging in ceremony in comparison to purging as a symptom of the ED	Vomiting that you engage in while experiencing ayahuasca is vastly different from any self-harming, bulimic behavior. You're very aware during the ayahuasca sessions, exactly what it is you are expelling from the body via vomiting, and it's very clear to you that this is a process in your body that's natural, that purging of the poison is something that needs to take place. Whereas doing so in a bulimic sense is an expression of an unconscious desire to reject something, or at least that's the way it was manifesting for me.	62.5
		I started thinking about the purging during ceremonies like a release and so it was like, well, no wonder, I kind of created my own ritual as release. So, I didn't know what to do with this energy that I was feeling. So, it helped me gain a lot of compassion for myself, and I think it took a lot of the shame away from it… And I think that is huge, especially with bulimia.	
	Improved the relationship with the body	I'm in love with my body. I've never been in love with my body. I've always been too tall, too this, too that, and there is a newfound relationship with my very own body. I look in the mirror, and I'm like "Hmm, I kind of like this."	50.0
		I've certainly learned to validate myself when it comes to my body. I can look in the mirror and look at myself and think that it's okay exactly how it is.	
	Improved the relationship with food and eating	I would tell you that I started working with the medicine in May, and it probably has cleared up since I started working with the medicine. Completely, where food and I are not arguing. You know either dieting or wanting to lose weight and being frustrated with obsessing over what I ate or didn't eat, what I should eat.	50.0

(continued)

Table 11.1 (continued)

Theme	Subtheme	Quote	Participant endorsement (%)
		When I am having a bad day, I will notice the tendency to crave foods or notice a tendency to catch myself picking at bad foods like I used to before when I was sick and being cognitively able to be "I'm having a bad day. It makes sense I'm eating this way," because it served a purpose and then being able to stop, whereas before I mistook it as actual hunger.	
	Regulation of weight	I'm doing well. I'm gaining weight. Fifteen pounds since last summer. It's a little hard to accept my body changing, but my health has improved.	31.3
		So, for me, last January, I got the plant telling me again, the intuition I have inside me telling me "Are you ready to go to, to start losing weight? You're going to start to walk. You can take stairs, in the subway, in your office, your work, whatever, you just take stairs. And then, next month, I want you to start taking the first four floors of your building, and then five floors next week… Ok, now I want you to start walking once a week to the office," which is half an hour. And I released 15 pounds.	
	Improved general physical health	I'm becoming healthier. My own cardiologist for the third time since I've come back from [the retreat] he tells me, "What's going on? You're just getting better. And your blood pressure, and this blood test and that … you're just getting better." And I said, "Well, for the third time, I'm doing ayahuasca, and also mindfulness every day."	18.8
		I've done ten ceremonies since May because it was like my body moves, it's changed, it's physically changed (participant with chronic fatigue).	

(continued)

Table 11.1 (continued)

Theme	Subtheme	Quote	Participant endorsement (%)
Relational effects and experiences	Improved relationships with family, children, and/or romantic partners	I definitely have more compassion for myself, for other people, my relationships with my family. They have really improved... I feel like whatever it is that has changed in me has allowed me to be more open with them, forgive, forgiveness. That ability to really genuinely connect and be loving.	62.5
		I guess one of the biggest things is my relationships with people. I don't experience conflict in the same way that I did anymore... I definitely can see now how imperfect relationships are.	
	Experienced intergenerational or relational visions/ insights in ceremony	I had an encounter with an ancestor. She showed me the lineage of the maltreatment or abuse, and she sat with me and held circle with me, and she helped me heal.	56.3
		In one of my visions I held myself as a child at the age of three. I just held hands with her for the entire ceremony, and so for some reason that vision or that picture sticks with me when I think about myself and when I think about when I shame myself, and it just kind of keeps me in check a little bit.	
Spiritual and/or transpersonal effects and experiences	Received awareness of spiritual/ transpersonal connections	I think that it helped me look at the world and at the Earth and the medicines that were put here as more sacred. I think it has overall confirmed a sense of spirituality that I have in me and an overall sense of sacredness about my body and the life that I have.	93.8
		I didn't really have [religion in my life], I put that all on pause in my mind... It just became not really important, and I didn't have the desire to really connect with that spirituality. And after I would say that it changed my desire. It made me feel more hopeful and it gave me a fresh start.	

(continued)

Table 11.1 (continued)

Theme	Subtheme	Quote	Participant endorsement (%)
	Experienced spiritual and/or transpersonal connections in ceremony	I saw like this: it was like mathematical; it was like what some people call the fifth dimension. So intricate, there's all these intricate lines, and there was so much space, so much depth... I felt like there was something beyond what we see with our senses. There's something completely beyond because like it blew my mind. I saw palaces, I saw worlds, and just colors, and it was completely something I could not fully describe. I don't know that I can say it could be possibly scientifically observable.	62.5
		I did have both visions of my body being like the Universe, so really expansive and big and also really, really small as a tiny part of the Cosmos.	
	Transformed spiritual or religious practices	I think it gave me a deeper appreciation for prayer.	56.3
		I had that satisfaction [with religion] before going to ayahuasca, but I guess, after doing ayahuasca in a ceremonial context, there was a greater sense of "this could be so much more," like going to church could be so much more... I go to church every week, like a muscle that I exercise, a spiritual muscle that I exercise. Church, it provides some community and so does the ayahuasca. It's an opportunity for me to put some time aside to commune with God or the creator.	

address specific issues, just to reflect on where I went and what it means to be stripped down to virtually nothing and also to understand the kind of language that I use with myself.

Participant 13 similarly describes: "I think, overall, the medicine confirmed the power of my internal resources and in my self-healing and collective healing abilities in terms of my own journey and my work with others."

Comparison of Ayahuasca with Conventional ED Treatment Interviews from a subset of participants ($n = 13$) who had a history of participation in both conventional ED treatment and ceremonial ayahuasca drinking were explored for comparison purposes. Five central themes were identified that related to the differences participants noted with respect to their experiences with conventional ED treatment and ceremonial ayahuasca use. They noted that, in comparison to conventional ED treatment, their work with ayahuasca (1) allowed for deeper healing, (2) provided a more effective and efficient process of healing, (3) helped process intense emotions and/or memories, (4) facilitated the embodiment of self-love and self-care, and (5) incorporated a spiritual component to healing and recovery (Renelli et al. 2020).

Participants described therapeutic advantages of ayahuasca-assisted treatment in comparison with conventional therapeutic practices for EDs. Some stated that participating in ayahuasca ceremonies allowed them to experience a more integral way of healing. Ayahuasca was considered an experiential form of therapy in contrast to other therapeutic interventions that focused more on addressing cognitive aspects of the psyche and associated behaviors. Participants also expressed that their experiences with ayahuasca were more effective and efficient at targeting their ED symptoms. In some cases, it allowed for therapeutic breakthroughs, where conventional therapeutic approaches had been less successful. Participant 6 stated: "It's by far the most healing experience I've ever had in my life."

Another perceived difference between ayahuasca and conventional ED therapy was that ayahuasca enabled better access to and understanding of profound emotions and memories. One participant, a medical doctor, suggests that ayahuasca works on an experiential level helping to process and work through intense feelings and recollections of the past that may have contributed to the development of the ED:

> I think it adds a dimension that is missing [to the Western model], which is this ability to really experience in a very embodied way all of the implicit material that played a role in leading into the formulation of the eating disorder, the wounding that happened... so I think ayahuasca adds to a large degree what can be added in a slower fashion maybe a less psychedelic fashion, that which somatic therapy could really add to standard Western therapy, which is just slowing down, working in body time, anchoring resource, feeling safe, and feeling the feelings that were buried in real time and giving them their time in acknowledgment. (P7)

Unlike conventional ED treatment, some participants described that ceremonial ayahuasca use led to spiritual experiences that were perceived to be therapeutically important. As participant 12 stated: "It tended to the part of recovery that is usually not included in treatment, like the spiritual recovery." Several participants also cherished the spiritual component of the ayahuasca experience, which they viewed was missing from conventional approaches. This was the case for participant 3: "To embody a connection with whatever you want to call it: God, greater intelligence, whatever word you want to use for it. And therapy could not."

A final theme expressed by the participants related to the cultivation of self-love. For one participant, although conventional ED treatment targeted self-love, it was through her experiences with ayahuasca that this lesson was embodied: "What it really communicated to me was self-love and I'd never, in over a decade, been able to identify something that's as effective as ayahuasca in doing this" (P2).

Challenges and Risks of Ceremonial Ayahuasca

Qualitative descriptive analysis using data from the total sample ($n = 16$) was utilized for the exploration of the perceived challenges and risks of ceremonial ayahuasca use (Sandelowski 2000). Specifically, whereas, in the previous sections, themes were identified based on frequency with which they were described by interviewees, in the following sections, all perceived challenges and risks identified by the participants, whether speculative, potential, or based on lived experience, are presented. One of the risks previously discussed in Lafrance et al. (2017) related to the preparatory diet. Specifically, two participants described that the food restrictions prior to the ayahuasca ceremony triggered some transitory ED thoughts and behavior patterns. Additional risks and challenges are discussed further.

Psychological Challenges Several participants ($n = 10$) described feeling challenged by the experience of acute emotional pain during the ayahuasca experience, including moments of what felt like overwhelming sadness, fear, anger, shame, and disgust. One participant likened this phenomenon to the concept of noble suffering, that is, suffering in the service of healing:

> It wasn't pleasant and it wasn't easy. It was kind of really horrid, it was horrible... like giving birth is really painful. It's the same; it's the same thing. Giving birth is very, very painful but... you don't look back at the pain, you're thankful to have done what you've done, and it's very much like that. (P16)

Two participants also described a sense of lingering emotional pain following ceremonial ayahuasca use. A participant described an ayahuasca experience that left her with a feeling of heaviness lasting for more than a month, which she attributed to a lack of therapeutic support offered at the retreat. Also, she cautioned against the use of ayahuasca for those with complex trauma, in particular in the absence of skilled integration or therapeutic support:

> I think the risk is to open up a whole can of worms, and if they have no way to integrate that, then it can be harmful. And people can have intense experiences and misinterpret them and that can be harmful. And that's why the integration is so important. And I also think, for certain people, it might not be the right thing; like, if somebody has a really severe and complex trauma and then they go and drink the medicine and, you know, it comes off, the default network is, you know, no longer there and people having intense experiences all around you and you've got somebody with a nervous system that's already so dysregulated from the trauma, the complex trauma they had, that may not be, that could be really terrifying and maybe not the appropriate thing. (P7)

Integration and Sharing Another challenge commonly reported by participants related to difficulties in sharing their ayahuasca experiences with family, friends, and colleagues ($n = 8$). Participants reported that, as a result, they often kept these experiences private or secret. Among those who did share with family and friends, they reported to have been met with responses that seemed to discourage further discussion. For some, this resulted in feelings of invalidation, frustration, and disappointment. As one participant stated: "There's a secrecy on my end for some people.

And also, I see the world in a very different way sometimes so, in terms of being able to relate to people, it has made it difficult" (P8).

Restricted Economic and Geographic Access Challenges associated with ceremonial ayahuasca drinking were attributed to the difficulties in the financial and geographic accessibility (*n* = 7) of ceremonial ayahuasca use. According to the participants interviewed, ceremonies ranged in price from approximately 75 to 350 USD (excluding travel), and participants attended a range of one to 30 ceremonies. Some participants indicated that their lack of financial resources limited their ayahuasca drinking. At the time of the interviews, all participants were living in North America, where ayahuasca is typically regarded as a preparation of a controlled substance (i.e., DMT), making it challenging to access retreats close to home. Participants either had to join underground circles or incur significant travel costs in order to drink ayahuasca legally, which for some included travel to remote areas of the Amazon. One participant believed that had she had been able to attend an ayahuasca retreat of longer duration, she could have recovered more fully from her ED symptoms:

> I feel like it took quite a few ceremonies for me to start digging deeper. I feel like if I could have had more ayahuasca ceremonies in shorter periods of time I wouldn't have struggled so long with my eating disorder. Like if I could have gone away somewhere for like six weeks or two months or something like that and just did a bunch of ayahuasca ceremonies it would've healed my eating disorder completely. (P14)

Legal Risks Under the legal regimes of most countries that are signatories to the 1971 Convention on Psychotropic Substances, a key component of the ayahuasca brew, DMT, is prohibited (Tupper and Labate 2014). This means that people who drink ayahuasca in these countries may be putting themselves at risk of arrest and criminal prosecution. Because of the potential legal risks, some participants (*n* = 3) expressed concerns about their own and others' involvement with ayahuasca:

> I have to be very careful about what I say and how I say it and who I say it to just to protect, you know because it is a Schedule 1, so I have to be able to protect the person serving the medicine here. You know, that's a bummer because it's just such a beautiful, wonderful, helpful, transformative tool. (P3)

Participants also expressed their disagreement with the legal status of ayahuasca from which they had received important therapeutic benefits, expressing hope that the legal situation might change:

> I think it's confirmed to me how important plant medicines are for the healing process and for healing in the eating disorder process. I hope to one day use ayahuasca in my own work as well as other plant medicines. Giving the political and legal climate that we live in today, I'm hoping that these things will change to allow for better use and integration of these medicines into health and wellness practices. I think it's given me a greater sense of hope. (P10)

Contextual Challenges Lastly, three participants spoke to the potential for serious risks in the event the ceremonial leader or other staff lacked adequate training, expertise, or integrity. As reported in Lafrance et al. (2017), one participant reported significant distress after having experienced inappropriate sexual contact by a support staff at the end of a ceremony. Additionally, a participant with a professional background in harm reduction commented on the risks related to the trust participants must place in the shaman's ability to safely prepare and administer the medicine:

> You don't self-administer it. I don't know how strong it is; I don't know anything about it. So, I want to have a shaman be able to administer it to me safely, but that is all out of my control, and I don't know how they assess that. And I think I did, in some ways, overdose on it that time. Overdose in the sense that I couldn't stand up and it's hard to know what component of that was psychological or what component of that was physical. (P16)

Queries Regarding Addiction Potential A question in the semi-structured interview invited participants to reflect on the addiction potential of ayahuasca. Although two participants described a type of fascination with ayahuasca, and one participant described that, at one point, her seeking of ayahuasca felt compulsive, none of the participants perceived ayahuasca as a substance that could lead to chronic dependent patterns of use nor did they feel concerned about its potential for addiction. Many participants noted that the foul taste was a factor in ayahuasca being an unlikely object of addiction, while others noted that, while many other psychoactive substances seem to numb emotional distress, ayahuasca brings one closer to the core of one's pain. That being said, one participant perceived that the feeling of connection and belonging that exists when people come together for an ayahuasca retreat could be a factor worthy of attention:

> I don't think people can become dependent on or addicted to ayahuasca as a substance. What I do think people can become dependent on is sort of the pseudo-culture of ayahuasca... because it is creating community, and community is something that's totally lacking in our culture; it's part of the pathology of a lot of mental illness that is present in our culture, as we don't have healthy communities. So, all of a sudden you have a community of people who are opening up and doing their healing work and, you know, having social experiences, and I've seen people return to the medicine, I think, for reasons that are more about being in the community than they are about drinking the medicine, but the medicine comes with it, so they do it. (P7)

As a follow-up question, participants were asked whether they would describe themselves as a "drug user" with respect to drinking ayahuasca. None of the participants endorsed this label; instead, 11 perceived ayahuasca as a medicine or form of therapy and not a "drug." Participant 1 likened the use of ayahuasca for the treatment of mental illnesses to the use of pharmaceuticals for the treatment of cancer: "I think it's about as accurate to call somebody who's undergoing chemotherapy a 'drug user'" (P2).

Discussion

Our findings suggest that, for some people along the continuum of ED recovery, ceremonial ayahuasca drinking may be therapeutically valuable. Not only did ceremonial ayahuasca use lead to a reduction of ED thoughts and behaviors, remarkably, a few participants reported experiencing complete resolution of ED symptoms following their ayahuasca experiences. They also reported and highly valued noted improvements across several other domains of psychological well-being; specifically, experiences with ayahuasca helped transform psychological, relational, and spiritual aspects of the self as well, including positive outcomes related to emotion processing and regulation, interpersonal relationships with close others, and spiritual and religious connectedness. The nature of ayahuasca healing can, therefore, be considered integrative and holistic.

Ceremonial ayahuasca use also appears to facilitate an internal process of healing of the various and sometimes disconnected parts of the self (Carhart-Harris et al. 2017). In fact, ayahuasca may act as a catalyst for the innate healing wisdom of the body and psyche to integrate the wounded parts of the self into an embodied whole. These findings are in line with recent investigations of psychedelic substances (e.g., LSD, psilocybin) as adjuncts to psychotherapy in treating certain mental illnesses (Tupper et al. 2015). Specifically, researchers studying MDMA-assisted psychotherapy for the treatment of chronic PTSD describe an "internal healing intelligence" that is defined as an individual's innate capacity to heal emotional wounds and trauma (Mithoefer 2013). The individual's inner healing intelligence, in tandem with the temporary non-ordinary state of consciousness induced by ingesting MDMA, allows for needed for healing and growth. In the same vein, ayahuasca has been described as a tool used to access internal healing resources of the self (Shanon 2002).

These findings are particularly compelling given that our sample included several individuals who had previously participated in conventional ED treatment programs in North America and who continued to struggle in their ED recovery. In fact, those participants with experience with both healing contexts claimed that their ayahuasca experiences were more significant for their healing process overall. Although many strengths were noted with respect to their experiences in the healthcare system, conventional treatment was perceived as lacking in integration in its focus, with too strong an emphasis on behavioral markers of ED recovery. Participants also spoke to experiencing more effective, efficient, and deeper healing experiences under the effects of ayahuasca. This was perhaps most evident when considering the findings relating to the healing of trauma. Historical trauma can be a risk factor for the development of an ED (Brewerton 2007). The majority of participants in our study experienced one or more significant episodes of trauma in the years prior to ED onset. Following their ayahuasca experiences, participants not only conceptualized historical trauma as being a root factor in the development and maintenance of their ED, they also expressed improved abilities to tolerate, regulate, and process challenging memories and emotions as a result of ayahuasca drinking. As such, it is possible that increased emotional regulation moderated participants' abilities to revisit and process root traumas. We also hypothesize that the temporary non-ordinary states of consciousness while under the effects of ayahuasca assisted participants in overriding psychological defense mechanisms that may have prevented access to this material in conventional ED treatment contexts, no matter the proficiency or expertise of the supporting clinician. As a result of the processing of root trauma and associated emotions, it is possible that individuals no longer had the same level of need for emotional regulation and avoidance strategies in the form of ED symptoms.

Given that etiological and maintenance factors of EDs are a complex interplay of biological, psychological, and social variables, our results suggest that recovery from an ED may be better achieved by an integrative treatment approach that focuses on a biopsychosocial-spiritual perspective. Our results also suggest that there may be therapeutic value for conventional treatment programs to integrate alternative healing modalities, including ceremonial ayahuasca use. Likewise, half of the participants recommended that ceremonial ayahuasca drinking for EDs should also be complemented with aspects of conventional ED treatment approaches, and, in particular, the focus on re-nourishment and psychotherapy support for integration, in order to facilitate a more comprehensive healing process. As such, we firmly believe in the bridging of conventional treatment methods (e.g., re-nourishment, psychotherapeutic support) with ceremonial ayahuasca use in order to capitalize on the strengths associated with each of the modalities in order to optimize the healing experiences of individuals with EDs.

In summary, ayahuasca appears to facilitate an integrative approach to healing an ED that encompasses simultaneous and integrative changes within the whole person: the physical, psychological, social, and spiritual. Moreover, ayahuasca may engage the innate healing resources of the self, promoting the embodiment of the healthy self. As EDs are notoriously difficult to treat, the positive outcomes expressed by our participants point to a need for the continued exploration and future development of progressive and integrative treatment approaches.

Challenges and Risks of Ayahuasca Drinking

Although the preliminary findings of this study show promise for the use of ceremonial ayahuasca along the continuum of healing from an ED, it is important to delineate the challenges and risks associated with ayahuasca drinking as well. A recent study found that disordered eating is equally prevalent across levels of income (Mulders-Jones et al. 2017). As some participants indicated that personal finances restricted their ability both to travel as well as participate in the number of ayahuasca ceremonies they would have liked, individuals of lower socioeconomic status (SES) are likely to experience barriers to accessibility. This suggests that individuals of lower SES may not have equitable opportunity to potentially benefit from this therapeutic modality.

In this study, some participants also described feeling intense and painful emotions during ayahuasca ceremonies. Although participants ultimately felt that they were an important cathartic component of the healing process, it is possible that, in the absence of appropriate therapeutic support, those who are regarded as most psychologically fragile or vulnerable could suffer unnecessarily. For those participants whose experiences were overwhelming or were perceived as threatening to their psychological well-being, this may have been a result of inadequately trained facilitators unable to contain these experiences appropriately or that comprehensive integration support for such deep emotional processes was not always provided. For this reason, at least for participants from communities where ayahuasca drinking is not an integral practice within the culture, we believe that it is important to offer support before and during ayahuasca retreats and establish therapeutic aftercare in order to ensure the psychological well-being of individuals with EDs participating in ayahuasca ceremonies. It may also be important to consider therapeutic support prior to the ayahuasca session to support the setting of intentions and to process possible challenges with the preparatory diet.

In terms of safety, it is well-known that ayahuasca tourism has contributed to the spread of pseudo-shamans who engage in inappropriate and even dangerous and harmful practices (Bauer 2018; Tupper 2009). These may include the admixture of other, sometimes dangerous, plants in the preparation of the ayahuasca brew, inappropriate dosing, and insufficiently trained and unethical practitioners. Some participants acknowledged their fears with respect to their personal safety associated with participating in ceremony with pseudo-shamans. Regrettably, one participant spoke about a traumatic experience in which a ceremonial facilitator made inappropriate sexual advances immediately following an ayahuasca ceremony (Lafrance et al. 2017). This speaks to the importance of not only being aware of the potential risks but also carefully researching the reputation of the retreats, shamans, and ceremonial facilitators. Thankfully, leaders in the field have developed a resource and set of guidelines for the awareness of sexual abuse in ayahuasca settings (Peluso et al. 2020).

Similarly, we believe that individuals with ED must obtain medical clearance from their primary physician in order to identify possible contraindications, such as electrolyte imbalance, cardiac arrhythmias, low blood pressure, gastrointestinal lesions, contraindicated medications, etc., in order to ensure that their body can handle the potential physiological stress sometimes associated with ceremonial ayahuasca drinking. In those instances where the individual is deemed too medically fragile for travel or for participation in an ayahuasca retreat, we are beginning to explore the potential for healing by proxy, where close others, such as parents or spouses, engage in ceremonial ayahuasca drinking with the intention of learning how best to support their loved one.

Study Limitations

The findings of this study should be regarded in the context of several limitations. Our study is not ethnically or gender diverse, mainly represented by Caucasian, North American women. A broader sample including a larger male population and a more ethnically diverse group of individuals could yield different perspectives. Also, given that the participants were a self-selected sample, it is possible that those who did not choose to participate may have experienced neutral or less than favorable outcomes related to their experiences with ceremonial ayahuasca. Furthermore, it is in the nature of our sampling procedure that we cannot definitively determine the generalizability of our findings to all individuals with an ED, as individuals who actively sought out ayahuasca were not compared to those that did not. With respect to the comparisons made between ayahuasca drinking and conventional treatment, it is also possible that participants' previous experiences with ED interventions may have initiated or deepened the process of healing experienced in the context of ceremonial ayahuasca and, therefore, colored these findings. Furthermore, while all participants reported ceremonial ayahuasca use, there can be great variability in practices across different communities and retreat centers, and therefore, it is possible that the data collected reflect different contexts of ayahuasca drinking. Future research should also explore further the perceived impact of the various ceremonial practices such as icaros (e.g., type, language, and method of delivery), rituals (e.g., the use of tobacco and perfumes), the number of practitioners in ceremony, etc. Despite these limitations, this exploratory study may provide hope for individuals struggling with an ED, especially if conventional methods have been less effective and opens a new line of scientific research into the therapeutic potential of ayahuasca.

Additional Considerations for Future Research

Studying the potential safety and efficacy of any new medical technology for eating disorders, including plant medicines, brings forth a number of philosophical issues to consider. This is especially relevant in the context of ayahuasca, given its deep cultural roots, and where plant can be separated from spirit, depending on the methodology used. As such, depending on one's field of study and conceptualization of illness, we first urge scientists to engage in a process of self-exploration to uncover potential biases and perhaps broaden their lens of inquiry.

There also continues to be stigma associated with the exploration of psychedelic substances within the field of mental health, and the subfield of eating disorders is not immune. The degree to which researchers internalize this stigma may therefore influence how they focus their efforts, which raises additional questions: Should early-career scientists potentially risk their reputation by publicly exploring this uncharted and sometimes controversial territory? Or should their mid- or late-career colleagues leverage their experience and perhaps even their status in order to clear the path forward?

When prospective studies are in development, it will also be important to ensure a robust process of screening and consent, especially given the consciousness-altering effects of ayahuasca. People struggling with eating disorders can be quite vulnerable, and those who have been struggling for years may feel desperate for relief. Therefore, it is possible that some individuals will be unable to weigh the risks and benefits of participation without additional support from family members, existing clinical supports, and the research team. Finally, until these studies are in progress and safety parameters have been more fully determined, we believe it is important to caution those suffering from an eating disorder about seeking out ayahuasca experiences on their own. Our research findings suggest that, though there are ceremonial ayahuasca settings conducive to healing from an eating disorder, the unique nutritional, emotional, and medical needs of many individuals may be overlooked or insufficiently addressed due to a lack of understanding on the part of the community or retreat staff.

References

American Psychiatric Association. (2013). *Diagnostic and statistical manual of mental disorders* (5th ed.). Washington, DC: Author.

Arcelus, J., Mitchell, A. J., Wales, J., & Nielsen, S. (2011). Mortality rates in patients with anorexia nervosa and other eating disorders: A meta-analysis of 36 studies. *Archives of General Psychiatry, 68*(7), 724–731.

Bauer, I. L. (2018). Ayahuasca: A risk for travelers? *Travel Medicine and Infectious Disease, 21*, 74–76. https://doi.org/10.1016/j.tmaid.2018.01.002.

Blinder, B. J., Cumella, E. J., & Sanathara, V. A. (2006). Psychiatric comorbidities of female inpatients with eating disorders. *Psychosomatic Medicine, 68*(3), 454–462.

Braun, V., & Clarke, V. (2006). Using thematic analysis in psychology. *Qualitative Research in Psychology, 3*(2), 77–101. https://doi.org/10.1191/1478088706qp063oa.

Brewerton, T. D. (2007). Eating disorders, trauma, and comorbidity: Focus on PTSD. *Eating Disorders, 15*(4), 285–304.

Carhart-Harris, R. L., Erritzoe, D., Haijen, E., Kaelen, M., & Watts, R. (2017). Psychedelics and connectedness. *Psychopharmacology, 235*(2), 547–550. https://doi.org/10.1007/s00213-017-4701-y.

dos Santos, R. G., Osório, F. L., Crippa, J. A. S., Riba, J., Zuardi, A. W., & Hallak, J. E. (2016). Antidepressive, anxiolytic, and antiaddictive effects of ayahuasca, psilocybin and lysergic acid diethylamide (LSD): A systematic review of clinical trials published in the last 25 years. *Therapeutic Advances in Psychopharmacology, 6*(3), 193–213.

Fassino, S., Pierò, A., Tomba, E., & Abbate-Daga, G. (2009). Factors associated with dropout from treatment for eating disorders: A comprehensive literature review. *BMC Psychiatry, 9*(1), 67.

Fotiou, E. (2016). The globalization of ayahuasca shamanism and the erasure of indigenous shamanism. *Anthropology of Consciousness, 27*(2), 151–179.

Hartmann, A., Zeeck, A., & Barrett, M. S. (2010). Interpersonal problems in eating disorders. *International Journal of Eating Disorders, 43*(7), 619–627.

Labate, B. C., Cavnar, C., & Gearin, A. K. (Eds.). (2016). *The world ayahuasca diaspora: Reinventions and controversies*. Abingdon: Routledge.

Lafrance, A., Loizaga-Velder, A., Fletcher, J., Files, N., Renelli, M., & Tupper, K. (2017). Nourishing the spirit: Exploratory research on ayahuasca experiences along the continuum of recovery from eating disorders. *Journal of Psychoactive Drugs, 49*(5), 427–435. https://doi.org/10.1080/02791072.2017.1361559.

Loizaga-Velder, A., & Verres, R. (2014). Therapeutic effects of ritual ayahuasca use in the treatment of substance dependence – Qualitative results. *Journal of Psychoactive Drugs, 46*(1), 63–72.

Löwe, B., Zipfel, S., Buchholz, C., Dupont, Y., Reas, D. L., & Herzog, W. (2001). Long-term outcome of anorexia nervosa in a prospective 21-year follow-up study. *Psychological Medicine, 31*(5), 881–890.

Mehler, P. S., Birmingham, L. C., Crow, S. J., & Jahraus, J. P. (2010). Medical complications of eating disorders. In C. M. Grilo & J. E. Mitchell (Eds.), *The treatment of eating disorders: A clinical handbook* (pp. 66–80). New York: Guilford Press.

Mithoefer, M. C. (2013). *A manual for MDMA-assisted psychotherapy in the treatment of posttraumatic stress disorder*. Santa Cruz: Multidisciplinary Association for Psychedelic Studies.

Mulders-Jones, B., Mitchison, D., Girosi, F., & Hay, P. (2017). Socioeconomic correlates of eating disorder symptoms in an Australian population-based sample. *PLoS One, 12*(1), e0170603.

Osório, F. L., Sanches, R. F., Macedo, L. R., Santos, R. G., Maia-de-Oliveira, J. P., Wichert-Ana, L., Araujo, D. B., Riba, J., Crippa, J. A., & Hallak, J. E. (2015). Antidepressant effects of a single dose of ayahuasca in patients with recurrent depression: A preliminary report. *Revista Brasileira de Psiquiatria, 37*(1), 13–20.

Peluso, D., Sinclair, E., Labate, B. C., & Cavnar, C. (2020). Reflections on crafting an ayahuasca community guide for the awareness of sexual abuse. *Journal of Psychedelic Studies, 4*(1), 24–33.

Renelli, M., Fletcher, J., Loizaga-Velder, A., Files, N., Tupper, K., & Lafrance, A. (2018). Ayahuasca and the healing of eating disorders. In H. L. McBride & J. L. Kwee (Eds.), *Embodiment and eating disorders* (pp. 214–230). Abingdon: Routledge.

Renelli, M., Fletcher, J., Tupper, K. W., Files, N., Loizaga-Velder, A., & Lafrance, A. (2020). An exploratory study of experiences with conventional eating disorder treatment and ceremonial ayahuasca for the healing of eating disorders. *Eating and Weight Disorders-Studies on Anorexia, Bulimia and Obesity, 25*(2), 437–444.

Sandelowski, M. (2000). Whatever happened to qualitative description? *Research in Nursing and Health, 23*(4), 334–340.

Shanon, B. (2002). *The antipodes of the mind: Charting the phenomenology of the ayahuasca experience*. New York: Oxford University Press.

Stephan, V. (2010). Singing to the plants: A guide to mestizo shamanism in the upper Amazon. UNM Press.

Thomas, G., Lucas, P., Capler, N. R., Tupper, K. W., & Martin, G. (2013). Ayahuasca-assisted therapy for addiction: Results from a preliminary observational study in Canada. *Current Drug Abuse Reviews, 6*(1), 30–42.

Tupper, K. W. (2008). The globalization of ayahuasca: Harm reduction or benefit maximization? *International Journal of Drug Policy, 9*(4), 297–303.

Tupper, K. W. (2009). Ayahuasca healing beyond the Amazon: The globalization of a traditional indigenous entheogenic practice. *Global Networks, 9*(1), 117–136.

Tupper, K. W., & Labate, B. C. (2014). Ayahuasca, psychedelic studies and health sciences: The politics of knowledge and inquiry into an Amazonian plant brew. *Current Drug Abuse Reviews, 7*(2), 71–80.

Tupper, K. W., Wood, E., Yensen, R., & Johnson, M. W. (2015). Psychedelic medicine: A re-emerging therapeutic paradigm. *Canadian Medical Association Journal, 187*(14), 1054–1059.

Waller, G. (2016). *Recent advances in psychological therapies for eating disorders*. F1000Research, 5, F1000 Faculty Rev-702. https://doi.org/10.12688/f1000research.7618.1.

Chapter 12
The Therapeutic Use of Ayahuasca in Grief

Débora González, Adam Andros Aronovich, and María Carvalho

Introduction

The word "ayahuasca" is derived from the Quechua language, which is variously interpreted as "dead person, spirit, soul, or ancestor," while *huasca* means "rope or vine" (Metzner 2006). Although, for some people, ayahuasca infers a direct relation between the plant and the realms of spirits – and the dead – few scientific studies have focused on its therapeutic potential for grief and bereavement. The first study compared the intensity of the suffering experienced after the loss of a loved one, over a maximum period of 5 years, between a group of people who took ayahuasca and a group of people who attended peer-support groups (Gonzalez et al. 2017). The researchers found that participants who reported having at least one meaningful experience with ayahuasca, in relation to their grieving process, presented grief that was less intense than the reported grief of the people who attended peer-support groups for an average period of 1 year. In order to gain a deeper understanding of the relation between the ayahuasca experience and the processing of grief, qualitative content analysis of the study participants' experience reports was performed. Three subthemes emerged that were identified within the analyzed narratives: emotional release, biographical memories, and experiences of contact with the deceased.

D. González (✉)
ICEERS—International Center for Ethnobotanical Education Research & Service, Barcelona, Spain

Fundación BeckleyMed, Barcelona, Spain
e-mail: debora.gonzalez@iceers.org

A. A. Aronovich
Medical Anthropology Research Center, Universitat Rovira i Virgili, Tarragona, Spain

M. Carvalho
Research Centre for Human Development, Universidade Católica Portuguesa, Porto, Portugal
e-mail: mccarvalho@porto.ucp.pt

© Springer Nature Switzerland AG 2021
B. C. Labate, C. Cavnar (eds.), *Ayahuasca Healing and Science*,
https://doi.org/10.1007/978-3-030-55688-4_12

The second study examined the course of grief over 1 year of follow-up in a bereaved sample that attended the Temple of the Way of Light ("the Temple" in what follows), a healing center located close to the city of Iquitos in the Peruvian Amazon (González et al. 2020). Situated at the intersection between the indigenous Shipibo medical practice and a variety of psychospiritual and integrative approaches, the Temple emphasizes ayahuasca as a vehicle for healing, individual transformation, and growth. Results of this study showed a significant improvement in grief after the Temple stay, being maintained over 1-year follow-up. In addition, a subgroup analysis performed in a representative subsample showed the improvement in grief symptoms were due to the effect of their stay at the center and not merely the passing of time.

This present chapter includes experiences from some of the Western subjects who participated in this last study. Accordingly, this chapter adopts a dual lens from which to examine the phenomena observed. From an *etic* point of view, and framed within current concepts and theory in the psychology of grief, we propose some possible therapeutic implications in relation to the emergent psychological content described by the grieving participants following their ayahuasca experiences. From an *emic* point of view, we present the voices of the participants as they share their first-person understanding and experiential knowledge regarding the benefits derived from their experiences. Furthermore, we share voice with four Shipibo ayahuasca practitioners (*onaya*[1]) who work at the Temple, with the intention of establishing an intercultural and inter-epistemic dialogue that enriches our understanding of grief and bereavement and further illuminates how ayahuasca can help us integrate and process painful human losses.

The reports that follow were gathered in the context of an observational study promoted by the International Center for Ethnobotanical Education, Research, and Service (ICEERS) and the Beckley Foundation and took place at the Temple of the Way of Light. The goal of the study is to evaluate the long-term effects that ayahuasca may have on personal well-being as well as its therapeutic potentials in relation to a variety of psychopathological symptoms as measured in specific and distinct samples of people who present symptoms of depression, anxiety, posttraumatic stress, and grief.

The reports presented in this chapter belong to the grief sample (n = 50) and were collected 15 days after attending a 9-, 12-, 13-day, or 3-week-long retreat at the Temple, where they participated in 4–9 ayahuasca ceremonies. Qualitative data for the present chapter was collected in a written, free-style form, prompted by an open-ended segment at the end of a battery of mostly quantitative measurements that subjects received at the aforementioned time intervals. For our follow-up chapters, we are also analyzing additional qualitative data collected through presential,

---

[1] Onaya (sometimes *onanya*), from the Shipibo *oni* (ayahuasca; also knowledge, to know) and the suffix -*ya* ("one who knows"), is the Shipibo auto-denomination for curandero, one of the three categories of healers in the Shipibo system, and specialized in the knowledge of plants and command of plant spirits. Throughout this chapter, the terms "onaya," "healer," and "curandero" have been used interchangeably.

semi-structured interviews, gathered by a team member on-site who is also a facilitator in some of the workshops.

Qualitative semi-inductive content analysis of these reports enriches and enlarges the concepts and categories previously identified in the study about the potential use of ayahuasca in grief therapy (González et al. 2017). To be included in the study, participants had to report going through a grieving process related to the loss of a loved one. Recognizing that the concept of "grief" is notoriously difficult to delimitate, in this study, we assume Faschingbauer et al.'s (1987) definition according to which the grief process presents the following components: (a) nonacceptance of a loss, (b) yearning or missing the deceased, and (c) crying or sadness. The reports that were subject to qualitative semi-inductive content analysis (Potter and Levine-Donnerstein 1999) resulted from the participants' answers to the question "Did you have an ayahuasca experience that you feel directly affected your grief?"; if the participant offered an affirmative answer, he received the following request: "Please describe your personal experience of how ayahuasca has influenced your grief. You can extend the description as much as necessary."

Psychological Processes That Could Mediate Grief Adaptation

Emotional Processing

People who undergo a grief process might suffer from feelings of yearning, loneliness, and profound sadness for varying periods of time. The crying that is provoked by these feelings has an adaptive function since, for example, the tears that result from emotional pain contain a substantial amount of stress-related hormones, such as prolactin or corticotrophin. Crying promotes their elimination from the body, hence diminishing the mourner's subjective discomfort (Frey and Langseth 1985). According to Hill and Martin, this hormonal release diminishes the emotional load associated with the cognitive contents that implicitly come with this pain, thus favoring the psychological reorganization in relation to the subject of pain (1997). Most participants in our study mention this type of emotional confrontation, which seems to lessen the pain and the psychosomatic load even after the dissipation of the ayahuasca's immediate effects. However, participants can understand the therapeutic effects derived from an emotional confrontation differently. For some people, confronting their pain is a way to gain relief from psychosomatic symptoms. A young woman who lost her father to suicide after being in an ambiguous relationship with him throughout her life expressed such an emotional release:

> Ayahuasca made very clear how much grief, and what kind, I was holding in my body from childhood and it forced me to sit with it during ceremony. And in sitting with it, I was able

to release a big portion of it. I am still having realizations of myself and the grief I had pushed away for so many years. Every day I feel a little clearer and lighter! (V.1[2])

For others, the emotional confrontation is valuable and meaningful as an act of love or as a way of honoring the deceased or the mourning process:
Ayahuasca helped me understand that the sadness I have felt right after… the suicide of my younger brother, and the sadness that I allowed myself to feel during the ayahuasca ceremony, is just another expression of the love I had for him, something that it is infinite and powerful. (V.15)

In one of the ceremonies, I asked Aya [ayahuasca] to help me feel my grief. Rather than looking and reviewing the stories associated with the many losses I've had in my life, Aya had me sit with my grief. I actually physically experienced in my body the emotion of grief as a heaviness, pressing my shoulders and back toward the earth, as an absence of any color except shades of grey. Even the visions were grey. Aya specifically told me not to go into the stories, as they were a distraction from the feelings. She [ayahuasca] also told me that by being with the feelings I was honoring the experiences, the losses, and myself. (V.3)

A 37-year-old woman who lost an intimate friend in a car crash found the emotional pain experienced under the effects of ayahuasca to be so intense that it helped her put her daily pain in perspective:
Five out of the seven ceremonies my grief was addressed with the ayahuasca. I went deep into the state of grief and shock and felt it very strongly… so in a way I feel like I felt my grief as hard as I am going to feel it. Time will tell. (V.2)

Another possible outcome, however, is the absence of emotional arousal despite increased contact with painful psychological content and the release of its attached psychophysiological reactions. A woman who lost her brother to a car crash describes her experience:
I dealt directly with the death of my brother during ceremony. The strange thing was the uncontrollable tears I cried in which I was an observer to my body's response to the grief. I sat there with tears flowing but felt very numb to any pain. (V.16)

In any case, the effects of ayahuasca cannot be considered independently from the ceremonial setting in which the experience occurs. For many participants, the emotional release is directly linked to the *icaros*, the healing chants of the Shipibo healers (see next section), while the release of psychosomatic tension is directly linked with the purging,[3] whether by vomiting or defecating: "Healers singing to me and feelings of sadness being pulled out of me. Purging and feeling packets of sadness, almost like losing the part of me that was causing sadness" (V.17). "I purged at least nine times and each time felt as if I was letting go of grief that had been trapped in my body" (V.3).

---

[2] "V" stands for "visitor." The number following was assigned to the participants to maintain confidentiality.

[3] Ayahuasca has emetic properties, and "purging," the act of vomiting, defecating, shaking, crying, laughing, and many other forms of psychosomatic release, is a primary dimension of its healing potential, according to the vast majority of both indigenous and Western practitioners.

Although intense emotions that arise with the loss of a loved one are inevitable, in some people the feeling of sadness may trigger defense mechanisms whose purpose is to allow the person to dampen or evade his pain. Circumventing sadness, however, does not contribute toward the resolution of grief, and masking or avoiding uncomfortable feelings could lead to emotional blunting and can pave the way for the emergence of feelings such as rage and bitterness that can cause future psychosomatic problems. In this scenario, cognitive-behavioral therapy uses emotional confrontation techniques such as "imagined revisiting" or "prolonged exposure" with the intention of increasing acceptance of the reality of loss (Boelen et al. 2006). For some people, it seems, ayahuasca – such as cognitive behavioral therapy (CBT) – may help the person fully experience their feelings, uncomfortable as they may be, without triggering defense mechanisms that may have hindered the grieving process through evasion or distraction.

In our study, we came across narratives of emotional confrontation that emerged in cases where experiences of loss may have been masked or avoided for prolonged periods of time. These were reported by women who had lost their mothers:

> The source of my grief,… my mother's suicide, was directly and clearly addressed in the ceremonies. This has been avoided for over years, and has shaped a great deal of my life. I feel as though a great deal of weight has been lifted off of me. I am very hopeful for my future now. (V.14)

> I had suppressed so many tears and I was finally able to release all of my grief so that it was no longer insurmountable. I was able to release emotions and feelings that I did not release when my mother passed away many years ago. Now I am free to love her and appreciate her life because my grief does not cloud me like it did (V.18).

Finally, one participant who lost a loved one during her childhood refers to a cathartic confrontation with the pain that remained hidden throughout her adult life. This 41-year-old woman arrived at the Temple with the intention of dealing with the loss of her mother and her father:

> I have had several people close to me die, many in sudden and unexpected ways, as a child. One was the suicide of my brother. Although I had drank ayahuasca before, I had been taking it to work with other deaths and other issues. My first night drinking at the Temple, I was taken to the period following my brother's suicide and had a cathartic cry and some time to explore the anger resulting from what happened surrounding this event. (V.13)

Making Meaning of the Past and Reconstructing Identity

The grieving process may affect an individual's identity and social environment. The persistent complex bereavement disorder, as formulated in the fifth edition of the American Psychiatric Association's (APA) *Diagnostic and Statistical Manual of Mental Disorders* (DSM-5) (APA 2013), includes "confusion about one's role in life or a diminished sense of one's identity" (p. 970) as one of the symptoms in the social identity disruption section. Grief may exert changes in a person's identity at various levels, including the personal, the social, the cultural, and the spiritual. For example, losing a long-term partner entails a change in marital status, as the person transitions from a married status to that of a widow or widower. Although contingent to culture and social status, such a life event has immediate repercussions and thoroughly transforms everyday roles. On the other hand, grief is never just about the loss of the other; losing a loved one necessarily entails the loss of a part of who we are, the part of our identity that we constructed in relation to that person, and the way that we see ourselves reflected through their eyes. In the timeless words of Vygotsky (Rieber and Carton 1987): "Through others, we become ourselves" (p. 161).

In our study, we have gathered numerous narratives that highlight the impact that the ayahuasca experiences can have on helping reconstruct and develop the identity of individuals going through a grieving process. One participant shares:

> Since the Temple, I realize the grief is more about the parts of myself that I had lost as a result of living with the grief…. I am really enjoying getting to know the parts of myself that were lost. All of my ceremonies contributed to this. The second ceremony was about strength, the third was about joy, the fourth was about dignity. All of these experiences have really helped me to move past feeling like I had no choices. (V4)

The ways in which we choose to reconstruct our life's histories, particularly through highly accessible and vivid personal memories, is closely related to the way in which we see ourselves and project our futures (Baerger and McAdams 1999). Narrative therapy, thus, emphasizes the importance of reassigning meanings to painful biographical episodes in the construction and inclusion of new scripts that could have been understated in the client's reports about their life history (Draucker 2003). This enables the process of reconstructing the life history while simultaneously helping the griever construct a more adaptive and empowering identity. However, in cases where the relationship with the deceased was an ambiguous one, additional challenges may emerge: the distorted reflection of who we are in the eyes of the deceased remains frozen, and an opportunity to evolve and recover that repaired sense of self is lost (Betz and Thorngren 2006).

Such an example comes from a man who lost his father, with whom he had had an ambiguous relationship throughout his life. He describes how ayahuasca helped him shed new light over meaningful biographical episodes that had affected his ability to accept and love himself:

> Ayahuasca helped me to see all of the situations in my life where these problems with self-love had stemmed from and how my thought processes during these times of distress shaped the person I was today. It opened my mind to see the troubles that I had created in my own mind over the years of my life and let me process these events properly. Ayahuasca shined a light on these things and showed me what I needed to see and enabled me to heal myself. (V12)

Another male participant, who lost his emotionally abusive mother to addiction, describes how ayahuasca helped him distinguish and reassign meaning to a previously unrecognized dimension of grief linked to a negative personal belief system that compromised his self-esteem:

> I think I misunderstood my grief as being solely related to external physical events, when, in fact, much of it was also tied to self-limited internal beliefs that separated me from my best self. The opportunity to witness those self-limiting beliefs, confront them with logic, reason, and compassion, and ultimately to change them, resolved such a strong store of grief within me. I am no longer separated from myself, or I am closer and more loving with myself than I was before. I believe I am beautiful, strong, a success, and good enough. I am no longer grieving those parts of me. Though I still have a journey to go on with my grief related to external events, I do so going forward with a new spirit. (V.11)

Reassigning meaning to a traumatic event is another approach to the process of identity reconstruction. The ability to assign new meanings to trauma may open new directions in ways to deal with future challenges, and it is associated with the promotion of resilience (Mancini and Bonanno 2006). As one of the participants writes: "I feel less pain and think of myself less as a victim of circumstance... I have more faith in the idea that there is a purpose to my trauma and life" (V19).

Lastly, we want to note that the opportunity to review biographical episodes under the effects of ayahuasca may increase empathy toward others who are suffering and increase forgiveness of oneself and others. Increased empathy and compassion for others have been described as signs of the personal growth derived from the grieving process (Lichtenthal et al. 2013). However, the role that forgiveness, both to self and others, may play in these kinds of transformative processes has barely been addressed in the scientific literature. Here are two examples from our sample: "During my first ceremony I was shown and experienced such incredible kindness, compassion, and love. I was also shown and experienced the pain of another person and therefore my sense of compassion has really expanded" (V.4).

> During ceremony, I was able to come to terms with everything that had happened with my ex-husband. I was able to forgive him for all the bad things he'd done, and I was able to forgive myself for the part I had played during our relationship. I realized how his death was somewhat like a rebirth for me and has allowed me to find my spiritual path and develop into a person that I'm a lot happier being, compared to when I was with my ex-husband. (V.6)

Continuing Bonds with the Deceased

Until recently, the psychoanalytical legacy of Freud, first expressed in *Mourning and Melancholia* (1924), dictated that the process of overcoming grief inevitably implied a detachment from the deceased, an act of letting go and saying goodbye forever. However, recent scientific research has evidenced the wide range of benefits associated with remaining attached to our loved ones, regardless of his or her physical absence (Klass and Steffen 2017). Not long ago, such a process might have resulted in the pathologization of the griever and the grieving process, as it contradicted the dominant discourse. Currently, however, the continuation of bonds with the deceased has become one of the major goals for a number of clinical models and approaches to the treatment of complicated grief (Neimeyer et al. 2006).

In our study, we find various narratives that describe experiences of contact or reunion with the deceased loved one. These experiences may represent significant therapeutic potential, as they allow seemingly unprecedented and exceptional psychological content to emerge. This has added value for the therapeutic application of ayahuasca in grief, as it may provide results that could hardly be achieved by other psychotherapeutic techniques, such as various forms of imaginal psychotherapeutic dialogues with the deceased (Shear 2010). When these techniques are used in psychotherapy, the griever intentionally projects the content that arises from this encounter. However, through the mediation of ayahuasca, the contact with the deceased may occur spontaneously, unrelated to the conscious expectations of the person. These experiences are often accompanied by a significant emotional intensity that may be partly responsible for the therapeutic impact of the experience. The following participants describe this:

> I mourned during some ceremonies. My dad came to mind and I cried a lot. I felt very close to him in one ceremony, like he was with me. I felt very connected. I also got a very clear message to appreciate what I do have, including my mother, who is still alive. (V.7)

> From the very first ceremony, one of the shamans sang me an icaro that felt like a funeral song, and I was able to tap into some deep grieving. The ceremony helped me feel a deeper connection with my deceased mother, her energy, her memories, her legacy, and helped me put death in general into context … This has really helped my grieving process and my ability to feel what I need to feel to move through difficult emotions. (V.5)

Kernberg defines an internal representation as "the parts of the self in the bond with the person, characterizations and thematic memories of the person, and the emotional states connected with the characterizations and memories" (as cited in Klass 2017, p. 3). In our sample, most narratives that describe these types of contact with the deceased seem to transform the internal representation the participants had of their loved ones. Accordingly, this transformation has a direct impact on the relationship that is later kept with the deceased. As far as we are aware, this process is yet to be described in the scientific literature and may result in new and unprecedented therapeutic approaches. The reports that follow refer to cases where the internal representation of the deceased is surpassed by a new archetypical image that widens and enriches the previous representation, imbuing it with novel and deeper meanings. A man who recently lost his grandfather describes:

> I was visited by my grandfather who passed away in the form of a golden light that spoke to me. He told me how proud of me he is, and how thankful he is that I was able to be there for his death. I feel at peace with his death in a way I never did before. (V.20)

While being concerned for the well-being of the deceased loved one cannot be framed logically within the boundaries of our Western, rational worldview, many grievers express an "irrational" worry about the departed, particularly in cases where the death event was painful or where death occurred before resolution of important aspects was achieved. So far, our data points out to the potential therapeutic effect achieved by "witnessing" the well-being of a deceased loved one following death and the subsequent reduction of the griever's concerns. The following fragment is from a young Buddhist man who lost his grandmother to a car crash and, 10 months later, lost his grandfather – the driver at the time of the accident – to what is known as "broken heart syndrome":

> On my second ceremony, the ayahuasca showed me my grandparents in spirit form. My grandfather was a proud, silent dragon; my grandmother a sweet little monkey. They stayed with me for a long time, maybe 30 to 45 minutes, while I cried for them without thinking of any complications from the time they died. (V.8)

This 56-year-old woman that lost her son in an accident describes:

> Ayahuasca was very loving and gentle with me. I have a direct contact with my son. I could talk with him again and he told me all the love and care he still has for me. He talked to me by being a star very big and bright. Now I feel he is part of me. The loss-feeling diminished because, now the separation does not exist. (V.21)

Lastly, this kind of experiences may also promote change in the representation of the space occupied by the deceased in relation to his or her former living space. Changes in the representation the griever has of the place the loved one was at the moment of their death can also impact and change the way the griever relates to the deceased. This 44-year-old woman, whose mother committed suicide 4 years ago, explains the relief she felt by experiencing her mother's spirit being able to move on:

> In my very first ceremony, Maestra Alicia saw a spirit of a dead person next to me and she helped that spirit to the light. I, at that very exact time, felt that she was sending my deceased mother to the light. I cried a lot because it felt like a big letting go, but I was also very relieved that my mom finally could go to the light. After the ceremony, I felt a big relief and ever since it feels much less dense and grateful when I think of her. (V.22)

Finding Existential Meaning

The search for existential meaning is a recurrent concern for people dealing with the loss of a loved one or when facing their own death. "What is the meaning of life?" "What happens when we die?" These questions encompass an additional burden when someone passes away before their expected time, seeming to violate our perception of what is just (Spillers 2007). In these situations, even religious people may get angry at God, asking questions such as "Where were you when this happened?" "If you are that powerful, why don't you bring them back?" Feelings of frustration, hopelessness, and vulnerability that arise from a lack of answers to these questions push the griever into an existential search.

The ability to make sense of life and death is the variable that better predicts adaptation during the grief process (Gillies and Neimeyer 2006). Presently, the World Health Organization recognizes the important role that spirituality and spiritual development play in our health and well-being (Chuengsatiansup 2003). The variety of ways in which ayahuasca enhances and promotes spiritual development and growth has been widely researched (Trichter et al. 2009; Tupper 2009) and is a recurrent theme in our data. The following excerpt belongs to the report of a participant who experienced great anxiety related to the fear of losing his loved ones and friends, a situation we could label as "anticipated grief":

> Ayahuasca helped me deal with the anxiety I'd get when worrying about my family and friends dying. One thing that helped me reduce my anxiety around death was seeing the spirit world and realizing our spirits never die. This was a rough experience, but I feel it helped me reduce my stress around the fear of death. (V.9)

The following excerpt is from a participant who lost his brother following a prolonged illness. Although secular and unattached to any religion, his experience resembles descriptions of the cycles of Hindu reincarnation (Martinez 2015):

> I was shown cycles of life, death, and rebirth. It no longer seems like an "end" to die but just another sort of beginning. Having faith in the universe restored has made my grief seem more like a part of a cycle, rather than the totality of my existence. (V23)

Another report offers an example of what could be considered a transpersonal experience: "Ayahuasca showed me the interconnectedness of the universe and all life. The experiences I had showed me that death is not to be feared" (V.10). Finally, as this other participant showss, these experiences can promote personal posttraumatic growth by enhancing a new way of relating with life, with death, and with existence: "It's hard to put into words, but the ayahuasca helped me have a larger, more expansive, and more dynamic relationship to life, death, and energy" (V.5).

Summary

To summarize, our approach, so far, has focused in juxtaposing a number of therapeutic concepts and approaches found in the literature with experiences that seem to have a direct relation to the grieving processes of the participants. These concepts and approaches relate to distinctive psychological theories that contribute to our understanding of the human experience of grief: from the cognitive-behavioral model to the narrative constructivist model, from the psychodynamic approaches to transpersonal psychology, among others.

Through our data, we conclude that the therapeutic potential of ayahuasca in relation to the grieving process can be related to experiences of emotional confrontation. We understand emotional confrontation to manifest variously as a simple expression of relief, as a way of honoring the deceased, as an opportunity to put daily pain into perspective, or as an opportunity to dissociate from pain by assuming a spectators' role toward one's own suffering. We also associate the therapeutic potential of these experiences with instances of emotional confrontation that bring insight into maladaptive patterns that are used as defense mechanisms to dampen or evade pain. Finally, we identify the therapeutic potential implicit in the opportunity to reconstruct one's identity, as it emerges from the chance to release obsolete personal narratives by assigning new meanings to traumatic life events, by preserving bonds with our deceased loved ones, or by exploring existential questions elicited by new spiritual insights.

Furthermore, we have encountered experiences that may shed light into beneficial psychological processes that, as far as we know, are yet to be described in the grief literature. Such seems to be the case when the internal representation of the deceased is transformed or the perception of the space occupied by him or her changes during the ayahuasca experience, providing an opportunity to continue the bonds with our loved ones in a different way, allowing that relationship to keep evolving, even in their physical absence.

To conclude this section, we would like to acknowledge the fact that the observed therapeutic effects of ayahuasca cannot be described solely in terms of psychological and intrapersonal dimensions, since they are embedded in a particular ceremonial and ethnomedical context. Many experiences reported in our data, such as experiences of emotional release, are reported *in relation to* the participant's perception of the role of other setting elements, such as the work of the Shipibo healers, the healing songs, the purgative qualities of the brew, or other environmental or relational factors. We believe that this invites an opportunity to share voice with some of these healers and present their perspectives on what it means to work with Westerners who are seeking help processing their grief.

An Approach to Grief and Grieving from the Shipibo Onaya at the Temple of the Way of Light

For the Shipibo, music is the spirit's language, and singing is the adequate mode of communicating with them (Illius, in Brabec de Mori 2012, p. 79). The purpose of this section is not to present a thorough analysis of the complex and eclectic medical practices of the Shipibo people – and, by extension, the plurality of Shipibo cosmologies – as this is beyond the scope of this chapter and has been done elsewhere (for a variety of perspectives, see Roe 1982; Gebhart-Sayer 1985; Tournon 2002; Brabec de Mori 2012, 2013). Instead, we merely want to complement the rich narratives offered in the previous section with the perspectives of four of the onayas who work at the Temple (two men, Soi and Segundo, and two women, Elena and Sara).[4] We believe that providing an approximation of the way that these four Shipibo healers perceive and address grief among Western participants will help elucidate some of the experiences described by the participants themselves, as they are the main social actors that attend to and are responsible for the ceremonial space from which these experiences emerge.

The data gathered comes from the experience of one of the authors of this chapter, who has been living at the Temple for the last 15 months, and from a focused conversation that he had with these four healers on the subject of grief. Even so, talking about the work of the Shipibo onaya requires an acknowledgment that any

---

[4]A more thorough interview with one of the onaya, Soi, is offered by Aronovich and Labate in this volume.

attempt to translate practices and phenomena that are rooted in Amazonian ontologies will always be accompanied by what Viveiros de Castro (2004) calls "equivocations": instances of fragmented communication where interlocutors may be sharing the same language to unknowingly refer to different things. However, as De la Cadena (2015) has shown, controlling for equivocation during the task of cultural translation allows us to avoid transforming the dissimilar into the same, maintaining the richness of the divergent perspectives while also making the excesses of each interpretive lens visible.

Making this last point explicit is important; most Western approaches to complex phenomena have a tendency to offer reductive explanatory systems that – whether psychological, biological, or cultural – would fail to account for the primary dimensions in which the Shipibo onaya situate their own practice: In our experience, these include the energetic, the aesthetic, the intersubjective, and the spiritual. By including their voices in this chapter, we also make space to honor the degrees of evidentiality in Shipibo ontology; as Brabec de Mori (2012) has shown, in contrast to naturalistic science where evidence can only be produced by third-person observation, for the Shipibo, it is the direct experience and expression in first person that commands the highest authority, particularly if that person is a respected healer.

In this volume, Aronovich and Labate present an in-depth interview with Soi, one of the Shipibo onaya who also contributed their voice to the present chapter. In the interview, Soi recounts how *oni* – the Shipibo name for ayahuasca – was given to his forefathers by the *chaikoni*, the mythical, invisible guardians of the forest and its people. Chaikoni narratives, as is often the case with oral history and myth, are varied and diverse. For Soi, the chaikoni were those wise ancestors who refused being conquered by the powerful "other," those who had foreseen what the European political and religious conquest would do to their people and their traditions. In Soi's narrative, ayahuasca, oni, was their gift to the bereaved, a remedy that was given to the native survivors to help them cope with the trauma and grief that resulted from the atrocities they endured.

Although the use of ayahuasca is popularly represented as being a millenary and uninterrupted pan-Amazonian tradition, there is very little evidence that its use was widespread in preconquest times (Gow 1994). Soi's account reinforces Brabec de Mori's idea that ayahuasca possibly reached the Ucayali River, the homeland of the Shipibo-Conibo and other Panoan groups, in the course of the last three centuries. This violent cultural exchange was likely a result of the waves of forced migrations and mass displacements of people from different regions and diverse ethnic groups who converged in new spaces regulated by the logic of the missions and, later on, the ruthless global rubber market (Brabec de Mori 2011). It is not farfetched to argue, as presented by Taussig (1987), based on his work on the Putumayo, and perhaps implicit in Soi's account, that the atrocities committed against native populations during these periods catalyzed the transmission and adoption of healing practices that became necessary to address individual and collective pain.

In our conversations with the *onayabo*, who lent their voices to this chapter, we asked them to describe the way in which they help, treat, and heal those *pasajeros* (the term used to refer to participants on a retreat, literally: "passengers") who have recently lost a very close and loved person. Soi responded that, in these cases, their job focuses on liberating and bringing out the sadness (*tristeza*) that is "blocking one's heart and mind." "Bringing all of that sadness out from the body, out from the spirit, the soul," added Elena, "It takes time. It's like a psychological trauma, a spiritual trauma." According to Soi, "when a person loses a loved one, that person's mind becomes ill, the body becomes ill. That person needs to seek help from a healer, so we can sing that sadness out." But what does it mean to "sing that sadness out?" How can grief be addressed through the medium of song, and how is that achieved? The response of the four onayas was unanimous and simultaneous: icaros!

Like the vast majority of Amazonian societies, Shipibo worlds are constructed upon an animist ontology, to use Descola's ontological approach (Descola 2013). In contrast to the hegemonic ontological discourses of modernity, animist constructions of the world are not rooted in a dualistic separation between nature and culture or subject and object. In animist societies, according to Descola (2013), the human and the nonhuman share the same internal nature; many plants and trees are conceived as persons, imbued with spirit, and, thus, fully able to communicate with humans. Because of this shared internal essence, nonhuman persons lead social and cultural lives, oftentimes, identical to those of men. They possess agency and intentionality. Emerging from an animistic ontology, where humans and plant spirits share a common culture and the subject/object duality is fluid, the Shipibo onaya relies on his ability to commune intimately with the spirits of the plants in order to diagnose and heal. The icaro – the magic song or chant – is the main and most important tool and skill, an embodiment of the power and the medicine that the healer has received and acquired from the different plants that he has "dieted."

"Dieting," in this context, refers to the initiation period and subsequent learning process in which the initiate goes into isolation, abstains from sexual activity, and observes special food restrictions, such as avoiding salt, sugar, fruits, meat, and fatty fish. Oftentimes, the person undergoing the "dieta" will ingest or bathe in a preparation of a specific tree or plant with which they aspire to commune, develop a relationship, and craft an alliance. It is during the initiation period, writes Luna (1986), "in which the novice is purified by the special diet and isolated from people, [that] he is open to the manifestation of the spiritual world. He receives from the spirits the tools of his future practice, in particular, the magic chants or melodies (icaros)" (p. 97).

"The *icaro*," says Soi, "is my connection to the plants." The spirits of each plant have their own icaros, so onayas will often keep "dieting" beyond their initiation period to gain more power by adding more plant spirits to their repertoire of allies. Through the icaro and the medium of sound, the onaya then becomes, transmits, directs, and guides that energy toward the patient. According to Segundo, "The icaros are medicine in its purest form." Furthermore, for Segundo, each plant and icaro hold a different form of medicine, particularly well-suited to treat different conditions, including grief and its physical, emotional, psychological, and spiritual manifestations: "This is why we diet many plants: With their help, we bring out those energies, that sadness and anguish. For this, we use the *tanti rao*,[5] and we also diet *piri piri*,[6] specially to treat the *susto*."

"Susto" is a Spanish word that translates as "fear" or "fright." It is a common experience of affliction throughout Latin America and an overarching diagnostic category in many non-hegemonic medical systems. It is included in the DSM-5 (APA 2013) as a culture-bound syndrome and described as "a cultural explanation for distress and misfortune," relating it to "major depressive disorder, posttraumatic stress disorder, other specified or unspecified trauma and stressor-related disorder, somatic symptom disorders" (p. 836–837). The symptomatology of susto is eclectic enough that Tousignant (1979) concludes that more than a syndrome, susto is an overarching strategy "to relate illness events to other levels of reality" (p. 347), with the condition being primarily spiritual in nature.

Susto, as commonly understood by the Shipibo and other Amazonian and Andean healers, is the experience of emotional and spiritual shock that occurs after a frightening or traumatic event, such as an encounter with a dangerous animal or a malevolent spirit. Susto can also result from an alarming illness or from the death of a loved one, and it often entails the loss of one's soul or part of that soul. As we wrote previously, Elena describes the experience of grief as a psychological and spiritual trauma that can lead to *susto*: "When you lose a close family member," Elena points out, "sometimes you can get *asustado* ('with susto')." She continues, "The death of a loved one is a terrible thing! It results in a lot of sorrow, in a lot of emotional and sentimental suffering. One gets susto, a very big susto! It creates an energetic blockage and one can get ill."

Like Segundo, both Elena and Soi agree about the wide-ranging reach of their command of plant spirits: "We have special icaros for different needs," says Maestra Elena. "We have *warmi*[7] icaros, we have medicine icaros… in this case, we would sing medicine icaros, especially the ones that cure susto, the ones that get rid of anguish." Soi adds, "We have icaros that we use specifically to bring the sadness out of the spirit and to help the person be open again to happiness and joy; we have special icaros for that too!" The icaros transcend and exceed their utilitarian value as a medical practice; they are a manifestation of spiritual connection as much as a sublime and refined form of creative and aesthetic expression. They are rooted in immediacy, occupying a space at the intersection between medicine and art: "Any of us can sing an icaro," comments Soi, "Segundo can sing his icaro, or Elena can sing hers. And tomorrow, you can come and ask us about that icaro we sung, but we don't remember it anymore! It comes from the moment, from our connection to that

---

[5] Jacques Tournon presents the category of *tanti rao* as a subcategory of the *isinma rao*, a native categorization of plants that are used by the Shipibo to modify human behavior. Tanti rao is a subcategory of plants that are used to calm the nerves and are part of the daily plant remedies that the Temple healers provide for workshop participants.

[6] Possibly *Cyperaceae* family. Tournon et al. identify a vast array of plants referred to as "Piri Piri" and so many different uses that it would collectively amount to a panacea (1998).

[7] Love magic is a big part of Shipibo practice. *Warmi* means "woman" in Quechua, and warmi icaros are sung in order to attract or secure the love of a romantic interest.

moment. Tomorrow, in another ceremony, it will be a different icaro." Elena agrees: "One time, a lady came to me after a ceremony. She said, 'I really enjoyed that icaro! Can you sing that for me again?" But I can't do that; I can't remember that! Because that icaro was lived, it was experienced in that moment alone… It was inspired!"

Inspiration is defined as the action or power of moving the intellect or emotions. According to the Merriam-Webster online dictionary, the word "inspire" (*inspirar*, in Spanish) means "to influence, move, or guide by divine or supernatural inspiration, to exert an animating, enlivening, or exalting influence." The word comes from the Latin *inspirare*, "breathe or blow into" (Merriam-Webster, n.d.). The icaro, as we understand it, is a collaborative, inspired, and in-spirited *Ars medica*. "This is an art," exclaims Soi. "An art, an inspired art!" follows Elena. "Inspired," concludes Segundo: "Inspired by the plants, by the spirits of the plants."

To conclude this section, we want to go back to the aforementioned notion of "controlled equivocations" (Viveiros De Castro 2004), with an acknowledgment that any attempt to translate the medicine of the icaros through our own understanding of plant spirits and our understanding of the onayas' experience of plant spirits will always be incomplete. Our goal is humble: to present the perspectives of four of the onayas whose medicine influenced the narratives presented in the previous sections.

Although ayahuasca tends to occupy the spotlight in the narratives of Western participants, for the Shipibo healers, ayahuasca serves mostly as a conduit or bridge facilitating the communion between the onaya and a multiplicity of plant spirits who are incorporated, transmitted, and guided through the voice of the healer to do the actual healing work. The main strategy to treat grief is, thus, to sing inspired icaros that make visible and then dissolve the sadness and the anguish and remove energetic and spiritual blockages caused by a susto originating in the traumatic experience of loss, allowing the numb person to feel that which has been suppressed and eventually becoming receptive to the experiences of happiness and joy. Although the folk psychology of the Shipibo medics and the insights of modern psychology and psychotherapy use radically different language and rely on different conceptual universes, it is our view that, by engaging both discourses in dialogue, they point toward the same core insight: Processing grief requires the capacity to feel; to allow the full impact of the sorrow, the sadness, and the loss to wash over us; and to be present to it and stay open to what is. Both CBT and the icaro-led ayahuasca session serve similar functions, as they help create a dedicated space where the grieving person feels safe to feel.

Conclusion

Our initial body of provisional theory-grounded data is constituted by the reports of 50 participants who attended workshops at the Temple of the Way of Light while processing grief over the death of a loved one. In order to organize and analyze the data, we identified a number of useful psychotherapeutic approaches, concepts, and theories in the literature on grief and juxtaposed them with the experiences of our participants. Our aim was to engage in an interdisciplinary and inter-epistemological dialogue that would further our understanding of the unique therapeutic potential that ayahuasca brings to the grieving process. In order to discover more facets and nuances of the extremely rich and complex phenomena observed in the field and evidenced in the reports of our participants, we combined available evidence and theory from the fields of psychology, psychiatry, and medical anthropology while also including the perspectives and knowledge of the four Shipibo healers. We believe all of these fields contribute to our understanding of the complex and unique dynamics of the ritual use of ayahuasca and its applications in the grieving process, within a syncretic context where Western practices and Amazonian medical systems meet.

Emerging from this exercise are partially novel and sometimes unexplored and uncharted concepts and approaches that await further scientific inquiry. Our data appeals to the need to further the development of a sound body of theory from which to build a model for the application of ayahuasca-based practices in the process of grief. However, we must recognize the limitations of our study; this is a work in progress, and we are far from a complete understanding of the complex interactions between the personal, the cultural, and the beyond-the-human (Kohn 2013) aspects that shape and contribute to the experience of the participants. Our work does offer some answers, or at least descriptions, as to *what* could be happening during an ayahuasca experience for people who are grieving. However, a lot remains unexplained in relation to *how* and *why* these processes are happening, in what personal and social context they are more likely to arise, and what aspects of the setting are more conducive to their emergence.

References

American Psychiatric Association (APA). (2013). *Diagnostic and statistical manual of mental disorders*. Arlington: Author.

Baerger, D. R., & McAdams, D. P. (1999). Life story coherence and its relation to psychological well-being. *Narrative Inquiry, 9*(1), 69–96.

Betz, G., & Thorngren, J. M. (2006). Ambiguous loss and the family grieving process. *The Family Journal, 14*(4), 359–365.

Boelen, P. A., Van Den Hout, M. A., & Van Den Bout, J. (2006). A cognitive-behavioral conceptualization of complicated grief. *Clinical Psychology: Science and Practice, 13*(2), 109–128.

Brabec de Mori, B. (2011). Tracing hallucinations: Contributing to a critical ethnohistory of ayahuasca usage in the Peruvian Amazon. In B. C. Labate & H. Jungaberle (Eds.), *The internationalization of ayahuasca* (pp. 23–47). Zürich: LIT-Verlag.

Brabec de Mori, B. (2012). About magical singing, sonic perspectives, ambient multinatures, and the conscious experience. *Indiana, 29*, 73–101.

Brabec de Mori, B. (2013). La transformación de la medicina Shipibo-Konibo: Conceptos etnomédicos en la representación de un pueblo indígena [The transformation of Shipibo-Konibo medicine: Ethnomedical concepts in the representation of an indigenous people]. In E. Sigl,

Y. Schaffler, & R. Ávila (Eds.), *Etnographia de America Latina, ocho ensayos*[Ethnography of Latin America: Eight essays] *(collecion estudios del hombre* [collection of the studies of man], *30)* (pp. 203–244). Guadalajara: Universidad de Guadalajara.

Chuengsatiansup, K. (2003). Spirituality and health: An initial proposal to incorporate spiritual health in health impact assessment. *Environmental Impact Assessment Review, 23*(1), 3–15.

De la Cadena, M. (2015). *Earth beings: Ecologies of practice across Andean worlds.* Durham: Duke University Press.

Descola, P. (2013). *Beyond nature and culture.* Chicago: University of Chicago Press.

Draucker, C. B. (2003). Unique outcomes of women and men who were abused. *Perspectives in Psychiatric Care, 39*(1), 7–16.

Faschingbauer, T. R., Zisook, S., & DeVaul, R., (Eds.). (1987). Biopsychosocial aspects of bereavement. In *The Texas revised inventory of* grief (pp. 111–124). Washington, DC: American Psychiatric Press.

Freud, S. (1924). Mourning and melancholia. *The Psychoanalytic Review (1913–1957), 11,* 77.

Frey, W. H., & Langseth, M. (1985). *Crying: The mystery of tears.* Minneapolis: Winston Press.

Gebhart-Sayer, A. (1985). The geometric designs of the Shipibo-Conibo in ritual context. *Journal of Latin American Lore, 11*(2), 143–175.

Gillies, J., & Neimeyer, R. A. (2006). Loss, grief, and the search for significance: Toward a model of meaning reconstruction in bereavement. *Journal of Constructivist Psychology, 19*(1), 31–65.

González, D., Carvalho, M., Cantillo, J., Aixalá, M., & Farré, M. (2017). Potential use of ayahuasca in grief therapy. *OMEGA-Journal of Death and Dying, 79*(3), 260–285.

González, D., Cantillo, J., Pérez, I., Farré, M., Feilding, A., Obiols, J. E., & Bouso, J. C. (2020). Therapeutic potential of ayahuasca in grief: A prospective, observational study. *Psychopharmacology, 237*(4), 1171–1182. https://doi.org/10.1007/s00213-019-05446-2.

Gow, P. (1994). River people: Shamanism and history in Western Amazonia. In M. Thomas & C. Humphrey (Eds.), *Shamanism, history and the state* (pp. 90–113). Ann Arbor: University of Michigan.

Hill, P., & Martin, R. B. (1997). Empathic weeping, social communication, and cognitive dissonance. *Journal of Social and Clinical Psychology, 16*(3), 299–322.

Klass, D. (2017). How continuing bonds got its name. In D. Klass & E. M. Steffen (Eds.), *Continuing bonds in bereavement: New directions for research and practice* (pp. 1–7). Abingdon: Routledge.

Klass, D., & Steffen, E. M. (Eds.). (2017). *Continuing bonds in bereavement: New directions for research and practice.* Abingon: Routledge.

Kohn, E. (2013). *How forests think: Toward an anthropology beyond the human.* Berkeley: University of California Press.

Lichtenthal, W. G., Neimeyer, R. A., Currier, J. M., Roberts, K., & Jordan, N. (2013). Cause of death and the quest for meaning after the loss of a child. *Death Studies, 37*(4), 311–342.

Luna, L. E. (1986). *Vegetalismo: Shamanism among the mestizo population of the Peruvian Amazon.* Stockholm: Almqvist & Wiksell International.

Mancini, A. D., & Bonanno, G. A. (2006). Resilience in the face of potential trauma: Clinical practices and illustrations. *Journal of Clinical Psychology, 62*(8), 971–985.

Martínez, E. R. L. (2015). Hinduismo: tradición y modernidad [Hinduism: Tradition and modernity]. *La Albolafia: Revista de Humanidades y Cultura, 4,* 29–42.

Merriam-Webster. (n.d.). Inspire. In *Merriam-Webster.com dictionary.* Retrieved February 10, 2018, from https://www.merriam-webster.com/dictionary/inspire

Metzner, R. (2006). *Ayahuasca: Sacred vine of spirits.* Rochester: Park Street Press.

Neimeyer, R. A., Baldwin, S. A., & Gillies, J. (2006). Continuing bonds and reconstructing meaning: Mitigating complications in bereavement. *Death Studies, 30*(8), 715–738.

Potter, W. J., & Levine-Donnerstein, D. (1999). Rethinking validity and reliability in content analysis. *Journal of Applied Communication Research, 27*(3), 258–284. https://doi.org/10.1080/00909889909365539.

Rieber, R. W., & Carton, A. S. (1987). *The collected works of LS Vygotsky: Volume 1: Problems of general psychology, including the volume thinking and speech.* New York: Plenum Press.

Roe, P. G. (1982). *The cosmic zygote: Cosmology in the Amazon Basin.* New Brunswick: Rutgers University Press.

Shear, M. K. (2010). Complicated grief treatment: The theory, practice and outcomes. *Bereavement Care, 29*(3), 10–14.

Spillers, C. S. (2007). An existential framework for understanding the counseling needs of clients. *American Journal of Speech-Language Pathology, 16*(3), 191–197.

Taussig, M. T. (1987). *Shamanism, colonialism, and the wild Mana study in terror and healing.* Chicago: University of Chicago Press.

Tournon, J. (2002). *La merma mágica: vida e historia de los shipibo-conibo del Ucayali* [The magical decline: Life and history of the Shipibo-Conibo of the Ucayali]. Magdalena del Mar: Centro Amazónico de Antropología y Aplicación Práctica.

Tousignant, M. (1979). Espanto: A dialogue with the gods. *Culture, Medicine and Psychiatry, 3*(4), 347–361.

Trichter, S., Klimo, J., & Krippner, S. (2009). Changes in spirituality among ayahuasca ceremony novice participants. *Journal of Psychoactive Drugs, 41*(2), 121–134.

Tupper, K. W. (2009). Entheogenic healing: The spiritual effects and therapeutic potential of ceremonial ayahuasca use. In J. H. Ellen (Ed.), *The healing power of spirituality: How religion helps humans thrive* (pp. 269–282). Westport: Praeger.

Viveiros de Castro, E. (2004). Perspectival anthropology and the method of controlled equivocation. *Tipití: Journal of the Society for the Anthropology of Lowland South America, 2*(1), 3–22.

Chapter 13
Healing at the Intersections between Tradition and Innovation: An Interview with Shipibo Onaya Jorge Ochavano Vasquez

Adam Andros Aronovich and Beatriz Caiuby Labate

Introduction

Every year, thousands of people from all over the world travel to the Peruvian Amazon to participate in ayahuasca ceremonies, workshops, and retreats run by one of the many centers offering "traditional" treatments. This relatively new phenomenon, conceived simultaneously as medical or shamanic tourism, a spiritual pilgrimage, psychonautical adventurism, personal development, or a legitimate quest for effective therapeutic options beyond the limits of the biomedical hegemony, among other possibilities, has changed forever the landscape of Amazonian *curanderismo*. The "traditional" merges with the psychotherapeutic, scientific positivism with the New Age or a variety of Eastern spiritual philosophies and practices. In this blend of afflictions, etiologies, diagnoses, and cures, the "exotic" becomes relative to the actor's own position, and "tradition" gives way to the cocreation of new emergent forms of sharing knowledge and practices, relating to the master plants and giving meaning to the experiences they offer.

Soi, or Jorge Ochavano Vasquez in his Castilian name, is a Shipibo *onaya* born in the native community of Nueva Betania on the Upper Ucayali. The community, which has about a hundred families, is located 4 h from Pucallpa by *peke peke*[1] or 45 min by speedboat. Soi is 36 years old and has spent more than 20 years working with ayahuasca, or *oni* in the Shipibo language. Despite learning from his own

---

[1] A small, relatively slow motorized boat, ubiquitous in the Peruvian waterways. The name is an onomatopoeia for the sound that the engine makes.

A. A. Aronovich (✉)
Medical Anthropology Research Center, Campus Catalunya, Universitat Rovira i Virgili, Tarragona, Spain

B. C. Labate
Chacruna Institute for Psychedelic Plant Medicines, San Francisco, CA, USA
e-mail: blabate@bialabate.net

© Springer Nature Switzerland AG 2021
B. C. Labate, C. Cavnar (eds.), *Ayahuasca Healing and Science*,
https://doi.org/10.1007/978-3-030-55688-4_13

grandfather how to work with the plants from a young age, Soi grew up at the intersection between tradition and Westernization. He belongs to a generation of young healers who, growing up in an increasingly urbanized, commodified, and Westernized rainforest, have built up their practice working mostly with a clientele of foreigners attracted to Amazonia by the growing popularization of ayahuasca.

The changes he has experienced in relation to the native way of life, driven by the ubiquitous neoliberal notions of "development," are significant: the construction of medical clinics, electricity, the arrival of industrial agricultural projects, and the introduction of intercultural education. Although Soi identifies these last evolutions as positive, he is also vividly aware of the long and violent history of exploitation and religious indoctrination that indigenous Amazonia suffered in the wake of European conquest, particularly the atrocities committed during the rubber-boom period. However, it is not suspicion nor resentment but rather hope and optimism that come forth through Soi's discourse. Curanderismo, he tells us, is a vehicle to unite people from all over the world under one common goal: healing.

Moreover, healing, in a broad sense, refers not just to the individual well-being of the thousands of relatively affluent Westerners able to afford a trip to the rainforest. Healing also involves forgiving, though not forgetting, the cultural and social traumas of the past and the present to encounter and foster new ways of relating and living with one another, driven by respect and a genuine curiosity. For Soi, healing resides in wishing to receive the message of the plants and share the teachings that they offer to anyone willing to listen: to be healthy, one must know how to live in harmony and reciprocity with other people and with our environment as a whole.

The following interview was conducted in November 2017 at the Temple of the Way of Light, a healing center offering ayahuasca retreats that works with Shipibo *onayabo*,[2] male and female. The center is located a few kilometers from the city of Iquitos and receives dozens of people each month who attend workshops lasting between 9 and 23 days, including from 5 to 7 ayahuasca ceremonies. Soi has been working at the Temple as part of the team of healers since 2010, treating hundreds of patients from around the world. Soi also attends to a new healing center that he built with his family in Nueva Betania, in the Ucayali region.

- **A: What does Soi mean, your Shipibo name?**
- S: Soi is like a design, something shiny, like a carved craftwork, something beautiful. It means something beautiful: So you can do anything, your art, and I say, "Oh, how beautiful!"

- **A: I thought Soi was the name of a bird.**
- S: A bird, a bird too. There are lots of meanings!

- **A: What is ayahuasca?**

---

[2] The plural form for *onaya*

- **S:** Ayahuasca is a very sacred plant. Its very name says it all: *oni*. In the Shipibo language, oni means wisdom, knowledge. I always explain that! Our wisdom comes from the beginning. In the past, there were no hospitals, clinics, or health centers in our communities. We Shipibo had none of these things before, and that's when the plant was discovered. It comes from the *chaikoni*.

- **A: Who are the chaikoni?**
- S: The chaikoni are invisible men, like spirits. They are our ancestors who, in the time of conquest, escaped into the forest. They didn't want to be conquered; they were afraid of what would happen to the Shipibo. They were very wise: they already knew everything, and they had the plant, oni. The chaikoni shared oni, wisdom, to help people during those times.

- **A: When did you start taking oni?**
- S: I started taking medicine when I was 13, just trying it. At the age of 14, I was already "dieting," understanding a little more. Because it isn't easy to understand and know our culture.

- **A: What is the diet?**
- S: The diet is what the plant wants to teach you and what you want to learn. The diet is like going to university. It's like enrolling (Laughs). For example, you might say, "I want to diet," and so you have to enroll, as they say, at a big university. Each plant is like a different course; it teaches something different. I began with small plants: a diet with *marosa* (*Pfaffia iresinoides*), followed by *piñon* (*Jatropha curcas L.*), then *sharomasho*. First, I dieted for a year and a half, then, continuously.

- **A: How much do you have to diet to become a healer?**
- S: The diet is a process that takes time. It requires patience. You have to advance little by little: you can't learn quickly. After the small plants come stronger, larger plants, like *toé* (*Brugmansia* sp.), which is a small tree. It's like working area by area. In that way, you rise in levels and rise, too, in levels of wisdom, levels of intelligence. It isn't easy, but it deepens over time. As you go deeper, you can diet on large trees, barks or stems, resins, leaves, everything, right? After many plants come the big trees: then you're going to see the real thing! As the diet advances, so does the level of knowledge, the intelligence of the plants: the wisdom of the plants, more than anything. Finally comes the "flying plant," *Niwe Rao*. This tree is the ultimate "mother of plants." The *meraya* (see below) of the past dieted not only on plants; they dieted on plants, animals, birds, fish, too... they could transform.

- **A: And when did you start working with patients?**
- S: I had already begun my work at the age of 17, treating people: the patients. That's where my path began. It was only then that I understood what my path was: healing and helping people. The connection with the plants and the visions of the *icaros* don't come quickly.

- **A: You mention the meraya. What different categories of healers are there in the Shipibo tradition?**
- S: There are three lines. There are *médicos*, healers (onaya), and meraya. All have the same medicine; the difference is that the meraya possess a much stronger energy. The meraya is the most "super" of them all, more super than the onaya. The meraya are connected to the chaikoni, they are the closest to the chaikoni. The meraya live purely on air, they feed themselves on air alone! During a ceremony, for example, a meraya can move the entire maloka[3] with their own energy.

- **A: How do you become a *meraya*?**
- S: In these times there are no more meraya, they no longer exist. There are only healers, onaya. Being a meraya is very difficult, the diet has to be strictly observed. The person must always be alone, without speaking: They have nothing to say because they are concentrating; that is their objective. Because if someone is dieting, their diet is aligned, it is highly channeled. Sometimes when you meet another person, a problem can arise, and afterwards you become blocked. That's why the meraya needs to be isolated, they shouldn't just wander off to chat wherever their friends are found.

- **A: So are you an onaya? How does an onaya heal?**
- S: I am healer, onaya. I take the medicine, oni, and I cure *pasajeros*[4] through the plants. My principal tool to heal is the *icaro*, to heal and provide protection. But when a meraya wants to treat a person, it is not through the icaro. They transmit themselves, they themselves enter the person's body. The meraya transform; they can enter a person and treat them from within.

- **A: What is an icaro? How does it cure?**
- S: For the onaya, the icaros are their connection with the plants. When someone is dieting a plant, an energy is left in their body. The diet, the plant, transfers its power to the healer. The healer afterwards uses this same energy to heal via their voice. That's the icaro, a transmission from the plant.

- **A: The plants give you their power?**
- S: To us, as healers, oni gives us power. The power is not ours alone, it is for all those who come to us, for the patients. So, the more you share this power with others, the more your own power decreases, diminishes, so that you are healing someone who is weakening you too. The power helps. How many people have been helped! Those who accumulate power have not treated anyone; the power is being hoarded. By contrast, here at the center, we share power in each workshop. Sometimes we become weak. To recover, you have to diet; the diet is the only way.

---

[3] Communal space, often where the healing ceremonies take place

[4] *Pasajero*, literally, "passenger," is a term deriving from the fluvial history of the Amazon where most foreigners and temporary guests traveled the waterways on ships. The same word is used pervasively today throughout the hospitality industry.

- **A: Did you learn everything through the plants?**
- S: My grandfather was my *maestro*, my teacher too; he taught me and told me just two things: "listen" and "tighten your belt." That's all he said. Listen! Sometimes, when I was just starting out, I would get annoyed with my teacher: I wanted to learn fast. But that's not how it works; first, you have to know how to listen.

- **A: What do you mean by "know how to listen"?**
- S: Icaros are born from plants. They are transmitted by plants. Icaros only come when you are already connected with nature, with the four elements, when you are already connected physically, spiritually, and emotionally. Then, when you take oni, the icaro comes. You are already connected to the energy of plants.

- **A: And what does "tighten your belt" mean?**
- S: (Laughs) As young people, when we want to diet, we get tempted by women. That's why my grandfather said "tighten your belt," right? A belt!

- **A: (Laughs) To keep your pants in place.**
- S: Yes, of course! The diet involves many trials, many temptations. Sometimes, if you don't stick to your diet, you may be left with nothing; or worse, you may be punished by the plant: You can become sick.

- **A: What do you mean by "stick to your diet"? Do these diets entail certain restrictions?**
- S: Of course. For example, when you are dieting on your plant and learning, you cannot have relations with a woman. You have to concentrate on your diet to connect with the plant. Your relation is with this plant. Women smell very strongly, for example. Their aroma is very strong. The plants also have their aroma, very strong, too. These can clash. Because they are also very strong energies, right? Some masters can manage these sexual energies; they can diet them as well. There are only a few masters who have practiced like that.

- **A: Aside from sexual restrictions, what other restrictions are important for someone who is dieting?**
- S: It depends on the teacher, the plant, the diet. Depending on the treatment, for example, some people may eat hot peppers and others not, or eat some fruits, but others not. But the most important [restriction] is to not have sex, not eat pork, and not consume alcohol and other drugs.

- **A: Why can't you eat pork?**
- S: Pigs eat anything; many things that the teacher himself has never eaten. It is a very dirty and fatty animal. To connect with the plant, you must be clean and eat little fat.

- **A: And alcohol and other drugs?**
- S: They can be medicines, too. But, sometimes, they are used only because the person is addicted, or they become addictions. Marijuana, for instance, is good for many things, like stomachache. You can also diet on this plant. But we don't

use it like so many people in the West use it, smoking it. The plant is not used like that. It is a healing plant. Its leaves are macerated in hot water. But in Western countries, it is used, more than anything, for fun; then things get completely crossed and it is the same plant that punishes you.

- **A: Many people smoke marijuana to relax and chill out.**
- S: But that doesn't heal you! If it calms you down a little, but doesn't heal you, then it's not medicinal. It becomes more complicated and blocks your mind.

- **A: We've talked a little about your learning process, your diets with plants. Can you tell me a little about your trajectory as a healer?**
- S: I began by working with my own family, the young children, the older women; people from my community who were sick, who came to me with susto[5]; some with more serious pains, bewitched, *daños* (harms): sorcery! With this kind of energy, I began to practice. Later, I spent 2 years working in Cajamarca, which is where I did my first work with foreigners. Then, I came to Iquitos, to Guillermo Arévalo's center, "Cosmic Anaconda." I worked there for 3 months, helping people from all over the world who came with psychological traumas, healing their spirit and their body. I gained a lot of experience there. Each maestro had their own ideology, their own way of working. It worked out for me, you know?

- **A: How did you end up working at the current center?**
- S: After I had finished working at the other center, I stayed in Iquitos and came to visit some friends who were working here. I asked them to invite me to work here too. I began in November 2010 and have been working for more than 6 years now as a healer. During my lifetime I have already worked with hundreds of people from all over the world, friends from many countries. I have worked much more with foreigners than with Amazonians.

- **A: Can you tell me a little about your spiritual life?**
- S: My spiritual life is my path as a healer: to help many people by doing good and learning a lot, with a good heart and good intention, and with awareness. There are some healers who use their spiritual connection to cause harm, cast spells, practice sorcery: I don't practice that. Above all, nature is my guide. Because nature provides everything, right? Food, life, humanity. Our path, our practice, is the intelligence and wisdom of the plants. We take oni and see what is now happening and will happen in the world. We take oni and we see the four elements connected: earth, air, fire, and water. There is a spirit that has created the world and has left here all the plants of the earth: the mother has left them here!

- **A: I see you are wearing a necklace with a cross. Is the Christian religion compatible with the path of the wisdom of plants?**

---

[5] Susto, a diagnostic category very prevalent in Latin American societies, is caused by an event that leaves a lasting impression or imprint on the experiencer. In the Shipibo healing system, susto is one of the main causes of illness among children and adults alike.

- S: I am religious, but I am not a Christian. It is the plants that guide me. Religions sometimes confuse people: there are some religious people who are Christians who also help people through prayer. But, sometimes, they do not want to heal with plants. There are Christian pastors who do not believe in that. They do not let us do it, and they hate us, don't they? They have a church, they preach, "this is going to happen." But they have no "vision," they make pure theory, they open their book, they read and read: "Oh, they are demons." But they are confusing people. We are not demons. We work physically, mentally, emotionally, and spiritually. We do not read, we have visions.

- **A: Is it very different for you to work with patients in your community and to work with foreign pasajeros?**

- S: It's very different! When patients from my community come to receive treatment, they don't take the medicine. Only the healer who is treating takes oni, and, through the icaro, the plants, we work deeply in the body and the spirit. With that, we cure. With that, we heal. This is the work material of a healer: oni, *mapacho*,[6] and *rao ininti*[7]; nothing more, because the healer is already prepared physically, spiritually, and emotionally. They are like a university professor. They already know, they don't clutch their books anymore, you know?

- **A: On the other hand, taking oni is a fundamental part of the workshops run for Westerners. Why?**

- S: Foreigners do drink oni; out of curiosity, more than anything else. Sometimes, they want to drink more and just want to look outside. They want to see. But some don't look deep inside, and first, you have to look inwards, because if you have a psychological trauma, or an emotional, spiritual, mental, or physical blockage, the vision is not going to open up.

- **A: What recommendations do you give to pasajeros who want to take oni?**

- S: Many pasajeros want to grow too quickly and see things before they put down roots. They want to grow in a night or two; but that's not how it works. Because you are born of your mother, right? Do you imagine you'll be able to walk just the next day? The work with plants is just the same. It takes time. First, you need to prepare. Cleanse the body. Some arrive addicted to liquor, alcoholism, a lot of drugs. Their energy has already penetrated the blood, the veins, the nerves, feelings, the mind. The *oni* works, but it is a process. It is like sowing a plant that is already almost withered. You have to take care of yourself: "Am I going to live or not?" When the plant has started taking root, the roots are like our nerves, right? The plant wants to rise up, it wants to become stronger, and, afterwards, when it [Soi makes a sound like a leafy tree stirring in the wind], then it's healthy!

---

[6] *Nicotiana rustica*, wild tobacco, processed very little: a master plant
[7] Flower perfume, generally based on commercially produced eau de cologne, used during ceremonies and healing rituals

- **A: Here at the Temple, what kind of preparation do pasajeros go through before taking oni?**
- S: They must diet, as we say, not eat pork, drink alcohol, or have sex for at least 2 or 3 days before and after. Before they take oni, we also give them a *vomitivo* (an emetic plant) and a steam bath. Vomiting washes out their stomach and purges them energetically, removing things that have happened in their life. The Shipibo know various plants to make you vomit: there are trees, plants, barks, stems, leaves. Here we use lemongrass (hierba luisa, *Aloysia citrodora*). Steam baths also help wash away negative energies externally. Vomiting and steam baths are essential before taking ayahuasca to clean both inside and out.

- **A: Do you think there are some people who shouldn't take oni?**
- S: Of course, because some pasajeros are already very sick when they arrive, feeling very strong pains.[8] Sometimes, women arrive feeling a lot of pain in their vagina, and the oni clashes with it. Someone who cannot even move, for example, must not take it. What would happen if they take it and the medicine opens up with those energies? They may scream and something may happen. What if they are taking pills or medicines!? Sometimes I get worried, taking measures to ensure that doesn't happen, because I've already had such experiences.

- **A: What experiences are you referring to? Where?**
- S: That was before I came here. It happened at the "Cosmic Anaconda." During a consultation, I heard that the healer asked the pasajero, and he replied, "I've already stopped taking my medicine," but this friend went to his *tambo* (cabin) and took his pills in secret, and then participated in ceremonies. And what happened? He died of poisoning. I don't know what pills they were because I had no idea he was taking them; he denied it! Oni is very strong. The pill is very strong, and the pain is very strong. That's how things became complicated. That's why, when people come with very strong pains, I always worry, I always ask them, and insist that they tell me the truth. Don't hide the fact they are taking other medications. I don't want to go through that again!

- **A: Do doses of oni vary a lot from person to person? Or do you give the same amount to everyone?**
- S: It depends on the person, the body. The first night, we give them all the same amount; very little initially. During this process, we start to open up and diagnose the person. Already after the first ceremony, we begin to calibrate the quantity to find the perfect dose for each person, because we're not all the same. Some people don't become *mareado*[9] easily, they are very strong, while other pasajeros are very sensitive. If they become very mareado, some vomit or want to "go

---

[8] The Temple of the Way of Light has a fairly comprehensive screening process, which excludes certain people from taking part in its workshops according to medical and psychological criteria. This process is supervised by professionals from diverse healthcare fields, so as to minimize the risk of adverse reactions.

[9] *Mareado*, literally seasick, used to describe the effects of ayahuasca

wild." That's why the optimal dose is when the pasajero is very mareado, but not too mareado, because then they no longer know anything nor remember what they have seen. The mind is simply blocked. When you have more experience, generally you need to take less, smaller amounts, because the oni has already penetrated your body. It is already connected. The oni rises alone!

- **A: How does oni cure? How does it help people?**
- S: It's not just oni. When we see a patient, the oni "opens": It helps us make a diagnosis, investigate their body, their *shinan*,[10] their spirit, everything, right? The oni also cures through the icaro. The oni opens, the healer enters, and the icaro heals. It heals spiritually through the transmission of the plants. The plants cure everything. Sometimes they draw energies from the mind, from the body, or any part of our organs, wherever there are pains, sadness, or frights.

- **A: And if you take oni alone?**
- S: If you take oni alone in your tambo, you can also be cured, but only if you know how to handle your mareación. If you can't handle that, drinking is just drinking: Anyone can drink ayahuasca! But to heal, you have to have a guide, a healer who transmits and guides the plants. The healer is just a guide, he tells the plants: "do it there, cure there."

- **A: Have you ever seen sick pasajeros cured over the course of a single workshop?**
- S: Yes, there have been lots cured, hundreds! Here a friend came, a pasajero, right? He hadn't walked for 8 years. He couldn't stand up. They brought him in a hammock. We cured him with the steam from different plants, and then icaros. He didn't drink oni, he couldn't. We healed him with plants and icaros alone. At the end, he walked out on his own. Many people are cured!

- **A: What types of illnesses have you cured with the help of *oni*?**
- S: Some people, for example, arrive with anemia, right? Others may have cancer or ulcers or prostate problems or urinary tract problems. With the help of the icaro, we can always extract the energy from these pains. In other words, we can *icarar*, or cure with the help of plants too. For example, you can chew the bark of the *ubos* tree (*Spondias mombin*) right? And how do you feel? It's bitter tasting, isn't it? You can bite a lemon. How do you feel? Acidic, right? That's the energetic side of the plant. Its energy fights sickness through the icaro, through the concoctions we give them to drink. With this energy, we extract, cleanse, and cure.

- **A: Can oni be harmful to some diseases?**
- S: More than any specific disease, oni can be dangerous for people who are taking lots of medications, because those medications, those chemicals, are still in their body. That's why we give an emetic first. You first have to extract all that energy from the body. Oni can also cause harm when your disease is already very

---

[10] *Shinan* or *Shina*, a Shipibo word that refers to a person's mind/spirit/consciousness

advanced, you know? Sometimes, it can weaken you even more, causing even more complications. The patient's spirit is unwell. It is very weak.

- – A: **What do you recommend in such cases?**
- – S: In those cases, we work with the icaros, or by blowing,[11] or some kind of plant concoction, so that the person recovers their energies. After the patient gets better, you can ask them: "How do you feel? How are you? How is your body? Is it weakening still, or not?" And, if the patient tells us, "Oh, I'm getting better," then you can give them oni too.

- – A: **What are the most common problems experienced by people taking ayahuasca?**
- – S: Strong, healthy people don't have any problems. For a normal man, who is not physically sick, the problems are few. There may be some psychological problems, but it's okay, you just have to keep working. Sometimes, the spirit reveals itself in the body itself or in your dreams. It's part of the healing process. There are also people who come with many addictions, such as alcoholism or stronger drugs. These penetrate deep into you: your veins, your mind, your *shinan*, your whole brain! For these people, it is very hard. That's why the treatment is lengthy. It takes time; a lot of time. We have to extract these energies from your body, from your spirit, right? I've had these kinds of experiences; it has also happened here. Some people have problems and decide to return home; they don't want to go deeper. They don't have faith!

- – A: **How you can help people who don't believe in the treatment?**
- – S: That's their decision, isn't it? Their own conscience. It doesn't come from outside; it comes from the same body, from oneself. It's not easy for those who don't believe. They lack cleanliness, lack healing, lack channeling of the mind. Alignment. Sometimes they have many "imaginations" that come to them, right? Thousands of "imaginations." In their life, they have done many things. These people are wrong, aren't they? Indecision. In their mind, they are thinking, "what am I going to do, this or that?" And, in the end, they close themselves, block themselves. In order to pass through these difficulties, the person needs to be healed by the plants.

- – A: **Have you ever seen ayahuasca cause someone more serious problems?**
- – S: This has happened to me with two people, two pasajeros, 3 or 4 years ago. This friend came to the ceremony, and we gave him four large glasses, four large cups of medicine, oni. Nothing! The next ceremony, again nothing! In the end, he did not believe in it, and on top of that he wanted to leave, he didn't want to be there any longer. And he went back to mess with his stash of heroin! Things got complicated. And this happens because they do not believe, and they fall back into addiction, you know? But what they need is time, time and more time.

---

[11] Blowing, *soplar*, is a basic Amerindian healing technique that involves blowing tobacco smoke, perfume, or other aromatic things on parts of the body to cleanse, cure, diagnose, or protect.

- **A: How did you help him?**
- S: They took him to hospital, washed him, and then he came back. Here, we made him vomit. We treated this case for 3 months. And, here, they took good care of him. The facilitators took care of him for a month. But this kind of person no longer has any sense; they're like a spirit, like a demon that slowly awakens. Sometimes, what we do to help cure them is very hard. What we do, what we seek, is to extract the demon they bring with them.

- **A: And what happened to him?**
- S: This friend spent a month in treatment with plants, icaros, and blowing. Finally, he completed 2 months and we gave him oni again: His mareación came slightly. He became just a bit dizzy. So, he still needed to purge more, to clear his body. After 15 days, we gave him oni again. Then he was mareado for almost an hour and a half, and again the mareación disappeared. That's the process, it's a process. After another fortnight: the same again. This time, he was almost normally mareado; he was equilibrated. All the energies contained in that poison were already coming out. Then, finally, when he had completed 3 months, he took oni three nights in succession. The second night he was very, very, very mareado! This friend was very mareado, well seated, concentrated, and not "wild." He had taken oni at eight or nine at night, and at dawn he was still mareado. Mareado all day! Because the oni was already inside, already at work. It wasn't until two in the afternoon that the mareación disappeared. Then he believed. Then he began to believe, right? And it was over. He was healed. This friend went away very calm.

- **A: Aside from the addictions, many people who visit the center suffer from depression and anxiety, and many have taken psychiatric medications. Is it good for them to take oni?**
- S: Here we always hear that some pasajeros take medications for depression. But they're just sedatives, right? They calm down for a moment, and then, when the medication leaves the body, the depression returns. Sometimes depression is caused by big sustos, big frights, or a lot of sadness. Things that happen to us, affect us, right? These things affect the heart and affect our brain. When the two don't work together well, it always gets complicated. That's when these things assume the form of an illness, like depression. When depression overwhelms us, a lot of anxieties surface, and the person can die. If they are taking medication when they take oni, it clashes. It clashes with their shinan, and everything, right? They can faint. That's why we spend all day on the treatment. We prepare medicinal plants. We give the emetic first, to cleanse first. A pasajero has to stop taking those medicines at home at least a week before coming here.[12]

- **A: Is ayahuasca good for pregnant women?**

---

[12] Guests at the Temple of the Way of Light are required to discontinue the use of certain pharmacological medications for longer periods of time, and special cases are reviewed by medical and mental health professionals.

- S: No, no, no, no! There are strong women, but some are not. They are more sensitive and may lose the baby. I seldom give oni to a pregnant woman.

- **A: People with epilepsy?**
- With people who have epilepsy, it is also very problematic, because ayahuasca absorbs everything, and the mareación affects everything: It can clash, and they may have a stroke. This kind of disease is very dangerous. I'm afraid to give oni to such people. That's why we first have to straighten them out, over time.

- **A: People with diabetes?**
- S: **Diabetes** is a very complicated illness. The treatment is with plants; afterwards, you can give icaros. You can continue the procedure with the icaros and plants; it gives strength, letting them recover their energy, because that comes from the blood. When the diabetes is mild, you can give oni. But when it is already complicated, very difficult, it can be dangerous, because ayahuasca also has acid, right?

- **A: Hepatitis?**
- S: It isn't good for hepatitis because the person has an inner fever, feels pain in their muscles, and the energies clash. It can be dangerous because the liver is already affected. You first have to treat it with plants.

- **A: And can women who are menstruating take ayahuasca?**
- S: Blood is strong! Blood has its own aroma. That's why some healers don't like it. Depending on the healer, there are some who don't like to give women ayahuasca while they are menstruating because it affects their energy. It depends on each healer's practice, how it affects their medicine. Sometimes some clash and vomit, then they wake up feeling weak because it has affected their body and their shinan. But for me, it is normal, menstruation for me is normal; it does not affect me.

- **A: Can oni be good for children?**
- S: Yes, for them too: babies over 2 or 3 months old. In the past, they would do so, crush a small plant and give the infant a few drops. Today, it is mainly done by those who have maintained the practice for generations, because it is our culture, allowing the baby to drink it and grow. They would give them a drop of oni with their mother's milk, they would suck it from the mother's teat, and the baby would grow. The baby would end up highly intelligent and wise. The baby would become a meraya!

- **A: Why do many people who drink oni vomit and purge?**
- S: Sometimes, they have stomachache or nausea because their stomach was filthy from the many things they have eaten. Many people eat fats, fruits, or ceviche, all of which are strong foods. They stay in the stomach. Everything they ate has penetrated their stomach. Ayahuasca doesn't like that, so it has to work to extract and vomit everything. That's why the ayahuasca cleanses; one has to expel and purge all the filth.

- **A: Is purging just a physical cleansing?**
- S: Purging is a cleansing of the stomach but also cleanses other energetic things, such as bad airs [*mal aires*]; bad air is physical, is it not? Sometimes people arrive who have a trauma. Sometimes they get dizzy in the head; that's the air they have in their brain. The oni extracts everything. Some are also addicted to eating, addicted to flavors. And, when you vomit, you purge with ayahuasca. How do you feel? Good. So that's why some of us like to vomit, for relief. Already, the next morning, some feel good; others don't feel so good, their head hurts, and they remain a little debilitated. These people still need to cleanse themselves.

- **A: How many times does someone have to take oni to be healed?**
- S: With ayahuasca it's not "how many times?" There's no minimum or maximum, you know? Anyone can take more and more and more. When your body is free or clean, that's it! When you are already healthy, you already know, you are already aware of your body, your mind, your spirit. Some like to take it, and they take it because they value the medicine: "Oh, that healed me, I'm going to carry on with this."

- **A: How often can you take oni? Can it be dangerous to take too much oni?**
- S: You can take it every day; hold a ceremony every night; but, sometimes, it affects us too, doesn't it? Sometimes, we don't want to drink it anymore, just the taste of it! Sometimes, it gets too much; it's not water! It's a sacred medicine. The medicine opens up; then, we see many things, have many visions. That also affects us. There are days when you become strongly mareado, there are very heavy days. You're not going to have the same vision every night. Everything changes; it always changes. Every time, it's very different. Some people want to go deeper, to continue going deeper. If you know how to practice with oni, it isn't dangerous. Most of all, when you deal well with the world of your mareación, physically and spiritually, it isn't dangerous.

- **A: You commented earlier that your work with Western pasajeros is very different from your work with people in your community. What are the main illnesses you cure among the Shipibo?**
- S: In my community, it's very different. The Shipibo often suffer from sadness, don't they? Sometimes, they get sick after being hurt by demons, evil spirits, which are strong. Because, sometimes, people walk in the forest and encounter virgin forests; they enter and the spirits of the virgin forests harm them. That's all we cure there: the damage inflicted by the spirits of the forests, because it's very strong! Others arrive with anxieties or stress from working too hard. That affects their organs and they become sick, and we have to try and help. Many are frightened; they come with susto.

- **A: What is susto?**
- S: Susto (fright) is when something happens to you and you're unwell, you're afraid. You can get frightened after an accident. Or some animal may also frighten you. Fright, fear, trauma, they're all along the same lines, aren't they?

Depression, too. All this comes from the mind, because the mind is connected to the heart. When you are unwell, how does your heart feel? When you're scared, you're unwell, your body is weak, just breathing hard, you're afraid, and no one can help you, you have to seek out a healer! When the person comes to us with fear, frightened, where do we see that? It's their mind that's not right. They are sad, wishing to escape, maybe: "Where shall I go?" Many pasajeros arrive here with fears, with sadness. They always arrive with that.

- **A: For some people, the world of ayahuasca is linked to sorcery. What is sorcery?**
- S: Sorcery is born in the mind of the human being, it is formed there. It is only what the human does, how the human handles it. The world of ayahuasca itself is not sorcery. Ayahuasca does not say "harm the person." It doesn't say that. It is the same human consciousness that shapes it, the same person who can use it to heal or harm. This is why many people who want to diet… Look, if you're going to diet, think hard about it! Because this path is the path of curanderismo, the path of healing, and you have to do it properly without thinking about other things. Someone who wants to do a job well directed, well aligned, well channeled; for them, their practice is pure light and pure medicine.

- **A: Are there many people who use ayahuasca to do harm?**
- S: Today, there are many sorcerers. It's a very complicated time. More than anything else, there's a lot of envy, a lot of envy in the world, they have it. For example, you may have something, things, valuables, a good house, work, and that affects you. That's where sorcery comes from, that's where it begins. That's why we have to be protected, to protect ourselves from the things humans do. We protect ourselves by taking ayahuasca, and a flower bath, or by preparing some plant.

- **A: You mentioned earlier that your Shipibo patients mainly suffer from** *daños* **("harms") caused by magic or sorcery. Do you also find daño among Western** *pasajeros*?
- S: Sorcery, less so, but magical harm, yes, of course! Because some read books, and books contain magic. People who have magic practices have to follow them like they were diets! Some who read books don't follow them, they don't perform their practice well, and suffer harm as a result. They block themselves. Those who practice yoga, it's the same.

- **A: What do you mean, "perform their practice well"? Is yoga compatible with oni?**
- S: Every practice is different, isn't it? You have to learn, deepen, relate. There are many who come here who practice yoga, practice meditation. Some come who practice other therapeutic fields, right? So, they come here and want to connect with oni. The oni is also a practice in itself. These are different practices, although they can be aligned. But, you have to take care of each practice: It's like a diet; if not, they block each other. Practices can be related; you only have to align them, you have to relate them.

- **A: Do you think ayahuasca can help cure any kind of problem?**
- S: Yes, I think so. Ayahuasca helps a lot with any problem. Family problems with your brother, or your dad, say. You can "be hated," right? Your dad hates you, doesn't love you, or doesn't want to see you. You can come and deal with that through oni; we can align the *shina* of your whole family with the icaros; we bring it together. And, by the end, the relationship with your family has returned. It's normal again. Ayahuasca can also help if you want to be with a woman. For example, you may love a woman, but the woman doesn't want you. Then, we make a "hitching." That's what we call *warmi*.[13] So, we give a *pusanga*,[14] right? (laughs) And now the woman will want you. That's also handled through oni.

- **A: I've heard some indigenous youths say they want to take ayahuasca to see their future. Can oni help them?**
- S: Yes, it can, because the medicine can "open up" to see the future, can't it? Also, their present and their past. But they have to be highly concentrated; that's why I always say, if you want to "see," you have to prepare yourself physically, mentally, spiritually, and emotionally. Because some young people come with many "imaginations," full of doubts. But then they don't just have to think; they have to think well, think clearly. If they concentrate well, they themselves can see: "that thing is going to happen, that's how the months, the year, are going to unfold."

- **A: Is oni different from other master plants?**
- S: Oni is the only master plant. It is the most sacred plant: the one that cures the most people. Because, of course, no other plant you take gives you the same mareación. You can take another plant, guayusa (*Piper callosum*, not to be confused with *Ilex guayusa*), and it doesn't give you a mareación. Some plants like piñón colorado (the black physicnut, or *Jatropha gossypifolia L.*) do cause mareación. Or you can take toé (angel's trumpet, or *Brugmansia sp.*), which can make you mareado too. Some who have dieted on toé know what its energy is like. They are strong plants; plants with many powers. Sacred plants. But ayahuasca is different from other sacred plants, its mareación is not "normal." The mareación given by oni, and *cahua* (*Psychotria viridis*, also called *chacruna*), enables you to see everything, everything! Oni is the only plant that controls all the other plants.

- **A: For many *pasajeros*, ayahuasca has a female essence. What's it like for you?**
- S: Male and female. Its very name says it: we Shipibo call ayahuasca oni, which is like *honi*: man! And the vine is mixed with chacruna, or *cahua* in Shipibo, which is female. We always work with both, because they are man and woman, they always work together. When they come together, oni, "wisdom," is born.

---

[13] In Quechua, "woman"

[14] Perfume or plant preparation used in love magic

- **A: Why do so many foreigners come to the forest to take ayahuasca today?**
- S: They come because the world today is globalized, modernized; now, everyone likes this plant, that's why they come. For me, it's good, because ayahuasca brings everyone together, of course. And the people who come receive messages, right? "Oh yes, I went to that place and the plant healed me," and the person shares that when they leave. And others then want to come: "I want to go too, I want to heal myself." That's what they think, and that's the message sent by ayahuasca. The message is healing, isn't it? Many people suffer from afflictions, psychological traumas, emotional, spiritual, and physical problems. Some people heal with medicine and what do they think? "I want it too. I want to practice it." They come afterwards and ask us: "I also want to be on that lineage, be on that path. I want to help my family and my friends." And there they diet and diet. And afterwards, they leave with that knowledge, with that wisdom.

- **A: What do you think about Western people working with ayahuasca in other countries?**
- S: Ayahuasca has its legend, its history; much of that comes from my ancestors, who told me about it and it exists still today. Today, this medicine has spread all over the world. Some people handle it well, others don't; some take it just to hallucinate or to have fun. That's why I always value my plant; because some people carry it around to perform their ceremonies with guitars, with anything, but that's not it! They don't value it fully. Ayahuasca is a sacred plant, principally for healing, you have to carry it well and treat it well, because they are beings: living beings! It's a very strong spirit.

- **A: Many people heal and learn with ayahuasca's help. What can we do to reciprocate and give something of value in return to the forest and its people?**
- S: Today, in this era, there is a lot of pollution, a lot of water is wasted. Today, big companies come to exploit the land, to damage the land, or logging companies enter and destroy nature, cutting down trees. A tree cut down by a chainsaw falls, it will never grow again. The tree is going to die: How many lungs are we killing? And how much oxygen are we killing? Every branch they cut is a life for me! For me, it's very sad, because this is my practice. Before, we were surrounded by animals; but now, the animals are moving away. That's what comes from big companies.

- **A: What can we do to help?**
- S: We all have to unite to defend our forests. We have to diet more to protect ourselves. We have to overcome our own energies, and change ourselves. We must plant ayahuasca, which I have seen is already becoming scarce. Many people are using it. Some prepare it and export it. How much ayahuasca leaves the forest every day?! Every day the plants die. They cut them down, crush them, and cook them. Yes, you have to plant it, of course! It has to be conserved.

Index

A

Acceptance and Commitment Therapy (ACT), 8, 53
Addiction, 144
Addiction disorder, 159
Addiction treatment
  bioenergetic therapy, 153
  cultural and cognitive approaches, 154
  doctoral dissertation, 154
  psychedelic movement, 153
  ritual healing, 154
  spiritual experiences, 154
  spirituality, 154
Adverse childhood experiences (ACEs), 110
  ayahuasca consumption, 100
  childhood trauma, 109
  childhood trauma-induced changes, 103
  child sexual abuse, 99
  chronic childhood trauma, 101, 102
  complexity of symptoms, 104
  cumulative impacts, 101
  DMT, 107
  Drug/alcohol dependence, 107
  economic burden, 100
  frontal and paralimbic areas, 108
  individual and societal costs, 100
  international data, 100
  interoception, 103
  neurological impact, 102
  original pathway, 102
  past traumas, 110
  pharmacotherapies, 104
  physical awareness, 108
  physical health issues, 107
  proprioceptive awareness, 103
  PTSD, 101, 104
  reconceptualization, 105
  self-awareness, 103, 109
  self-care, 103
  serotonin and dopamine systems, 107
  therapy and other spiritual practices, 106
  trauma, 101
  traumatic childhood experiences, 99
  in the United States, 100
  well-being effects, 109
Alzheimer's disease, 3
Amazonian shamanic context, 166
Amsterdam Resting State Questionnaire (ARSQ), 27
Amygdala, 4
Antecedents
  addiction, 155
  cognitive alterations, 156
  dopaminergic reward pathway, 155
  psychological and transcendental experiences, 155
  subjective experiences, 156
  therapeutic applications, 155
  transcendental hypothesis, 155
Anterior cingulate cortex (ACC), 23
Antidepressant effects
  ayahuasca, 25–29
  depression, 23–29
  financial and psychosocial stressors, 34
  ketamine, 33
  LSD, 33
  MDD, 21, 22
  neurobiological bases supporting, 29–33
  neuroimaging, 23, 24
  personality disorder, 34
  sleep, 24, 25
  treatment-resistant depression, 34

© Springer Nature Switzerland AG 2021
B. C. Labate, C. Cavnar (eds.), *Ayahuasca Healing and Science*,
https://doi.org/10.1007/978-3-030-55688-4